FUNCTIONAL NEUROSCIENCE

Volume 2

NEUROMETRICS

NEUROMETRICS

Clinical Applications of Quantitative Electrophysiology

E. ROY JOHN

Routledge
Taylor & Francis Group
LONDON AND NEW YORK

First published in 1977 by Lawrence Erlbaum Associates, Inc.

This edition first published in 2022
by Routledge
4 Park Square, Milton Park, Abingdon, Oxon OX14 4RN
605 Third Avenue, New York, NY 10017

Routledge is an imprint of the Taylor & Francis Group, an informa business

British Library Cataloguing in Publication Data
A catalogue record for this book is available from the British Library

ISBN: 978-0-367-75066-4 (Set)
ISBN: 978-1-003-18026-5 (Set) (ebk)
ISBN: 978-0-367-75398-6 (Volume 2) (hbk)
ISBN: 978-0-367-75402-0 (Volume 2) (pbk)
ISBN: 978-1-003-16237-7 (Volume 2) (ebk)

Publisher's Note
The publisher has gone to great lengths to ensure the quality of this reprint but points out that some imperfections in the original copies may be apparent.

Disclaimer
The publisher has made every effort to trace copyright holders and would welcome correspondence from those they have been unable to trace.

FUNCTIONAL NEUROSCIENCE VOLUME 2

NEUROMETRICS:

Clinical Applications of Quantitative Electrophysiology

E. ROY JOHN

Brain Research Laboratories
Department of Psychiatry
New York University Medical Center

LEA LAWRENCE ERLBAUM ASSOCIATES, PUBLISHERS
1977 Hillsdale, New Jersey

DISTRIBUTED BY THE HALSTED PRESS DIVISION OF

JOHN WILEY & SONS

New York Toronto London Sydney

Lawrence Erlbaum Associates, Inc., Publishers
62 Maria Drive
Hillsdale, New Jersey 07642

Distributed solely by Halsted Press Division
John Wiley & Sons, Inc., New York

Library of Congress Cataloging in Publication Data

John, Erwin Roy.
Neurometrics.

(His Functional neuroscience ; v. 2)
Bibliography: p.
Includes indexes.
1. Evoked potentials (Electrophysiology)–Mathematical models. 2. Brain–Diseases–Diagnosis. 3. Cognition disorders–Diagnosis. I. Title. [DNLM:
1. Electrophysiology. 2. Nervous system diseases–Diagnosis. WL102 R979 v. 2]
QP395.J63 vol. 2 [RC386.6.E86] 612'.82s
ISBN 0-470-99272-7 [616.8'04'754] 77-23967

Printed in the United States of America

To Josh, Sasha, Suki, Sanyi,
Penny, Andy, and Mir,
with love

Contents

Foreword

It is a unique priviledge to write a Foreword to this second volume of the distinguished two-volume work by E. Roy John and R. W. Thatcher entitled *Functional Neuroscience*. The scope, quality, and insightfulness of this volume is unusual and of such a pioneering nature that it is difficult to do it justice within the confines of a foreword. Hence, I will simply indicate its significance to clinicians and scientists in ophthalmology, otolaryngology, pediatrics, geriatrics, neurology, neuropsychology, electroencephalography, neurophysiology, psychiatry, and special education. Few scientific-clinical workers can match the courage, persistence, and ingenuity of the author. It is rare to have the background and breadth of knowledge necessary to deal with the complexities of brain function in a manner which extends so many horizons, from neurodynamic to psychodynamic dimensions, from sensation to perception and cognition.

In the early chapters we find discussions of the methods and techniques employed in the area of interdisciplinary effort and specialization designated by Dr. John as *Neurometrics*. This term, in itself, is in the nature of a breakthrough, opening avenues for objective, quantitative evaluation of brain-behavior relationships, especially with regard to learning and brain dysfunction which result in disturbed cognitive processes.

Neurometrics, as an approach, as a field, tells us that though subjective readings of electrophysiologic recordings are useful, if we are to make progress it is essential to up-grade procedures for obtaining and understanding data on brain functions. The studies presented in this volume represent ways in which this can be done. This may be the most generally significant contribution of this work. It not only shows what can be done but it establishes an objective, quantitative basis for identification and definition of brain function and dysfunction in relation to developmental maturation, normal and abnormal. In so doing, it clarifies what is happening when children learn, as well as when they do not learn. There are implications for special educators concerned with instructional programs for children who have learning disabilities, are mentally retarded or emotionally disturbed, as well as for those who have childhood aphasia and dyslexia.

Moreover, the data also elucidate what occurs in the brain when people pass through middle life and become elderly, as well as what disturbances occur when individuals encounter a variety of brain diseases. In each instance, the data are scrutinized for accuracy, for agreement with other techniques, and for the comparative findings presented. There are suggestions for new training programs, perhaps in regional settings, which would focus on the development of personnel who are qualified to use the neurometric approach. These programs would extend knowledge and encompass the applied, clinical aspects that already are showing promise for bettering understanding of those who have disabilities.

The neurometric approach, with emphasis on cluster analysis of electrocortical functions, is meaningful—not only in its use of numeric indices but in its inclusion of psychologic processes, such as sensation, perception, and cognition; because it relates neuroelectric and psychologic processes so effectively, its value and usefulness is extended immeasurably. Such knowledge is more comprehensive when we see that certain evoked potentials are basic to sensory responses, others are characteristic of perceptual responses, and still others characterize those in which the individual engages in meaningful, representational interpretation.

It is in this respect that the discussion is noteworthy. Dr. John does not report statistical data alone and omit the complex problems confronting clinicians, educators, and psychologists. He presents data within a behavioral frame of reference and analyzes the results in terms of these basic hypotheses, consistent with the highest requirements of neuroscience, of work relating brain and behavior. Those who are responsible for developing instructional and other types of remediation programs will be appreciative. In fact, in a somewhat different sense, the author has extended the use of the EEG to a greater understanding of learning and cognition. We envisage studies of maturation and aging, from infancy to adulthood, from middle age to senility. The studies reported in this volume already demonstrate that neuropsychological deficits will be more readily recognized and that persons with significant deviation will receive needed attention before the onset of debilitating consequences.

Three major studies are presented in later sections of this volume; neurometric findings for those who have sustained various neurologic diseases, for those with senility, and for learning-disabled children are discussed. Cluster analysis of neurometric indices revealed subgroups within each of these populations, previously assumed to be homogeneous. Such studies also could be made of the mentally retarded, the emotionally disturbed, the deaf, and the blind. The within-group clusters disclosed varying patterns and differences in relation to how previous experience is used in making judgments. These results reflect the brain regions participating when representational, symbolic significance is involved.

There is remarkable potential for greater comprehension of the true character of language disorders–aphasia, dyslexia, and dysgraphia. The individual's response comprises more than a physical stimulus. His response involves both short- and long-term memory, the influence of experience, and anticipatory attitudes. These data disclose the possibilities underlying the neurometric approach, in that they are derived from the assessment of cognitive processes by use of figure-ground test, cross-sensory interaction, and match-mismatch techniques, and from clarifying the role of memory and meaning in learning.

Simultaneously, the anatomical integrity of the brain is being determined because the procedure yields data both for ongoing electrocortical functions and for averaged evoked potentials. This is of consequence to those concerned with learning disabilities in children, because a metric ratio is obtained, reflecting the maturational development of each of the major cortical regions. This in turn makes it possible to follow growth patterns and deviations, as well as the effect of treatment programs over critical periods. The neurometric approach also is fundamental to the development of large scale screening programs in hospitals, mental health centers, and schools.

The results for the geriatric group opens avenues for recognizing in new ways the various needs of these persons, because now they can be more precisely diagnosed and defined. Five different neurometric clusters were revealed in this population, each with its own characteristics and, presumably, with largely varying needs for treatment and remediation. Numerical taxonomy clearly separates the senile from the normal. In those with senility, there was low interhemispheric symmetry, a discrepancy in response of the homotopic regions of the hemispheres to afferent input. The responses from the two hemispheres were poorly coordinated, with selective cognitive integration being disturbed; the hemispheric responses were inconsistent and contradictory, thus poor awareness and misunderstanding ensued.

We see too that factor analysis and multivariate statistical treatment of EEG data ensures new insights into subtle, complex brain dysfunctions. Insights hitherto unavailable introduce us to a new frontier, a new field that combines neuropsychology and medical sciences in a manner that expands the all-embracing horizon of neuroscience.

HELMER R. MYKLEBUST

Professor, Urban Education Research
University of Illinois, Chicago

Preface

For over twenty years, I have been studying the neurophysiology of learning and memory. Those studies, primarily electrophysiological in nature, convinced me that the electrical activity of the brain can reflect cognitive processes as well as responses to sensory stimulation. Although the initial observations that led me to that belief were obtained from cats (John, 1963)[1], it was not difficult to construct analogous experimental situations for human subjects. In a series of experiments with Sutton and other colleagues[2], it was possible to confirm that the waveshapes of average evoked responses from humans were not fully determined by the physical characteristics of the stimuli but also depended upon such subjective factors as set, prior expectations, and the meaning attributed to the stimulus. Those experiments provided the impetus for an active and still expanding line of research into a set of phenomena variously referred to as "late positive components" or "$P_{\overline{300}}$ processes," and the related phenomena of "readout processes" or "emitted potentials."

Although there still exists considerable disagreement as to whether these "endogenous processes" reflect the operation of brain mechanisms involved in cognitive processes or actually provide information about the content of the operation, there is by now a general consensus based upon a large body of evidence that such phenomena do in fact reveal a great deal about a person's

[1] John, E. R. Neural mechanisms of decision making. In W. S. Fields & W. Abbot (Eds.), *Information storage and neural control*. Springfield, Ill.: Charles C Thomas, 1963.

[2] Sutton, S., Tueting, P., Zubin, J., & John, E. R. Evoked potential correlates of stimulus uncertainty. *Science,* 1965, *150,* 1187–1188; Sutton, S., Tueting, P., Zubin, J., & John, E. R. Information delivery and the sensory evoked potential. *Science*, 1967, *155*, 1436–1439; John, E. R., Herrington, R. N., & Sutton, S. Effects of visual form on the evoked response. *Science*, 1967, *155*, 1439–1442.

reactions to and interpretations of present events in the context of previous experiences.

In retrospect, it seems surprising that I was so slow to realize that these electrophysiological phenomena were potentially of great practical clinical utility for the evaluation of people with cognitive disorders. They could be used as tools to probe such brain mechanisms as those concerned with focus and maintenance of attention, the memory of recent events and their use to generate expectations about the future, and the identification of meaningful information within the sensory barrage from the environment. Perhaps this lag was due to my long preoccupation with basic research issues, which may make one relatively oblivious to practical implications.

I think it was in large part the war in Vietnam that made me more aware of those practical implications, as I sought ways to become more relevant to the world around me. The mathematical and computational methods my colleagues and I had developed to quantify electrophysiological observations made obvious the need for and advantages of numerical treatment of such data. Advances in electronics made it possible for us to design and build special-purpose computing circuits that would extract critical features from electrophysiological measurements. My collaborative relationship with the group of Cuban investigators, headed by Dr. Thalia Harmony, at the Centro Nacional de Investigaciones Cientificas in Havana, gave me the initial opportunity to test the clinical utility of the neurometric approach. As those collaborative studies confirmed that the approach was practical, I decided to change the orientation of my personal research. While maintaining an active interest in basic neural mechanisms of learning and memory, I set a higher priority on the attempt to develop concrete methods to use quantitative electrophysiology for the diagnosis of brain dysfunctions in general and cognitive dysfunctions in particular.

Since my experience in clinical neurology and electroencephalography was totally inadequate, it was necessary to review the current status of knowledge about electrophysiological assessment of brain dysfunction. At the same time, I began systematic development of a system of neurometric instrumentation and techniques. This book describes the results of both of these endeavors and their application to a variety of problems.

This period of reorientation has been both intellectually challenging and rewarding. Much to my surprise and delight, I have learned that "applied" research can be just as difficult and just as "basic" as basic research. Perhaps I've learned more in the last few years than in any comparable period of my scientific development.

I owe numerous people thanks: to my friend Thalia Harmony for having enough trust in these ideas to devote untold effort to their evaluation and improvement, to my colleagues Paul Easton, Bob Laupheimer, Dan Brown, Bernie Karmel, Hansook Ahn, Frank Bartlett, Bob Thatcher, Alfredo Toro, Herb Kaye, Leslie Prichep, Eric Schwartz, and Bill Corning for their numerous creative contributions to this work, to George Brosseau for concern, interest, and help far above and beyond the call of duty of an NSF program manager, to Michaela Lobel for endless painful hours producing an accurate manuscript, and to my wife Miriam and my children who patiently endured the frequent preoccupation, fatigue, or irritability that overwork so often produces in me.

E. ROY JOHN

Brain Research Laboratories
Department of Psychiatry
New York University Medical Center

The bulk of the original research reported here was conducted while the author was affiliated with the Brain Research Laboratories of the Department of Psychiatry at New York Medical College.

1
Introduction

I. A DEFINITION OF "NEUROMETRICS"

In Volume 1 of this work, we attempted to utilize neuroanatomical, neurochemical, neurophysiological, neuropharmacological, and neuropsychological data from fundamental research in order to construct an integrated functional view of the brain mechanisms which mediate higher processes ranging from arousal and attention to subjective experience and psychopathology. In this volume, we will attempt to show how the existing state of knowledge and technique in neuroscience can be effectively applied to a variety of practical clinical problems that are presently dealt with less than adequately. Traditionally, clinical electroencephalography has been one of the major techniques by which our knowledge of neuroscience has been brought to bear upon these problems. The utility of this technique has been sharply limited and constrained by reliance upon qualitative interpretation of electrophysiological observations. In contrast, the approach which we propose is based upon quantitative measurements of salient features extracted from electrophysiological data which reflect various aspects of brain function related to sensory, perceptual and cognitive processes as well as to the structural and functional integrity of different neuroanatomical systems. We will call this quantitative approach "neurometrics."[1]

Much of our discussion of this new neurometric approach depends upon thorough familiarity with facts, methods and concepts presented in Volume 1. Wherever possible, we have endeavored to refer the reader to the most relevant

[1] We wish to acknowledge our indebtedness to our colleague, Dr. B. Z. Karmel, for independently coining this apt term to describe this approach during one of our many discussions together about our joint studies (see Chapter 8).

sections of Volume 1. We have been constrained from presenting overlapping discussions in both volumes by practical reasons and apologize to the reader if it is sometimes necessary to refer to the other volume for complete understanding of the presentation.

II. FUNCTIONAL INSIGHTS AVAILABLE FROM SCALP RECORDINGS

A great variety of information about brain functions can be obtained by analyzing the weak electrical activity that can be recorded from the intact human scalp. Not only does this electrical activity contain diagnostically valuable information about neuropathology but it reveals many phenomena related to sensory, perceptual, and cognitive processes that are otherwise extremely difficult to measure in an objective manner. This information is reflected in two major aspects of the electrical activity: (1) the spontaneous fluctuations of voltages generated by the brain comprising the electroencephalogram, or EEG. The physiological origin of the EEG has been described in Chapter 2 of Volume 1; and (2) the evoked fluctuations of voltages occurring after presentation of a sensory stimulus reflect the responses of different brain regions to the afferent input. The physiological origin of evoked potentials (EP) has been discussed in Chapters 3 and 9, and some of the insights provided by EPs have also been presented in Chapters 9 and 10 of Volume 1.

In this Introduction to Volume 2, we will examine a variety of clinical applications of the information derived from these two kinds of electrical activity. It is our intention neither to provide a manual for interpretation of the conventional EEG nor an exhaustive review of the literature on sensory evoked potentials. Such reference works of excellent quality already exist, and will be cited where appropriate. Rather, it is our intention to provide the reader with a general overview of the kinds of information available, how they can be obtained, and the uses to which they are generally put, so that he will more fully comprehend the basis of the quantitative computer methods presented in later chapters, as well as their practical utility.

The clinical utilization of electrophysiological measures has become the almost exclusive prerogative of the neurologist who evaluates the EEG to assess neuropathology. Such clinical practice still depends overwhelmingly on the experience of the skilled electroencephalographer who diagnoses a variety of brain disorders by visual assessment of particular features of the EEG. In the last fifteen years, there has been a great expansion of our knowledge about the electrical activity of the brain, particularly as a result of the development of minicomputers and their application to the precise analysis of electrophysiological processes. Although a vast amount of information has been accumulated during this period, such information has had little impact on clinical practice. Not only has there been a lag in the application of more objective and precise methods of EEG evaluation to such traditional problems as the assessment of neuropathology, but there has been little attempt by psychologists, psychiatrists, pediatricians, and those physicians concerned with visual or auditory function to take advantage of the insights into sensory, perceptual, and cognitive processes which can be obtained by electrophysiological measurements. Such insights can be invaluable in the diagnosis and treatment of certain kinds of patients and are often difficult or impossible to obtain if one relies upon the verbal cooperation of the patient.

All too often, the pediatrician, psychologist, psychiatrist, or other practitioner is forced to draw inferences about subtle nuances of sensory, perceptual, and cognitive function by interpreting the behavior of the patient, the observations of family members or teachers, or the results of psychometric tests that emphasize product rather than process, often rely upon language, and are severely "culture-bound."

III. LIMITATIONS OF QUALITATIVE ANALYSIS OF BRAIN ELECTRICAL ACTIVITY

There are many reasons to be dissatisfied with this situation. The number of skilled neurologists and encephalographers in every country of the world is grossly inadequate. For example, there are about 3,000 neurologists, including only 250 child neurologists, in the United States, which has a population of about 65 million children aged 15 years and under. Outside major cities, most locales are completely unable to benefit from electroencephalography. Patients presenting symptoms suggesting the presence of neurological diseases are often not submitted to electroencephalographic examination until those symptoms have persisted for some time and have become sufficiently distressing to occasion referral to an electroencephalographer or neurologist located in the nearest major community. The attendant delay may well have adverse consequences for the patient, if not in ultimate prognosis then at least in terms of distress in the intervening period.

Proper evaluation of the EEG takes not only skill but a substantial amount of time; therefore, even in major cities competent EEG examinations are not readily available. Furthermore, because of differences in experience and judgment, not all electroencephalographers evaluate EEGs in the same way. Most serious practitioners of other specialties are forced to depend upon the neurologist for direct assessment of the adequacy of brain function, but the methods of the neurologist are not optimally devised for the objective assessment of subtle nuances of information process in the brain.

IV. THE NEUROMETRIC ALTERNATIVE

With the advent of minicomputers and the development of several new mathematical techniques, it has become possible to obtain rapid, objective, and detailed quantitative analysis of electrophysiological phenomena related to many aspects of brain function. This neurometric methodology promises a number of important advances:

1. Assessment of neuropathology can become much more readily available, and perhaps significantly more sensitive, than has hitherto been the case. Assessment of the amount of damage and prognosis of recovery based upon repeated evaluations after short intervals may be of major utility in decisions about the treatment of patients after strokes or head injury.
2. Development abnormalities reflecting such factors as perinatal trauma, inadequate sensory functions, or the absence of various capacities expected to emerge with maturation can now be recognized early and objectively. Such early identification of brain dysfunction often enables interventions which minimize psychological trauma to both child and parent, and provides valuable bases for guidance of pharmacotherapy, remedial treatment, and rehabilitation;
3. Alterations in brain states caused by centrally active drugs can be described in precise mathematical terms. A quantitative nomenclature for psychoactive drugs can be envisaged, which would provide a measure of the changes in excitability levels and reactivity to stimuli in different brain regions by analysis of the EEG and evoked response changes resulting from medication with such agents. Such a catalog of the magnitude and locale of drug effects might provide a preliminary bioassay method for drug screening by comparison of the profiles of effects of new versus familiar compounds.
4. Changes in brain states related to particular diseases can be quantified. Such assessments may reveal characteristic features of specific disorders, and may permit selection of drugs most appropriate for treatment of the individual patient. Reexamination after initiation of a course of medication might permit adjustment of dosage or indicate the advisability of changing the initial treatment. Such procedures might offer a very useful adjunct to the pharmacotherapy of psychiatric and metabolic disorders. Neurometric assessment may also permit early identification of certain disease processes, permitting more effective intervention. This might be especially useful in detection of brain changes related to cognitive impairment resulting from chronic brain syndrome or senile deterioration.

It can be expected that as practitioners in the diverse fields in which neurometric evaluations would be valuable become cognizant of their availability and potential utility, large service centers will come into existence which will provide the necessary analyses automatically and economically. Computer centers for automatic assessment of brain functions reflected in electrical processes will become an important adjunct to practice in a number of specialties in addition to neurology itself. Recognizing this, plans for such centers are already being elaborated in certain countries, for example, in Sweden (Petersén, 1973). Small general purpose minicomputers and special purpose diagnostic computers already are available which provide the necessary measures in a fashion adequate for routine use. Personnel to operate such centers, to interpret the results of computer assessment of brain function, and to implement treatment based upon such evaluations will be needed. It is our hope that this book will provide an impetus for the training of such neurometricians.

2

Diagnostic Electrophysiology

The purpose of this chapter is to provide an overview of the present status of knowledge about the diagnostic utility of qualitative electrophysiological measures. Later chapters will deal with the techniques of rapid automatic computer implementation of quantitative neurometric indices and results obtained by applying neurometric techniques to a variety of clinical problems.

I. EEG ASSESSMENT OF NEUROPATHOLOGY

A survey of the value of the EEG in various fields of medicine can be found in Volume 1 of the *Handbook of Electroencephalography and Clinical Neurophysiology* (Rémond, 1971). It has long been recognized that examination of the patterns of electrical voltages recordable from electrodes affixed to the human scalp by conductive paste or inserted into the scalp as needles can provide a wealth of information about the integrity of the brain generating those voltages. Such examination is based upon the recognition of particular features of waveform and frequency composition characteristic of normal function as well as of particular neurological disorders. The interested reader can find exhaustive reviews of various aspects of the interpretation of the EEG in numerous volumes, for example, Hill and Parr (1963), and atlases of the waveshapes characteristically found in adults (Gibbs & Gibbs, 1964) and in children (Kellaway & Petersén, 1964) are available. Only a general introduction to principles of EEG analysis will be provided here.

A. EEG Recording Methods

1. Electrode Placement

Since the EEG electrode on the scalp is primarily responsive to the electrical activity of the subjacent cortex and its transactions with other brain regions, it is obvious that the information available from an EEG examination will depend substantially upon the loci selected for electrode placement. Many systems for electrode placement have been proposed, each with its own rationale. Some years ago, it was considered essential to define a standard set of electrode placements so that results obtained in different clinics and laboratories could be compared. Accordingly, the so-called international 10/20 system for EEG electrode placement was devised (Jasper, 1958).

The goal of this system is to define a set of relative electrode placements whose position will be determined by the dimensions of the head of each individual patient and which will nonetheless place electrodes in comparable positions on heads of differing dimension. Basically, this system constructs a set of arcs of "latitude" and "longitude" upon the surface of an approximately spherical cranium. Arc lengths from nasion to inion and from left ear over vertex to right ear are measured. Electrodes are then placed at points 10, 20, 20, 20, 20, and 10% along each of these arcs, from which the name of the 10/20 system derives. Using the points thus defined, additional archs are located and electrodes are placed at particular intersections and positions along these arcs. If the arc lengths are measured with a flexible tape and the 20% interval then set on a pair of blunt tipped calipers, the appropriate loci can be marked with remarkable rapidity by an experienced technician. Application of the full 10/20 electrode set can be completed in about 10 minutes.

In principle, this system enables one to place electrodes over corresponding brain regions in patients with heads of various sizes and shapes. In practice, although it has the merit of unequivocally defining an electrode location on the scalp, it neither accomplishes the goal of reproducible placement over particular brain regions nor does it necessarily place electrodes where anatomic considerations might suggest optimal information might be obtained with minimum redundancy.

2. Recording Method

Electrodes are applied in a variety of ways, with subcutaneous needles, collodion, or electrode paste. We find needle electrodes distressing to many patients, particularly children, and the use of collodion is time-consuming and accompanied by odors offensive to many subjects. Therefore, we have found it most convenient and least objectionable to place electrodes over a small mound of thick electrode paste (a variety of commercial pastes are available which serve excellently) which is sufficiently adherent to hold the electrode firmly while providing good electrical contact.

With modern EEG amplifiers, electrode–skin impedance as high as 20,000 Ω can be tolerated. If the oil on the scalp is removed with a cotton swab soaked in alcohol, little difficulty is encountered in obtaining impedances below 5,000 Ω. With some pastes, this impedance actually drops lower after the electrode has been in place for a few minutes. High impedance contacts increase the amount of 60-Hz "noise" introduced into the EEG record by environmental electrical fields. Most modern EEG amplifiers permit rejection of most or all of this noise by a combination of high "common-mode" rejection ratios and built-in 60-Hz notch filters, so that artifact-free records can be readily obtained without the need for special electromagnetically shielded recording chambers that were previously required.

Whenever possible, it is highly desirable to obtain electromagnetic tape recordings of EEG data both for subsequent "off-line" computer analysis and for a permanent record which can be dynamically reproduced for purposes of longitudinal comparisons of the development of features of interest. Such tape recorders, which generally utilize FM recording techniques to permit accurate registration of the low-frequency EEG signals while minimizing electrical noise, are now available in economic and compact cassette form.

3. Recording Derivations or "Montages"

Two basic methods of recording exist:

1. *Referential* or *monopolar* recording, in which the EEG voltages detected by an electrode placed over some brain region are compared to a distant, electrophysiologically inactive reference electrode. Linked earlobes, linked mastoid processes, or the tip of the nose are among the commonly used reference loci. Although each of these suffers from its own peculiar shortcomings in terms of muscle artifacts, volume conducted potentials, and movement artifacts, linking the earlobes is usually preferred. When in doubt as to whether the reference electrodes are truly inactive, especially in average evoked potential studies (discussed below), it is desirable to compare two reference loci directly.

The advantage of referential recording is that the record, given sufficient amplification, will reveal the presence of spontaneous EEG or evoked activity originating a substantial distance from the location of the active electrode. It is also sensitive to potentials of local origin. The major disadvantage of monopolar records is that they offer little utility for precise localization of an electrographic phenomenon.

2. To overcome this shortcoming, *bipolar* recording is used. In bipolar recording, voltage fluctuations occurring in phase at two active electrodes are excluded by common-mode rejection; that is, a so-called *push–pull* or differential amplifier is responsive to the difference in voltage between the two electrodes, and does not respond to voltage oscillations that occur in common at both electrodes. Since this ability depends upon the accuracy with which the two halves of the differential amplifier are balanced, it is conventional to evaluate EEG amplifiers in terms of their *common-mode rejection ratio*, which is a measure of the output ratio between a signal out of phase at the two electrodes and a signal which is in phase. In general, the higher the common-mode rejection ratio, the less will the amplifier be susceptible to artifact from ambient electrical fields and the better will be the information it provides about electrical events occurring differently near one or the other recording electrode.

Therefore, *bipolar* recordings offer the advantage of more precise localization of electrophysiological events. They suffer from the shortcoming that true voltage swings taking place under both electrodes at the same time will be obscured or not detected at all. For this reason, both monopolar and bipolar recording arrays, or *montages*, are conventionally examined. The usual EEG scan, or "run," consists of a systematic examination of monopolar electrodes, of bipolar pairs oriented in the sagittal (anterio–posterior) plane, and of bipolar pairs oriented in the coronal (left–right) plane. Since the 10/20 system includes 19 electrodes, a total of 57 derivations constitutes an exhaustive scan of the array. Usually, a far lesser number of derivations is examined because of the limited number of amplifier channels available.

Recordings are conventionally obtained from patients seated in a relaxed fashion or reclining, with eyes open and with eyes closed. Data are obtained of the spontaneous EEG, of the EEG during repetitive photic stimulation or "driving" intended to provide an estimate of reactivity and to provoke epileptiform discharge, and during and after a period of hyperventilation, which frequently evokes certain characteristic signs of *petit mal* epilepsy.

B. Principles of EEG Analysis

In order for the reader to understand the basic principles of automatic computer analysis to be presented in later chapters, a brief summary of the major features upon which visual analysis of the EEG is based will be presented. In visual analysis, the factors evaluated include: frequency composition of the EEG signal, voltage (amplitude), brain region from which the particular signal features are obtained, peculiarities of waveform, interhemispheric coherence (symmetry and synchrony of amplitude and frequency), duration and envelope of recurrence of series of waves at particular frequencies, relations between voltage and frequency, and reactivity. Ideally, automatic analyses must take all of these parameters into account. In practice, some of these parameters are of significant utility far more often than others. By this we mean that a few of these electrophysiological parameters are markedly disturbed by head trauma, space-occupying or irritative lesions, and cerebrovascular disease or accidents, which constitute the great majority of neurological disorders (Gibbs & Gibbs, 1964; Merritt, 1973). These parameters are the frequency distribution or spectrum of the EEG signal characteristic of particular brain regions, the relationship between frequency and voltage, the absolute and relative amplitudes of the EEG, and the coherence or symmetry of amplitude, waveshape, and phase between bilaterally symmetric or homologous electrode placements. Features such as the specific waveform of potentials, the rate of change of voltage or frequency, and recognition of specific electrographic patterns or sequences, while of indisputable value in differential diagnosis, are of decisive importance in the identification of a relatively small percentage of abnormalities. While these latter parameters are potentially within the scope of automatic computer recognition, those parameters which are most critical for the diagnosis of the majority of neurological disorders producing clear EEG abnormalities are already substantially detectable by automatic quantitative computer analysis, as is discussed in Chapter 5.

C. The Accuracy of EEG Detection of Neuropathology

1. Incidence of Major Neurological Disease

It is difficult to obtain an accurate assessment of the relative incidence of the different neurological disease. Percentages vary substantially depending upon whether autopsy statistics or diagnostic statistics of a typical neurological service are utilized. The data in Table 2.1 are a composite taken from several sources, particularly Merritt (1973) and NINDS (1973) and can only be regarded as a rough approximation.

a. Head injury. The single greatest cause of neurological disorder seems to be cerebral trauma caused by automobile accidents, estimated to cause head injury in more than 3 million persons each year, with about 50,000 fatalities. With traumatic head injury of almost any sort, there is usually a change in the electrical activity of the cerebral cortex. These changes have been systematically reviewed by Courjon (1972). The electrical activity gradually returns to normal with recovery from the damage, after going through a phase in which there is generalized slowing and increase in voltage. Areas of focal damage may show evidence of abnormal activity (localized slowing) for many weeks or months after injury, with epileptiform-like spike activity often appearing. Thus, quantitative assessment of the EEG might provide an index of the severity of cerebral trauma in auto accidents or other head injury, and might serve as a useful measure of the rate of recovery or deterioration in such patients.

b. Paroxysmal disorders. (i) Migraine Headaches.. About 5 million persons in the United States suffer from migraine attacks, generally considered to arise from traction or distension of the cerebral blood vessels, or inflammation or pressure on the cranial or cervical nerves. Although the etiology of migraine headaches is not understood, the possibility that such headaches are associated with a developing brain tumor requires careful evaluation to exclude that possibility. Only rarely are migraine headaches accompanied by clear abnormalities in the EEG in the interval between headaches (Niedermeyer, 1972).

(ii) Epilepsy. Another paroxysmal disorder of the brain, but one with clear and often unequivocal EEG features, is epilepsy. Epilepsies are sometimes divided into symptomatic epilepsy, for which a known causative agent can be identified and idiopathic epilepsy, in which no lesion or other plausible agent can be demonstrated. A more common classification method is to separate patients into different groups according to the symptoms manifested.

TABLE 2.1
Incidence of Major Types of Neurological Disorder[a]

Etiology	Approximate incidence (cases/year)	Common EEG signs
Cerebral trauma	3,000,000	Diffuse or focal slow waves
Migraine headache	5,000,000	No characteristic EEG signs
Epilepsy	2,000,000	Spikes, excess slow or fast waves
Developmental, congenital, and degenerative defects	6,250,000 children (ages 1–10)	Range from normal to generalized or focal slow waves, no consistent signs
Infectious disease	1,000,000[b]	Sometimes accompanied by slow waves during acute phase
Cerebrovascular disease	575,000[c]	Diffuse or focal slow waves
Intracranial tumors	140,000	Diffuse or focal slow waves

[a]Based upon H. H. Merritt, *A textbook of neurology* (5th ed.). Philadelphia: Lea & Febiger, 1973; and *Neurological and Sensory Disabilities,* DHEW Publication No. 73-152, Information Office, National Institute of Neurological Diseases and Stroke, 1973.

[b]This is a rough estimate because of wide fluctuations in use of antibiotics, variability in local statistics, and sporadic occurrence of epidemics, especially of measles, rubella (or German measles), mumps, and influenza, all of which may be accompanied by an encephalomyelitis.

[c]Note that this is an obvious underestimate of the rate of occurrence of cerebrovascular disease, if it is compared with the 25% rate in autopsy findings. It is an estimate of the incidence of debilitating cerebrovascular disease.

Draft board medical statistics indicate that about .5% of the United States population, or about 1,250,000 persons, suffer from epilepsy. Other estimates go over 2 million (NINDS, 1973). In 77% of the cases, no apparent cause can be identified; 5.7% are associated with a history of head injury, 5.6% with a history of birth injury, and the remainder with histories including developmental defects, infectious diseases, tumors and abscesses, degenerative diseases, metabolic disturbances, and intoxication with a variety of substances. Four major forms of epilepsy have been defined, and they may be encountered singly or in combination in the same patient: (1) *grand mal*, characterized by generalized convulsions; (2) *focal seizures*, characterized by a systematic anatomic progression of convulsion or by a focal onset leading to a generalized seizure; (3) *petit mal*, characterized by brief muscle spasms, arrest of activity for a brief period, and momentary "absences", and (4) psychomotor attacks, characterized by the domination of particular alterations in consciousness, complex feeling states, and absences or partial amnesia. A fifth form is infrequently encountered, often referred to as "atypical" or psychic equivalent attacks, characterized by prolonged periods of automatic behavior and mental cloudiness with amnesia for the whole period. Approximately 70% of all epilepsies are classified as grand mal, 7% as focal, 10% as petit mal, 10% as psychomotor, and 3% as epileptic equivalents. A very high percentage of epilepsies are accompanied by marked EEG changes detectable in the interval between overt seizures (see Table 2.3, adapted from Gibbs & Gibbs, 1964).

c. Developmental and degenerative defects. About 5 million children in the United States of elementary school age or younger suffer some form of developmental defect involving the brain. The largest single group of these children suffer from so-called minimal brain dysfunction, or MBD, discussed more extensively in Chapter 8. Estimates of the incidence of MBD in the United States range from 5 to 14%. There are about one million mentally retarded children in this age range. Congenital, hereditary, and degenerative diseases, of which Down's syndrome (sometimes called mongolism) and cerebral palsy are perhaps the major entities, account for approximately 750,000 more neurologically impaired children. Children suffering from these diseases often show severe intellectual deficit, speech difficulties, and impairment of movement. There are a total of about 500,000 cases of multiple sclerosis, a disease characterized by the appearance of numerous areas of demyelination in the brain accompanied by a variety of neurological symptoms. It is especially noteworthy that the incidence of intracranial hemorrhage and subdural hematoma resulting from complications in obstetrical delivery and postnatal trauma is estimated to be between 2 and 3%. Merritt (1973) states that this is obviously an underestimate. It is interesting that the incidence of subdural hematomas found among autopsies of psychotic patients was 8%. Many hereditary, congenital, and perinatal diseases are accompanied by marked EEG abnormalities, recently reviewed by Dreyfus-Brisac and Ellingson (1972).

d. Cerebrovascular disease. Cerebrovascular disease is found in about 25% of all autopsies. Estimation of the fraction of those individuals who suffered significant functional impairment from this condition is impossible. About 300,000 persons are disabled by cerebrovascular accidents and an additional 210,000 die from this cause in the United States each year. Roughly 70% of cerebrovascular disease is due to occlusion (thrombosis) of a cerebral vessel, 20% is due to intracerebral hemorrhage, and 10% is due to embolism. Patients are usually asymptomatic until the occlusion occurs, although transient ischemic episodes are sometimes noted.

A high percentage of victimes of cerebrovascular accidents show abnormalities in the EEG (see review by Van Der Drift, 1972). The two most consistent electrophysiological indicators of cerebrovascular abnormalities are: (1) asymmetry of EEG recorded from homologous derivations (Epstein & Lennox, 1949; Jones & Bagchi, 1951; Lavy & Bental, 1964; Omae *et al.*, 1969) and (2) increased energy of spontaneous EEG in the slow frequency range (Birchfield *et al.*, 1959; Cohn *et al.*, 1948; Lavy *et al.*, 1964; Rohmer *et al.*, 1952; Roseman *et al.*, 1952). Specific electroencephalic phenomena associated with pathology are spike and dome foci (Cohn *et al.*, 1948; Omae *et al.*, 1969) and large delta waves (Marguardsen and Harvald, 1964; McDowell *et al.*, 1959) which are frequently related to the location of the disturbance.

Even though there is agreement that these two categories of phenomena are correlated with pathology there is considerable confusion regarding the reliability, sensitivity, and the ability of these two measures to discriminate between various forms of pathology. For example, with standard EEG measures the percentage of patients with confirmed vascular lesions exhibiting normal EEG ranges from 60% to 13% (Birchfield *et al.*, 1959; Carmon *et al.*, 1966; Lavy & Bental, 1964; Van Buskirk & Zarling, 1951; Strauss & Greenstein, 1948; Zfass & Hoefer, 1950). The inconsistency of subjective EEG measures prevents accurate discrimination between the different types of pathologies. For instance, the percentage of cases showing slowing of EEG or focal electrical abnormalities in patients with occlusive vascular lesions varies from 100% (Jones & Bagchi, 1951) to 60% (Zfass & Hoefer, 1950) to only 37% in a study by Harvald and Skinhoj (1956). In nonocclusive cerebrovascular lesions, disturbances in the EEG vary from 60% of the patients by Frantzen *et al.* (1959) to 20% in a similar study by Harvald and Skinhoj (1956).

Frantzen *et al.* (1959) and Carmon *et al.* (1966) found that the incidence of diffusely disturbed EEG was highest among patients with diffuse arteriosclerotic changes. Unilateral slowing or depression of cortical activity is frequently correlated with hemispheric infarction or hemorrhage (Carmon *et al.*, 1966; Cress & Gibbs, 1948; Lavy *et al.*, 1964; Martin, 1953; Rohmer *et al.*, 1952). However, correlations between subjective EEG indicators and angiographic analysis are poor (Carmon *et al.*, 1966).

The above studies demonstrate that even though clear EEG changes are frequently associated with cerebral vascular lesions, the degree of false negatives is high while the ability of the EEG to discriminate various lesions responsible for stroke is limited. These limitations are due, in part, to the different criteria used in the evaluation of the EEG as well as an absence of quantitative measures sensitive enough to detect subtle electrophysiological changes. Quantitative frequency analysis may serve as an index of functional impairment of the brain due to occlusion or hemorrhage in cerebral vessels, may indicate the extent of residual function available for rehabilitation, and may serve as a prognostic index for the rate and amount of recovery. Further, sufficiently sensitive measures may detect gradual onset of ischemia due to atherosclerosis or decrease of the internal diameter of cerebral vessels, as well as asymptomatic mini-strokes, that is, cerebrovascular accidents with no lasting or significant functional signs. Such measures may identify the development of conditions making a person susceptible to cerebrovascular accidents and thereby serve for the initiation of preventive therapy.

e. Intracranial tumors. Intracranial tumors are found in about 2% of all autopsies. Tumor cases constitute a large proportion of admissions to the typical neurological service than any other diseases of the brain with the exception of cerebrovascular disease and acute infections. About 10–20% of all brain tumors are metastatic in origin, usually (~80%) from lung cancer. It is estimated that about 22,000 people die in the United States each year from brain tumors. The annual incidence of brain tumors in the United States is difficult to ascertain.

As with cerebrovascular disease, quantitative EEG analysis may permit early detection of brain tumors, with a substantial improvement of the effectiveness of brain surgery. Many tumors cause significant changes in the EEG, particularly in the form of increased slow activity (Magnus, Van Leeuwen, & Cobb, 1961). As discussed later in this chapter, in the section on the utility of evoked potentials for clinical diagnosis, in some cases averaged evoked response (AER) techniques have demonstrated tumor-related abnormalities before the onset of clinical symptoms or EEG signs.

f. Infectious diseases. A wide variety of infectious diseases can disturb brain function. The major types of brain infections are acute and subacute meningitis, infections of the veins and sinuses, brain abscesses, virus infections including a number of types of encephalitis, and syphilis. EEG signs of brain involvement are increasingly clear with rise in temperature. Generally, these signs subside as the acute phase of the disease passes. In general, the EEG is a good indicator of infection or inflammation in the vicinity of cortical neurons – Gibbs and Gibbs (1964) believe that if the EEG remains normal during an acute febrile illness in childhood, the prognosis for complete recovery without residual deficits in brain function is excellent. Spread of the infection to the brain substance is usually accompanied by slowing of the EEG and occasionally by the appearance of spike activity. However, Gibbs and Gibbs note that a high percentage of children (68%) who showed persistent EEG abnormalities after an acute infectious disease subsequently behaved in a fashion that elicited concern from parents about their intellectual development or behavior. Vague though this report may be, it strongly suggests the utility of quantitative EEG examination for any child after an episode of acute infectious disease, especially if accompanied by high fever.

2. *Incidence of EEG Abnormalities in Major Neurological Diseases*

Few neurological services have compiled and published a sufficiently large body of case material to permit the construction of reasonably accurate estimates as to the relative incidence of different brain diseases among cases examined in an EEG clinic, the incidence of false negatives as well as correct diagnosis of abnormality for each different etiology, and the incidence of false positives in a large control population. Unless such data are taken from a single clinic, where the same group of electroencephalographers develop and use similar diagnostic criteria, variabilities between criteria might invalidate any conclusions. The three-volume *Atlas of Electroencephalography* (Gibbs & Gibbs, 1964) is perhaps a unique source for such information. By collating data from different sections of those volumes, it was possible for us to construct Table 2.2. It must be appreciated that the validity of the subsequent analysis rests entirely upon the clinical accuracy of the diagnosticians whose conclusions form the basis for the classifications of clinical etiologies and EEG abnormalities.

In Table 2.2, we have first listed the absolute number and percentage of each of the major etiologies in the Gibbs' diagnostic classification (adapted from Gibbs & Gibbs, 1964, Volume 3, Table 1, page 9). We then estimated the percentage of EEG abnormalities found in each of these categories by studying the detailed results of examinations on these various subgroups as reported in that volume. The incidence of false positives was similarly obtained. By multiplying the percentage of incidence of each disease by the percentage of incidence of EEG abnormality for that disease, we obtained an estimate of the relative contribution of different etiologies of disease to the total incidence of EEG abnormality.

Several interesting "facts," or first approximations, emerge from this exercise: first, the incidence of EEG abnormality in 38,082 cases on Gibbs' Neurological Service was 61.5%, with 39.5% false negatives; second, the incidence of false positive findings of ostensible EEG abnormality among a clinically healthy population of 3,476 individuals was 12% (note that about two-thirds of these false positives displayed so-called "14 and 6 positive spikes per second," a feature about which there has been much controversy since many electroencephalographers do not consider it abnormal); third, 38% of 1,228 children with apparently uncomplicated childhood diseases showed an EEG abnormality.

Inspection of Table 2.2 reveals an apparent discrepancy: the incidence of EEG abnormality in the major diseases obtained by calculating the incidence for each separate disease and summating (23.5%) is markedly less than that reported by

TABLE 2.2
Incidence of Disease and EEG Abnormality, for Major Etiologies[a]

Etiology	N	Incidence of disease (%)	Incidence of EEG abnormality (%)	EEG normal (%)	Total EEG abnormalities related to particular etiology (%)
Cerebral trauma	5,851	15.4	31	69	4.7
Chronic & acute encephalitis	2,714	7.2	20	80	1.5
Perinatal trauma	2,027	5.3	80	20	4.0
Febrile illness	2,020	5.3	38	62	2.0
Cerebrovascular disease	1,458	3.8	85	15	3.2
Premature birth	908	2.4	70	30	1.7
Miscellaneous neurological diseases	867	2.3	60[b]	40	1.3
Intoxications	847	2.3	75[b]	25	1.8
Cortical tumors	405	1.2	90	10	1.0
Tumors	345	0.8	55	45	0.5

Developmental defects	696	1.8	34[b]	66	0.6
Poliomyelitis	450	1.2	41	59	0.5
Meningitis, acute & chronic	480	1.3	50	50	0.6
Miscellaneous medical & metabolic	350	0.9	13[b]	87	0.1
Unknown disease (see Table 2.3)	18,664	49.1			
Total:	38,082		61.5	39.5	23.5
Uncomplicated childhood diseases[c]	1,228		38	62	
No significant disease	3,476		12[d]	88	

[a]Based upon Gibbs & Gibbs, *Atlas of electroencephalography* (Vols. 2 & 3). Addison-Wesley, 1964.

[b]Approximate average across conditions taking incidence into account.

[c]Mumps, measles, chicken pox, scarlet fever, rubella with slight to greatly elevated rectal temperatures.

[d]Includes 8.3% of normals who displayed 14 & 6 positive spikes/sec, a finding considered normal by many electroencephalographers.

Gibbs and Gibbs for the full group of 38,082 patients (61.5%). This discrepancy probably arises from the fact that 18,664 patients with "unknown diseases" are included in Table 2.2 but not in our calculation. Examination of the table in Gibbs and Gibbs (1964), from which these data were compiled, shows that about 25,000 of the 38,082 patients had epileptic seizures or were considered at risk for epilepsy. Most of those patients are in the "unknown etiology" category.

Table 2.3 presents the data for incidence of EEG abnormalities in all forms of epilepsy. Multiplying the percentage of incidence of each form by the percentage of incidence of EEG abnormality for that form, and summing the resulting values, it is possible to reach the estimate that 65.2% of all forms of epilepsy are accompanied by EEG abnormalities, which are apparent in the sleep record, while 28.7% can be detected in the waking record.

These findings justify the assertions that (1) the visual inspection of the EEG is of great utility in the diagnosis of a wide variety of brain diseases; (2) even if EEG diagnosis of the outstanding caliber represented by this team was available in every community, the incidence of false negatives permits a great deal of improvement; (3) the incidence of false or equivocal positives among the normal population is undesirably high; and (4) the incidence of EEG abnormality among children with common childhood diseases is high enough to warrant routine EEG surveillance of children as soon as possible after such diseases, to establish whether they are at risk for significant brain damage or dysfunction. If we take these conclusions into consideration, and if we realize from Table 2.1 that over 15 million cases of neurological disorder exist in the United States in a typical year, whereas only about 2 million EEGs are administered (based upon statistics published by the American Hospital Association, 1972), then it becomes clear that there exists a significant need for the development of automatic computer centers for the evaluation of EEG data. Such centers would both improve the accuracy of conventional EEG evaluation and make quantitative assessment of electrophysiological activity reflecting various aspects of brain function as well as disease processes available to the great majority of persons who might benefit from such assessments and presently have no access to them.

II. DETECTION OF NEUROPATHOLOGY USING SYMMETRY OF AVERAGED EVOKED RESPONSES (AER)

When a sensory stimulus is presented to a human subject, a transient oscillation of voltage occurs in the EEG recorded from electrodes over responsive brain regions, which often is obscured by other ongoing activity. This oscillation, or evoked response, represents the response of the brain to the sensory stimulus and occurs at a latency determined by the central transmission time of the sensory system that was stimulated. Basic mechanisms involved in generation of the evoked response, and the relationship between the evoked response and the EEG, were discussed in Chapter 2 of Volume 1. By use of an averaging

TABLE 2.3
Incidence of EEG Abnormalities in Various Forms of Epilepsy[a]

		Abnormal EEG (%)		Abnormality due to each type (%)	
Clinical signs	Relative incidence (%)	Awake	Asleep	Awake	Asleep
Grand mal	48	22	46	10.5	22.1
Grand mal & psychomotor	15	32	85	4.8	12.8
Grand & petit mal	6	75	94	4.5	5.6
Psychomotor	6	32	88	1.9	5.4
Jacksonian	5	35	65	1.8	3.3
Focal	4	38	70	1.5	2.8
Petit mal	4	84	90	3.3	3.6
Other	12	45+	80[b]	5.4	9.6
Total: 100				28.7	65.2

[a]Compiled from Gibbs & Gibbs, 1964.
[b]Based upon average of all other categories.

technique, often implemented by a special-purpose average response computer, the details of the waveshape of the voltage oscillations that are time-locked to the delivery of the sensory stimulus can be ascertained from the *averaged evoked response* (AER). The waveshape of the AER reflects the anatomy of the responding systems, the characteristics of the stimulus, and certain dynamic factors to be discussed later.

Since the advent of the average response computer, numerous investigators have discussed the characteristics of the AER in various types of neuropathology, and have drawn inferences about brain function based upon deviations of certain components of the waveshape from some expected normal contour. That such AER methods might be used for diagnostic purposes have won remarkably little acceptance in routine clinical practice, however, perhaps because assessment of the AER waveshape depends on visual recognition of patterns by the practitioner, and thus the technique suffers from some of the same shortcomings as qualitative evaluation of the EEG. Vaughan and Katzman (1964), Ciurea and Crighel (1967), Oosterhuis *et al.* (1969), and Jonkman (1967) have demonstrated significant correlations in the asymmetry of evoked potential (EP) amplitude and component latencies in patients with cerebral vascular lesions. Ciganek (1961), Kooi and Bagchi (1964), Kooi *et al.* (1965), Vaughan *et al.* (1963), and Gastaut *et al.* (1964) have demonstrated differences, particularly in regard to the latency, presence, or absence of the primary response of the EP in tumor and cerebral vascular lesion (CVA) patients. Unfortunately, peaks I and II are absent in 15 to 20% of normal people (Ciganek, 1961; Oosterhuis *et al.*, 1969; Vaughan *et al.*, 1963). False positive judgments may therefore arise if evaluation is restricted to the early components. A more reliable component to discriminate neuropathologies is wave III of the EP (Jonkman, 1967; Oosterhuis *et al.*, 1969). However, even this wave is unreliable since wave III is known to change with age (Oosterhuis *et al.*, 1969). Furthermore, intersubject variability is such that only rarely can a single AER be identified as abnormal (Oosterhuis *et al.*, 1969).

Evaluation of the symmetry of the AER from homologous derivations partially circumvents the issue of how to define a "normal" waveshape (although we shall return to this problem in our discussion of factor analytic and pattern recognition techniques in later chapters). Bilateral asymmetry provides a far less equivocal way to assess neuropathology because each individual serves as his own control. Many disease processes cause unilateral changes in AER waveshape. A number of workers (Bergamini & Bergamasco, 1967; Crighel & Poilici, 1968; Jonkman, 1967; Oosterhuis *et al.*, 1969; Vaughan & Katzman, 1964) have reported that asymmetry of latency or amplitude in averaged evoked responses to light (visual evoked responses, or VER) simultaneously computed from homologous derivations is an extremely sensitive index of brain damage due to tumor or cerebrovascular disease. Positive results have been reported in some patients in the presence of tumors which, because of their nature and localization, had not yet produced subjective distress, clinical signs, or EEG changes

(Jonkman, 1967). Asymmetric VERs can be the result of space occupying lesions restricted to nonspecific cortical areas (for example, frontal tumors) or to subthalamic and thalamic nonsensory-specific structures (Bergamini & Bergamasco, 1967; Crighel & Poilici, 1968) as well as to portions of the visual system itself. Thus, the visual evoked response reflects in its amplitude and morphology an extensive set of transactions between structures of the visual system and many other regions of the brain. Localized damage or otherwise abnormal processes in a particular brain region may therefore be reflected as an asymmetrical VER. This seems intuitively plausible if one realizes that a large percentage of neurological disorders, especially tumors, cerebrovascular accidents, and epileptic foci are unlikely to involve both sides of the brain in a symmetric way. Although this discussion focuses primarily upon the VER, the same conclusions undoubtedly apply to auditory and somatosensory evoked responses as well.

There is general agreement that the detection in homologous structures of VER components with latency discrepancies greater than 10 msec or amplitude discrepancies of 50% indicates pathology. Similarly, absence of a component from one side which is present homotopically is abnormal. Probably the most systematic study of the VER in neurological patients has been conducted by Jonkman (1967). In 130 patients with assorted diseases, he achieved better than 90% detection of tumors and CVAs, and called attention to the potential utility of this method to facilitate early detection of tumors because of its high sensitivity. His study was deficient in several respects, however. The control group was relatively small ($N = 14$), primarily parieto-occipital and occipital derivations were studied, the number of patients with any single pathology was limited, and the assessment of similarity of waveshapes was qualitative.

III. LOCALIZATION OF LESION SITE BY AER METHODS

The assessment of AER symmetry, as described in the preceding section, permits evaluation of the probability that a patient is suffering from a variety of neurological disorders, but provides relatively little information about the localization of the abnormal process. This is implicit in the fact that VERs, for example, can become asymmetric as a result of space occupying lesions in widespread nonvisual structures due to transactions between those regions and the visual system. However, this limitation holds true for assessments made with binocular stimulation only, and using flash presentation of spatially unstructured visual fields. By using a variety of stimuli in an attempt to evaluate different functions of visual pathways, by using stimuli in other modalities, by examining responses to monocular as well as binocular stimuli, and by combining VER with electroretinogram (ERG) recordings, these limitations can be overcome in a number of conditions.

For example, Vaughan and Katzman (1964) have pointed out that analysis of VER and ERG data permits visual disorders to be classified into lesions of retina, optic nerve, optic chiasm, retrochiasmal projection, or visual cortex, as follows: (1) unilateral retinal involvement is accompanied by unilateral changes in ERG and VEP; (2) optic nerve involvement is characterized by VER absence when the eye on the afflicted side is stimulated, with no ERG changes; (3) chiasmal involvement produces VER asymmetries indicating bitemporal hemianopia; and (4) geniculocalcarine involvement is accompanied by VER asymmetries indicating hemianopic defects in either both left or both right half-fields. Further, visual problems related to pregeniculate lesions produce complete loss of all cortical responses to flash stimulation, while lesions in the geniculocalcarine projection block the early response (25–35 msec) while longer latency components persist. Presumably, these long latency responses reflect arrival of afferent influences via the midline nonsensory-specific projection system.

The issue of localization of the site of a lesion has been thoughtfully analyzed by Regan (1972), who provides an excellent review of the applications of evoked potential methods in psychology, sensory physiology, and clinical medicine. Regan points out that investigations of the utility of AER methods for lesion localization have thus far focused mainly upon the analysis of EP waveforms, and have only used spatially unstructured visual fields. The use of colored stimuli permits the distinction between bundles in the optic nerve coming from macular and peripheral regions of the retina. The topographical distribution of VERs evoked by changes of brightness in visual stimuli with spatial structure, as well as the morphology of such potentials, are markedly different from those elicited by unstructured flashes, suggesting that different neuronal populations and anatomic sites may mediate responses to these different stimuli. In our own unpublished studies, we have often observed that VER waveshapes found to be perfectly symmetric to unstructured visual stimuli have become markedly asymmetric when patterned stimuli were presented. Such asymmetry seems to have an unusually high incidence in children with learning disabilities and may be related to abnormality of processing visual information. These observations will be further discussed in Chapter 6.

IV. ASSESSMENT OF CHANGES IN BRAIN STATE

In a series of papers (discussed in Chapter 12 of Volume 1 on neurophysiology of mind), Livanov (1967) and his colleagues have provided evidence that the alteration of brain state in certain psychiatric diseases is accompanied by changes in the electrical activity of the brain. These studies showed that the relationships between spontaneous electrical rhythms in frontal and more posterior brain regions reflect certain cognitive processes and are drastically altered by psychoses, as well as by the action of some psychotropic drugs. A number of other workers have explored the possibility that the changes in reactivity related to psychopathological states might be detected in alterations of the electrical responses evoked by test stimuli in different sensory modalities. In order to minimize the contribution of nonfunctional factors to interindividual differences in the absolute form of the evoked response, such studies have tended to focus upon relative differences between successive determinations of response, under circumstances devised to illuminate functionally relevant alterations. One line of investigation has examined changes in the "recovery cycle"; to make recovery cycle determinations, paired "conditioning" and test stimuli are delivered and, in successive measurements, the interval between the two members of a stimulus pair is varied. When brief intervals are used, the response to the second stimulus is smaller than to the first, presumably because of refractoriness. As the interval becomes longer, the second response becomes more similar to the first, indicating recovery of responsiveness. Gastaut and his colleagues (1951), who first measured such functions using visual evoked responses, interpreted their observations as the "cycle of cortical excitability." Using somatosensory stimuli, Shagass and his co-workers (1972; Shagass & Schwartz, 1961) have reported abnormal recovery functions in a variety of psychiatric disorders. Other workers have confirmed these findings using visual and auditory as well as somatosensory recovery functions, but have found little evidence for diagnostic specificity (Floris *et al.*, 1968; Ishikawa, 1968; Satterfield, 1972; Speck *et al.*, 1966).

Evidence was also found indicating that the remission of psychopathology was associated with normalization of the recovery function (Heninger & Speck, 1966; Shagass & Schwartz, 1961; Shagass *et al.*, 1962). As might be expected, changes in the morphology of the evoked response and in recovery functions have been found after administration of a variety of pharmacotherapeutic agents (Heninger, 1969; Saletu *et al.*, 1971a, b; Shagass, 1972). John *et al.* (1972, 1973) have described a neurometric method for the use of factor analysis of evoked potentials to provide a quantitative description of the action of drugs upon the brain (see Chapter 3).

Evidence has been provided by other workers showing that schizophrenic patients process information in an unusual fashion which is reflected in evoked response measurements (Callaway *et al.*, 1965; Jones *et al.*, 1966; Jones & Callaway, 1970; Lifshitz, 1969). These findings suggest that such patients possess perceptual dysfunctions that seem to be related to the structuring of figure ground relations in the environment, as inferred from the waveshape elicited by various stimuli.

V. AER EVALUATION OF SENSORY ACUITY

The most extensive clinical application of AER techniques has probably been in the evaluation of sensory acuity. These methods are unique in their capability to provide objective estimates of visual and auditory function in children and in patients from whom reliable verbal cooperation cannot be obtained. It is particularly difficult to decide whether the failure of an infant or young child to show behavioral response to sound is due to peripheral damage, brain damage, mental retardation, or psychosis, such as autism. Failure to provide hearing aids for children at the earliest possible time can cause severe difficulties in language acquisition and in emotional adjustment and maturation.

A. Evoked Response Audiometry

Evoked response audiometry is the term applied to the estimation of the threshold for auditory sensation by ascertaining the sound intensity required to elicit an auditory evoked response. The electrode configuration used for this determination is usually either the temporoparietal versus mastoid or the vertex versus mastoid. It is important to pay close attention to details of recording procedure when attempting to obtain a reliable threshold estimate. Stimuli should be delivered at low repetition rates to minimize effects of slow recovery cycles. Interstimulus intervals of 2–3 sec are usually utilized. The presentation of stimuli at irregular rather than regular intervals helps to defeat the effects of habituation (see discussion later in this chapter). Evoked potential components used to define the auditory evoked response are usually the excursion between the first positive component, P_1, (vertex positive process between 50–60 msec) and the first negative component N_1 (vertex negative, 90–105 msec), or the excursion between P_2 (vertex positive, 160–200 msec), and N_2 (vertex negative, 250–300 msec). It must be pointed out that the latency of these components is variable with age and brain state and identification of a particular component in an individual patient may be quite difficult. $P_2 - N_2$ is most often used for infant assessment because it is more constant. Interindividual differences in morphology further complicate unequivocal identification. Some of the computer methods described in Chapter 3 were devised to circumvent these problems.

In spite of these reservations and complications, it has been well established that careful EP audiometry provides results corresponding well with those obtained by pure tone audiometry (Cody & Bickford, 1965; Keidel & Spreng, 1965; Suzuki & Taguchi, 1965). H. Davis and his colleagues (1966, 1967) compared the detection threshold obtained by EP audiometry and behavioral methods in children up to 10 years of age and found that the usual difference between the two types of measures was somewhat less than 10 dB, with EP thresholds consistently higher. Rapin and Graziani (1967) evaluated EP audiometry in infants under one year of age, with comparable findings. There seems to be agreement among users of this technique that EP audiometry seldom *underestimates* the severity of peripheral hearing loss. It should be emphasized, as Rapin and Graziani took pains to point out, that the existence of clear auditory EPs merely establishes the functional integrity of the afferent auditory pathway, but is not equivalent to a demonstration that the auditory input can be behaviorally utilized; that is, integrity of auditory sensation as reflected by an auditory EP does not imply intact auditory perception. Methods of estimating perceptual processes are discussed below.

B. AER Assessment of Visual Acuity

Evoked potential techniques for the estimation of visual acuity are beginning to be applied in routine clinic practice in some installations and undoubtedly will become substantially more common as the methodology is simplified. Electrode derivations that have been used include occipital (O_1/O_2) versus linked earlobes, occipitoparietal (O_1P_3/O_2P_4), or occipital versus vertex. (Symbols are in international 10/20 nomenclature.) In our clinic we usually use both referential occipital and occipitoparietal derivations. Thorough evaluation of visual acuity by EP methods should include use of both unstructured (blank) and patterned visual fields. Copenhaver and Perry (1964) showed, in patients with normal vision in only one eye and lesions peripheral to the chiasm in the other eye that stimulation of the fovea with a small unstructured field elicited much smaller EPs from the affected eye than from the normal eye. In contrast, subjects with a defect in imaging resolution (refractive errors) on one side showed little difference between the EPs elicited from monocular stimulation of either eye with such stimuli. This maneuver provides a preliminary evaluation as to whether poor vision is due to a neural lesion or a defect in the imaging structure.

In contrast to the results obtained with small unstructured fields, numerous reports have established that the sharpness of the retinal image produced by patterned stimuli or the change from unpatterned to patterned stimuli dramatically alters VER amplitude or waveshape (Eason, White, & Bartlett, 1970; Harter & White, 1968, 1970; Spehlmann, 1965; White & Eason, 1966). An admirably clear example of the change in amplitude of a component, between 90–100 msec, caused by defocusing retinal images by a series of increasingly concave or convex lenses, was provided by Harter and White (1968) and is reproduced in Fig. 2.1.

As can be seen from the figure, the minimum point on the curve, generated by plotting change in amplitude versus power of the lens, corresponds to the refractive correction required by the subject. The method used in this and similar studies was to measure the effect of systematically degrading the image by successively placing a series of different lenses in front of the eye. This method, although accurate, is relatively slow and cumbersome. An observation reported by several authors suggests a simpler way to approximate the same measurement: when EPs to a blank stimulus field are compared with EPs to a high contrast patterned field, a clear change in polarity of a prominent wave at 100–150 msec of latency can be observed (Harter & White, 1968; Rietveld, Tordoir, Hagenouw, & van Dongen, 1965; Spehlmann, 1965). This effect is illustrated in Chapter 6.

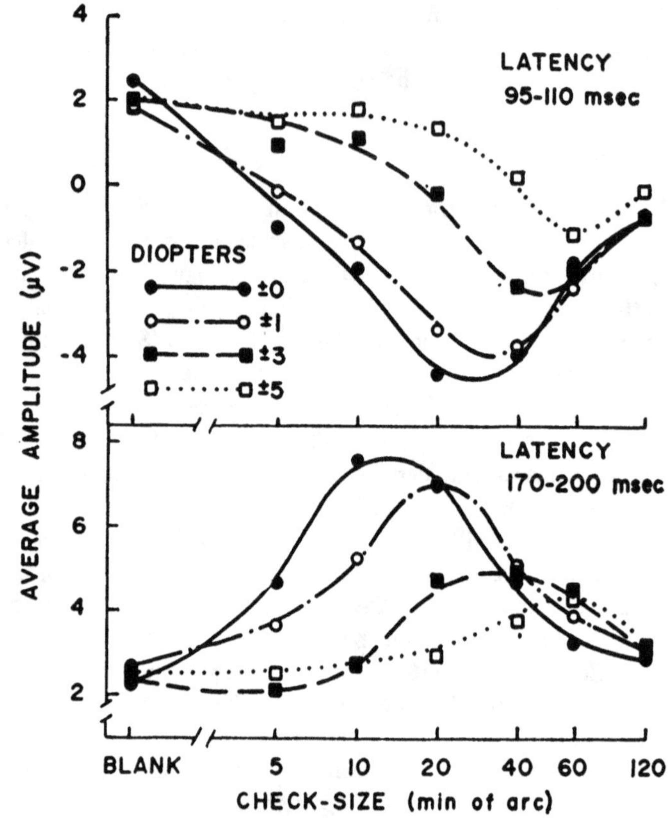

FIG. 2.1 Amplitude of evoked responses at a latency of 95–110 msec (surface negative deflection) and 170–200 msec (surface positive deflection) in reference to base line (see text) as a function of induced refractive error (diopters) and check size (minutes of arc). Data have been combined across subjects, replications, and diopter sign (– and +). Each plotted point is a mean based on 12 averaged responses, except 0 diopter conditions which were based on 6 responses. (From Harter & White, 1968. Copyright © 1968 by Pergamon Press Ltd. Reprinted by permission.)

Suppose that one has constructed a series of checkerboard patterns with a range of spatial frequencies (check sizes). If the white and black squares are of equal size, transillumination of any of these patterns results in 50% transmission; that is, they are equated for stimulus intensity. Presentation of a pattern with a spatial frequency too high to be resolved by a particular imaging system will result in the perception of a gray flash without structure. As the spatial frequency is reduced, the polarity reversal at 100–150 msec associated with pattern perception will appear when the coarseness of the checkerboard reaches the threshold for pattern resolution by that imaging system. The check size at threshold for this effect seems to be about 2–3 min of visual subtense. For a rapid visual screening test, a check size of about 10 min produces a strong effect. Although we have made no effort to correlate the threshold for polarity reversal with acuity of different subjects, we have explored the utility of this measure for screening, using a check size that we calculated could be resolved by a person with 20/70 vision (John, 1974b). The details of this neurometric method are discussed in Chapter 6. An excellent review of studies about the effects caused by presentation of structured visual fields, including studies relating check size to evoked potential features, can be found in Regan (1972).

C. AER Assessment of Color Vision

A number of reports indicate that different stimulus colors elicit evoked potentials with different waveshapes (Clynes, 1965; Clynes & Kohn, 1967; Clynes, Kohn, & Gradijan, 1967; Shipley, Jones, & Fry, 1965, 1966). These changes in waveshape as a function of wavelength of the visual stimulus are illustrated in Fig. 2.2.

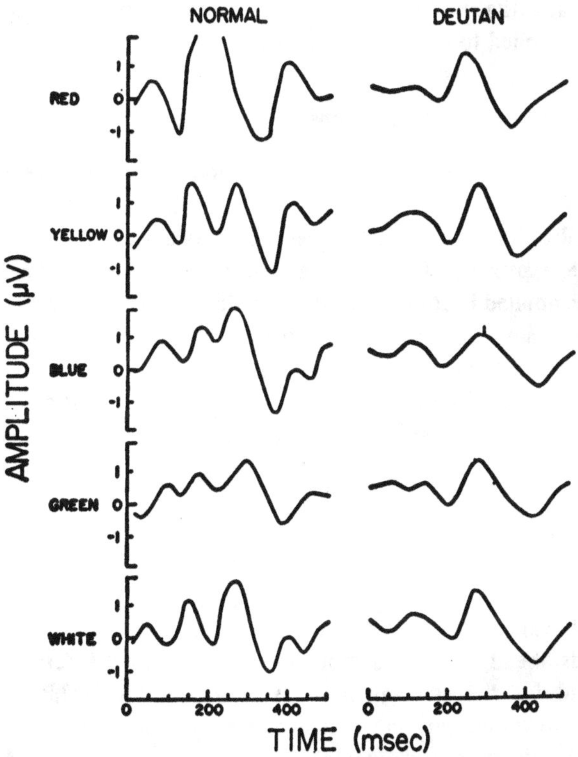

FIG. 2.2 Representative chromatic occipitograms for a color-normal and a deutan observer. (From Shipley *et al.*, 1965. Copyright © 1965 by the American Association for the Advancement of Science.)

Potentially, detection of differences in EP waveshape between red, green, and blue stimuli matched for intensity would constitute an objective test of color vision. Shipley and his colleagues have demonstrated that color-blind individuals (deutans) fail to show waveshape differences under these conditions. A neurometric technique was devised for assessing color vision by utilizing this phenomenon (John, 1974b) and it is discussed in Chapter 6.

VI. AER ASSESSMENT OF OTHER ASPECTS OF BRAIN FUNCTION

A. Indices of Development and Maturation

A substantial literature exists describing systematic maturational features of the EEG and AER. As yet, such information has not been systematically utilized in the management of clinical problems, in spite of its substantial potential utility. Chapter 6 will provide a review of our knowledge about developmental changes in the electrical activity of the brain and will describe neurometric methods we believe can be applied to a variety of clinical problems.

B. Perceptual and Cognitive Functions

A large body of literature describes AER phenomena related to perceptual and cognitive processes. As with the indices of development and maturation, no systematic utilization of such techniques has yet occurred. Later chapters review that literature (especially chapter 6) and describe neurometric methods that we believe can be applied in routine clinical practice. Illustrative findings in children with learning disabilities and elderly patients with intellectual deterioration are discussed in some detail in Chapters 7 and 8.

We have not included a detailed discussion of the topics just mentioned in this chapter, preferring to confine ourselves here to those methods which are already well recognized.

VII. SUMMARY

The EEG and evoked response literature reviewed in this chapter establishes that these methods already possess demonstrated utility for the detection of neuropathology and for the assessment of sensory functions. This utility can be enhanced by providing automatic, neurometric methods that will increase the accuracy of the electroencephalographic practitioner, permit the rapid screening and evaluation of far more patients per unit time in any existing clinical facility, and make electrophysiological evaluation available to the vast majority of potential patients who now have essentially no access to these assessments. Not only the electroencephalographer and the patient but a wide variety of other clinical practitioners, in psychiatry, psychology, pediatrics, and general practice, will derive benefits from such a development in that information essential for their treatment of their own patients will be more readily available. In many instances, such practitioners may elect to obtain such data themselves, since the neurological specialist is so overloaded or geographically remote as to be in effect inaccessible.

We shall turn now to a discussion of the principles of neurometric evaluation of electrophysiological data (Chapter 3) and the technology of automatic computer assessment (Chapter 4). In Chapter 5, we will see the effectiveness of these neurometric methods applied to the conventional problems of identification of patients with neuropathology. In Chapter 6, we will review the literature and discuss methods and findings related to the possibility that electrophysiological techniques can be extended to provide valuable information about developmental, perceptual, and cognitive processes. Such methods have thus far been of little value in this domain. However, neurometric techniques might here fulfill a unique need because of the great difficulty entailed in obtaining meaningful and reliable insights with the time-consuming and costly psychological and neurological techniques presently utilized. Results will be presented from our current studies of elderly patients with cognitive impairment (Chapter 7), and of learning disabilities in children (Chapter 8) demonstrating the sensitivity of neurometric indices of brain functions related to cognitive processes.

3

Principles of Neurometric Analysis of Brain Electrical Activity

I. ANALYSIS OF THE SPONTANEOUS EEG

The major features which must be taken into consideration for visual analysis of the EEG were outlined in the previous chapter. Many workers have recognized the need for more precise, objective criteria for EEG diagnosis in order to exploit more fully the potentialities of clinical and research electroencephalography. Achievement of this goal requires that impressionistic evaluation of the EEG be replaced by numbers and that subjective descriptions be replaced by mathematical characterizations, as pointed out by Kellaway (1973) who outlined a sequence of visual analyses of the EEG which he felt could be replicated by computer methods, or by "numerical taxonomy." Ultimately, neurometric analysis must consider all of the factors enumerated: (1) frequency distribution, (2) voltage, (3) locus of the phenomenon observed, (4) waveform, (5) interhemispheric coherence (i.e., symmetry of voltage, frequency, and waveshape at homologous placements), (6) character of waveform occurrence (random, serial, continuous), (7) regulation of voltage and frequency (abrupt or *paroxysmal* changes are usually abnormal), and (8) reactivity (changes in an EEG parameter with changes in state). Kellaway provides a discussion of each of these factors, outlining what he considers the features which would be considered abnormal for each separate factor. It is not our intention to provide such details here, since the interested reader can readily gain access to them. What is relevant for our purposes is that none of the proposed criteria present insuperable obstacles to computer implementation. Quite the contrary, most of them appear quite straightforward.

Furthermore, the relative diagnostic utility of these different EEG features remains to be demonstrated. It may well develop that certain features, such as the distribution of slow waves in different loci as a function of age, are of major utility for detection of a wide variety of neurological disorders (see Chapter 2), while other features, such as particular waveforms or features of regulation, are primarily of utility in differential diagnosis. In our opinion, the first goal of a strategy for neurometric analysis of the electrical activity of the brain should be to accomplish the most powerful, rapid, and economic separation of individuals who probably have neurological disease from those who probably do not. The method should strive to minimize false negative findings because the consequences for the patient can be quite serious. Translated into practical terms, the first purpose of the neurometric approach is to increase the proportion of neurologically diseased patients who reach the stage of physical examination by the neurologist or other practitioner, excluding as many healthy individuals as possible. Doubtful cases should not be excluded.

A. Frequency Analysis

Many methods for neurometric analysis of the EEG have been proposed, most of which have concerned themselves primarily with objective determination of the frequency distribution, or spectral analysis, of the EEG. The major methods have been of two basic types, those implemented with special electronic circuits and those utilizing high-speed digital computation, and have included spectrum analysis by narrowband electronic filters (Walter, 1943) or by wideband electronic filters corresponding to conventional EEG bands, such as delta (0.5–3.5 Hz), theta (3.5–7 Hz), alpha (7–13 Hz), and beta (13–25 Hz) (Kaiser, Petersén,Selldén, & Kagawa, 1964; Kozhevnikov, 1958); measurement of amplitude distributions (Drohocki, 1939); measurement of the distribution of time intervals between two successive zero crossings (Stein, Goodwin, & Garvin, 1949); pattern recognition of relations between amplitude and frequency distributions (Kaiser & Sem-Jacobson, 1962); computation of auto- and cross-correlation functions (Brazier & Barlow, 1956; Brazier & Casby, 1952); and computation of the fast Fourier transform (FFT) for spectral analysis (Walter, 1963). These methods have been reviewed by Bickford (1961), Walter and Brazier (1969), and most recently by several authors in the volume, *Automation of Clinical Electroencephalography*, edited by Kellaway and Petersén (1973).

Examination of the literature shows that many practical and accurate methods of computer analysis of frequency and other features of the EEG already exist. Nonetheless, these techniques are seldom applied in routine clinical examinations. Many workers have concluded that the problem now is not so much to develop new analytic methods as to simplify the representation of critical features of the EEG by computers in order to facilitate interpretation of the EEG by individuals who are not electroencephalographers. The rest of our discussion of methods of computer analysis of the EEG will be focused on some of the representations devised for this purpose.

1. Compressed Spectral Array

In a series of papers, Bickford and his colleagues have proposed that the vast amount of information in the EEG can be effectively presented to the clinician by a procedure which they term the compressed spectral array, or CSA (Bickford et al., 1971, 1973). The CSA presents the results of the whole EEG examination as a picture that can be viewed as a single page report. The CSA is generated in several steps: first, the record is divided into successive 4-sec segments; second, each segment is subjected to power spectral analysis using the FFT; third, successive spectra are displayed sequentially, one above the other, using the technique of "hidden line suppression." This technique ensures that no subsequent line spectrum crosses a previously plotted spectrum, and provides a three-dimensional display. The procedure for generating the CSA is illustrated in Fig. 3.1.

The dynamic qualities of this display are dramatically evident when changes in state are studied. For example, Fig. 3.2 shows the change in response to photic stimulation as the rate of a light flash was progressively increased from 1 to 16 Hz, and the frequencies quantified in the spectral analysis. Notice the major diagonal band of response peaks at the fundamental frequency of the flashing light, and the divergent series of smaller peaks below the main peaks, representing the second harmonic response (from Bickford *et al*., 1973).

The final step in the CSA representation consists of arranging the CSA from single EEG channels in an array corresponding to the relative positions from which they were sampled on the head, yielding a display of the distribution of EEG frequencies related in space and time. An example from a normal subject is shown in Fig. 3.3. Note the fundamental symmetry of the array and the clear, consistent alpha peak.

An example of CSA from an infant with a congenital cyst in the left temporal region is shown in Fig. 3.4. Note the clearly localized increase in delta activity on the left temporal and the absence of any alpha peak, since the patient was an infant.

2. "Neurometric" Displays

a. The "canonogram." Although the CSA permits clear visualization of the results of spectral analysis, it does not provide a numerical or quantitative evaluation of the local features of the array. Thus, it is not truly neurometric. Although the task of the clinician has been simplified, qualitative judgments and impressions are nonetheless necessary to interpret the CSA, even though each line represents the result of a precise quantitative computation. Several methods have been proposed to provide a quantitative topographic representation of the EEG.

Gotman *et al.* (1973) extracted what they considered significant "channel features" from each derivation. A channel feature was defined as the ratio of

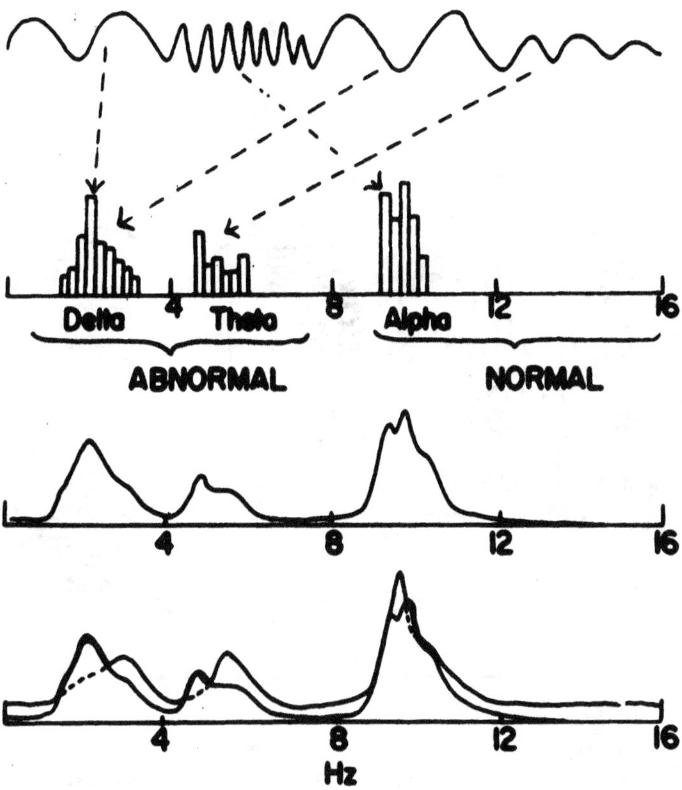

FIG. 3.1 Row 1 shows the original EEG tracing with mixed pathologic and normal frequencies. These are resorted during power spectral analysis (row 2), which tends to separate abnormal from normal frequencies since the former tend to be at the lower end of the spectrum in the waking state. The power spectrum is now smoothed (row 3), in preparation for the plotting process shown in row 4. Here spectra from successive 4-sec periods of primary data are compressed sequentially down upon each other using the technique of "hidden-line suppression." This technique ensures that no subsequent line spectrum crosses a previously plotted spectrum and thereby provides the display with three dimensions for easy comprehensibility. In the CSA type of display, time rises vertically in 4-sec spectral increments. (From Bickford *et al.*, 1973.)

activities in various frequency bands within the channel, appropriately weighted. The general form of such a feature was

$$\frac{(a \times \text{delta}) + (b \times \text{theta})}{(c \times \text{alpha}) + (d \times \text{beta})}.$$

The weighting coefficients *a, b, c, d* were 2.0 for "low delta" (.7–1.1 Hz), 4.0 for "high delta" (1.1–3.9 Hz), 5.0 for "low theta" (4.3–6.3 Hz), 1.0 for "theta" (6.7–7.1 Hz), 1.5 for "alpha" (7.1–1.3 Hz), and .5 for "beta" (13–25

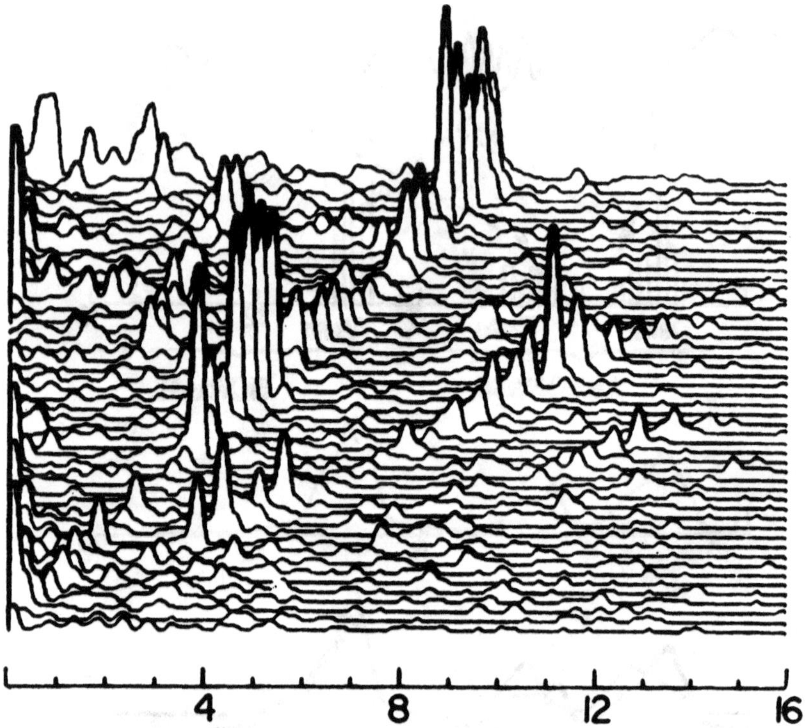

FIG. 3.2 "Photic sweep." CSA of a C_Z–O_Z EEG recording of the response to photic flash in a normal subject as the flash frequency was swept slowly from 1–16 Hz. Response at the flash frequency is shown by the diagonal series of main response peaks. A divergent series of peaks can also be seen below the main peaks which represent the second harmonic response. (From Bickford *et al.*, 1973.)

Hz), for all but frontal electrodes. Different values were used there, to compensate for artefacts due to slow eye movements. Once the channel feature was specified and calculated, it was multiplied by a symmetry coefficient. This coefficient reflected the ratio of the corresponding feature between the two sides. These channel features, weighted by the appropriate symmetry coefficient, were then summed to provide a global estimate of pathology which was a single number. This neurometric estimate was moderately effective in separating normal patients from abnormal ones, as seen in Fig. 3.5 (top).

These results are of only preliminary interest, in our opinion, because they were obtained from records preselected to contain a high incidence of marked and consistent slow wave abnormality. Of greater interest was the "canonogram," which Gotman and his co-workers devised to represent the results of performing these feature extraction operations upon the EEG, as illustrated in

Fig. 3.5 (middle and bottom). This display is easily interpretable, and accomplishes a substantial amount of "data compression."

Maulsby, Saltzberg, and Lustick (1973) propose a very similar representation, in which four head diagrams are used to represent the total spectral power in the delta, theta, alpha, and beta frequency bands. One symbol (X) represents analysis of records taken with eyes closed and another symbol (O) represents data obtained with eyes open. The number of symbols at each anatomic site represents the spectral power recorded at that site in the corresponding frequency band. Examples of the analytic displays from a normal and an abnormal subject are illustrated in Fig. 3.6.

b. The "age dependent quotient." Although the methods of Gotman and Maulsby and their colleagues move in the direction of further quantification, while preserving the advantageous features of topographic representation of the EEG frequency composition as provided by the CSA, yet they must be considered as only "semineurometric" in that they still fail to provide an unequivocal numerical index of abnormality.

Many of the investigators involved in the attempt to implement quantitative computer analysis of the EEG have realized that no matter which of the various

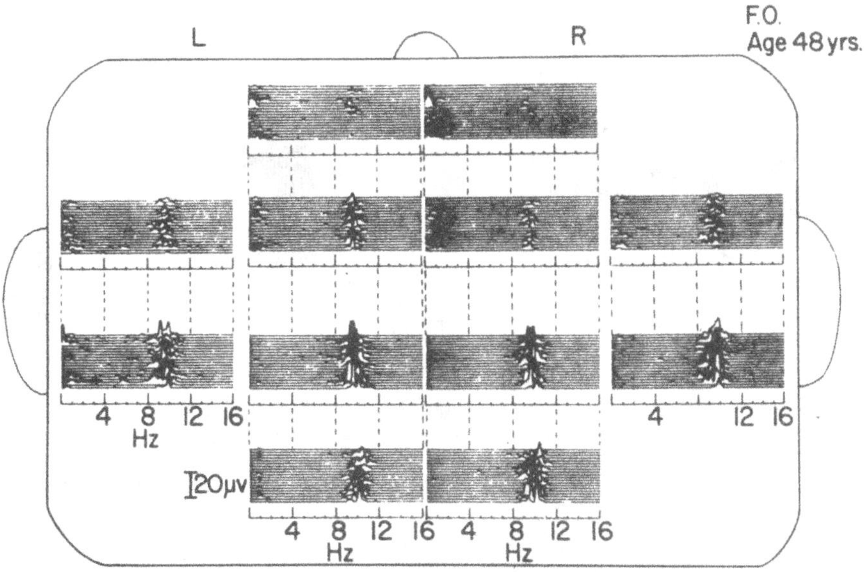

FIG. 3.3 The compressed spectral array from a 12-channel EEG of a normal subject (bipolar montage). Note the symmetric alpha rhythm in the lower six arrays and an asymmetry of the mu-like rhythm in the motor regions. There is no appreciable slow activity present. (From Bickford *et al.*, 1973.)

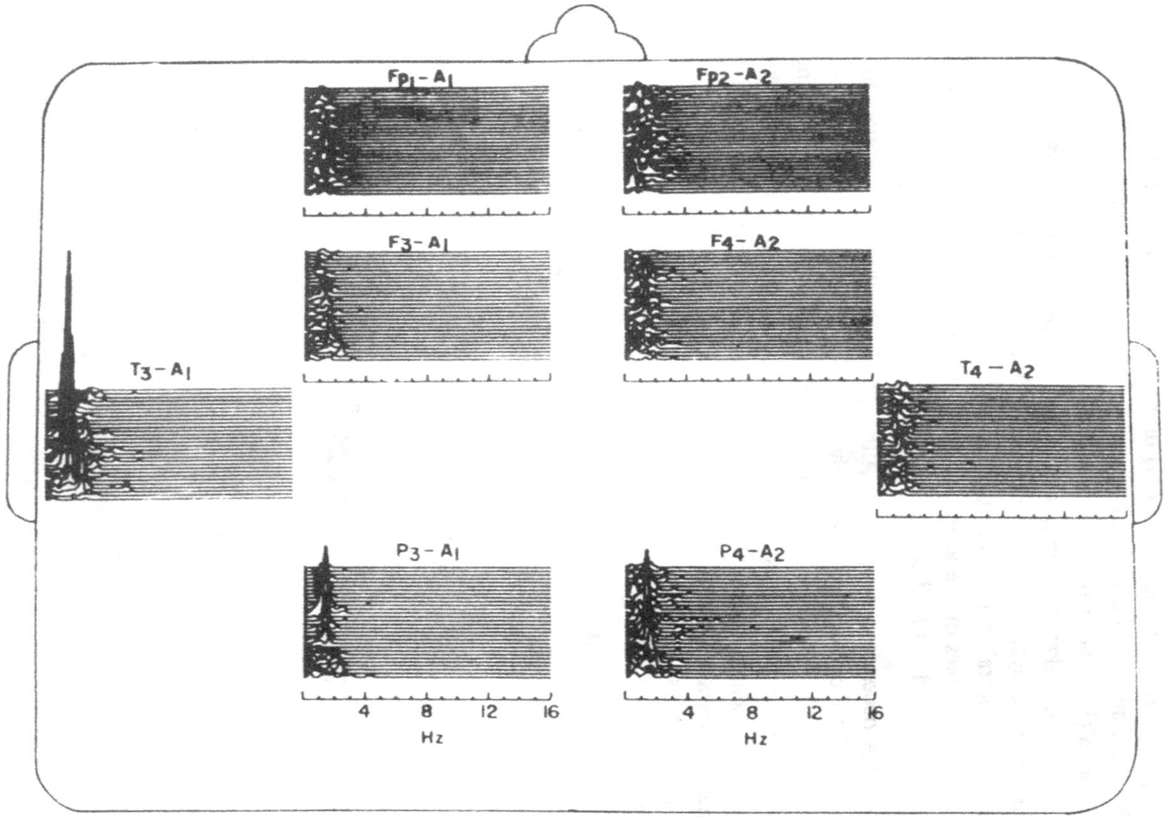

FIG. 3.4 CSA on an infant with a congenital malformation (cyst formation) in the left temporal area. Note the clearly defined slow wave delta discharge appearing from the left temporal region. No adult type of alpha peak is seen since the patient was an infant. Bipolar montage. (From Bickford *et al.*, 1973.)

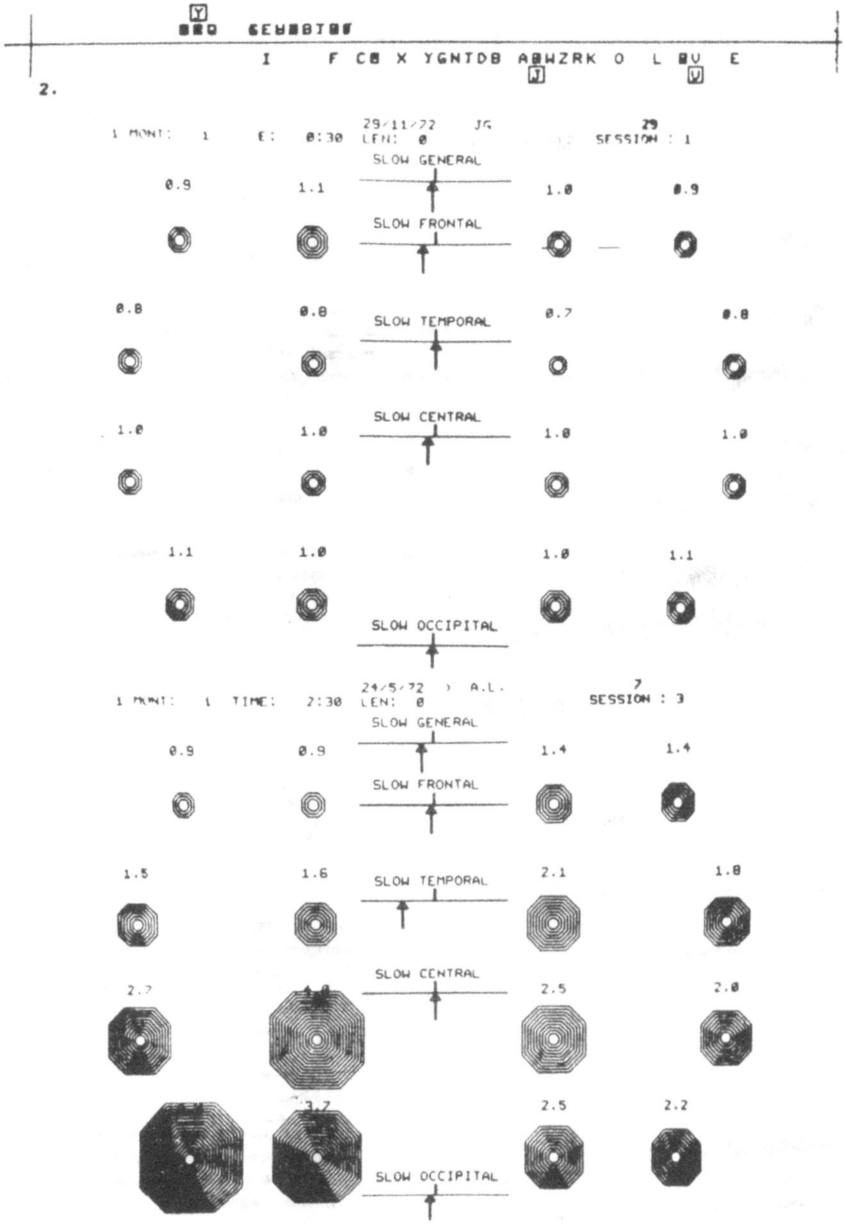

FIG. 3.5 Top: Values of global features for 26 normal subjects (A–Z above horizontal line) and 26 patients (A–Z below horizontal line). Two subjects with the same value of global feature appear on top of each other, but may be singled out (normal subject "Y," patients "J" and "V"). Middle: Canonogram of normal subject. The polygons are arranged in a topographic pattern which corresponds to the position of the channels on the subject's head: frontal on top, occipital on bottom. The number of rings in each polygon is proportional to the slow wave abnormality in that channel. The position of the arrows indicates the degree of slow wave asymmetry. The first arrow at the top, labeled "slow general," indicates the overall slow wave asymmetry between the two hemispheres. The other arrow

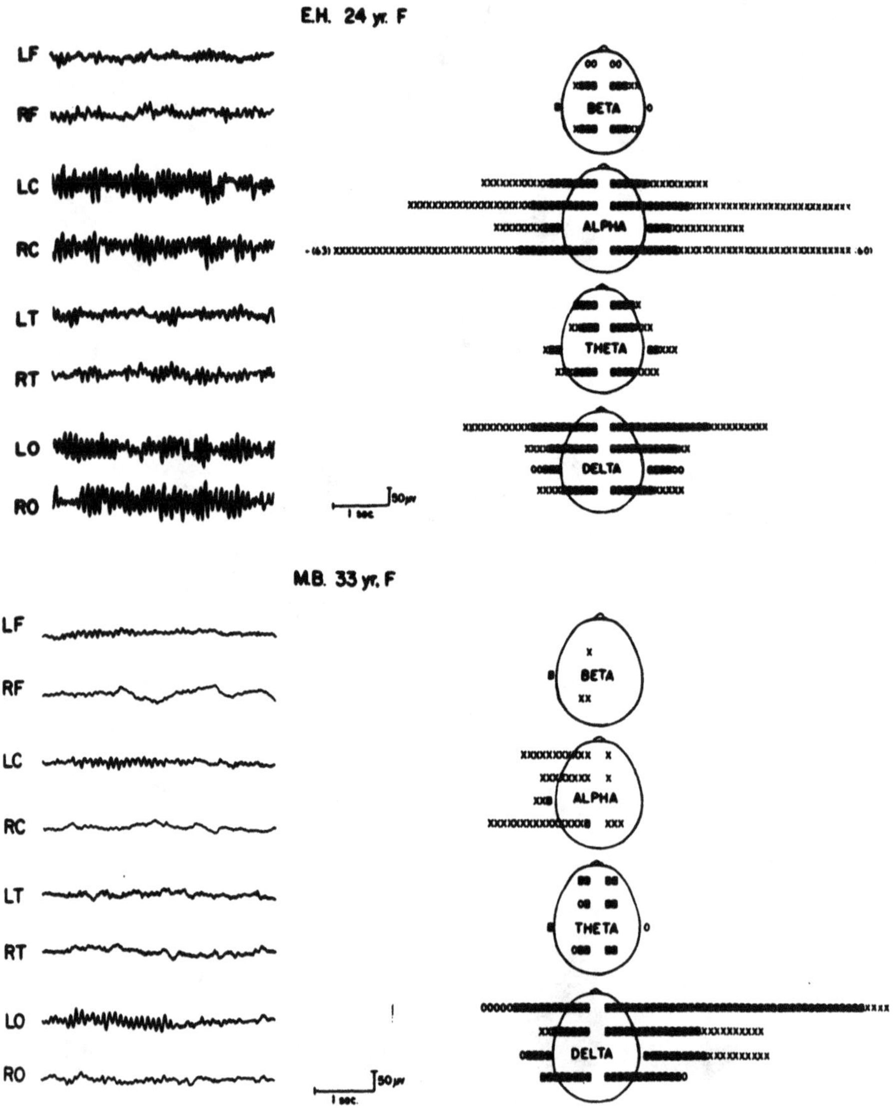

FIG. 3.6

FIG. 3.5 (*contd.*)
below indicates the slow wave asymmetries in the frontal ($F_{p1}-F_3$, $F_{p1}-F_7$, $F_{p2}-F_4$, $F_{p2}-F_8$), temporal (F_7-T_3, T_3-T_5, F_8-T_4, T_4-T_6), central (F_3-C_3, C_3-P_3, F_4-C_4, C_4-P_4), and occipital channels (P_3-O_1, T_5-O_1, P_4-O_2, T_6-O_2). Bottom: Canonogram of patient with left occipital lesion. (From Gotman *et al.*, 1973.)

methods already available was utilized to quantify features of the EEG, such analyses would be of limited value until a normative data base was established. Faced with this realization, Matoušek and Petersén (1973a) undertook the heroic task of obtaining analyses of the average amount of activity in frontotemporal (F_7T_3/F_8T_4), central (C_3C_Z/C_4C_Z), temporal (T_3T_5/T_4T_6), and parietooccipital (P_3O_1/P_4O_2) regions of groups of normal subjects from 1 to 22 years of age. Means and standard deviations of the amplitudes in microvolts were calculated for frequency bands defined as follows: delta (1.5–3.5 Hz), theta (3.5–7.5 Hz), alpha 1 (7.5–9.5 Hz), alpha 2 (9.5–12.5 Hz), beta 1 (12.5–17.5 Hz), and beta 2 (17.5–25 Hz). Sharply tuned broadband filters were employed. These invaluable data are reproduced in Table 3.1, with the permission of the original authors.

Using these normative data, Matoušek and Petersén (1973a) devised an extremely useful neurometric descriptor, the *age dependent quotient* (ADQ), which will be discussed in greater detail in Chapters 5 and 6. Of direct relevance to the present discussion is their use of this metric to obtain a numerical evaluation of the extent to which a particular EEG tracing deviated from normal. Their normative data showed a systematic diminution of slow (delta and theta) and a systematic increase in fast (alpha and beta) activity as a function of age, with different curves characterizing this process in different head regions. This EEG evolution seems to stabilize in late adolescence, with relatively stable values thereafter. Matoušek and Petersén calculated multiple regression equations between various parameters of the EEG and chronological age, and found impressive correlations between these variables. The best single correlation was between the ratio of theta activity to alpha activity plus 8 (an arbitrary value found to maximize the regression) and chronological age. This correlation was found to be .72. Equations with up to 20 variables were constructed.

Reasoning that abnormality would cause deviation from EEG features characteristic for a given age most marked in the particular head region involved, Matoušek and Petersén defined the ADQ as the ratio of the average values for a particular head region in a group of normal individuals of a particular age to the values obtained for the corresponding head region in a patient of that age. The

---

FIG. 3.6 Top: EEG and display of spectral results from a normal subject. Each head diagram on the right represents the total spectral power in the frequency band indicated, during three 10-sec epochs with eyes open and three 10-sec epochs with eyes closed. The numbers of figures at each anatomic site are proportional to the spectral power recorded from that site in the specified frequency band. Results during eyes open (os) are superimposed upon those with eyes closed (Xs). At a glance, one can see that alpha activity is maximum posteriorly and that there is more of it with the eyes closed (Xs) than open (os). The large amount of delta activity in the front of the heat is believed to be eye-movement artifact. Bottom: EEG and analytic display from a patient with prove arteriovenous malformation involving the right frontal lobe. Note the marked asymmetry of beta and alpha activity and the presence of excessive delta activity over the right side, maximum in the right frontal area. (From Maulsby *et al.*, 1973.)

TABLE 3.1
EEG Activity in Various Frequency Bands Related to Age (Means and Standard Deviations of Amplitudes in Microvolts)[a]

Age group	1–	2–	3–	4–	5–	6–	7–	8–	9–	10–	11–	12–	13–	14–	15–	16–	17–	18–	19–	20–	21–22
n=	19	18	18	22	26	18	26	41	25	31	24	29	29	49	26	26	32	21	29	25	27
F7-T3 and F8-T4																					
delta	23.6	23.6	19.2	18.9	19.9	16.6	15.0	14.7	14.2	14.6	12.6	11.2	12.0	10.2	8.7	7.8	7.5	7.1	7.5	7.4	6.8
	5.3	7.3	3.9	5.2	7.3	3.0	5.6	4.1	4.4	4.5	3.5	2.9	4.0	3.9	2.5	2.3	2.2	1.9	2.1	2.8	2.3
theta	14.5	16.9	13.8	15.0	15.1	12.5	11.1	10.2	11.2	10.8	9.9	8.5	8.8	7.4	6.2	4.9	5.2	4.4	4.7	4.3	4.2
	5.1	5.2	3.5	4.7	5.2	3.3	3.6	2.3	3.8	4.3	2.8	2.6	3.3	3.5	2.0	1.6	2.8	1.3	1.6	1.5	2.0
alpha 1	6.3	7.1	6.8	7.7	8.4	7.3	6.5	6.0	7.4	7.1	6.3	6.0	6.8	6.3	5.1	4.5	4.6	4.1	4.7	5.0	4.2
	1.9	2.0	2.5	2.3	3.1	2.4	1.8	1.7	3.6	3.8	2.6	2.6	3.6	5.0	2.4	2.4	2.8	2.0	3.2	3.0	2.2
alpha 2	4.8	5.1	4.8	5.3	6.0	5.4	5.2	4.9	5.3	5.4	5.1	5.3	5.5	5.4	5.4	3.8	3.9	4.1	4.0	4.6	4.3
	1.8	1.3	1.2	1.5	2.4	1.9	2.3	1.6	1.8	3.0	2.8	2.7	2.5	3.3	2.3	2.3	1.7	1.9	2.3	2.4	2.3
beta 1	8.4	7.4	6.5	6.7	6.9	5.9	5.5	5.3	6.3	5.5	5.2	4.9	5.5	4.8	4.6	3.9	4.0	4.5	3.7	4.1	3.4
	3.5	2.3	1.6	2.2	3.2	1.5	1.4	1.4	1.8	2.4	1.4	1.8	3.3	2.6	1.4	1.4	1.3	1.8	1.0	2.3	1.2
beta 2	11.7	8.6	7.5	7.7	7.9	6.2	5.9	5.6	6.4	6.4	5.2	5.0	6.4	4.9	4.6	4.7	4.4	5.6	4.2	4.9	4.5
	5.9	3.4	2.5	3.5	3.9	1.4	1.9	2.0	2.1	3.4	1.8	2.1	4.7	2.9	1.4	2.6	1.6	3.2	1.1	2.9	2.8
CO–C3																					
delta	24.4	24.0	17.7	17.1	17.7	15.9	14.1	13.5	13.6	11.9	10.7	11.1	10.5	8.2	6.7	6.0	6.1	4.7	4.8	4.2	4.8
	7.5	8.8	4.8	3.2	6.8	3.9	7.3	3.7	4.1	3.7	2.5	3.7	3.0	2.8	2.0	1.5	3.5	1.3	1.1	1.3	2.0
theta	18.2	19.4	16.8	17.5	18.0	16.6	14.0	13.6	14.8	12.2	12.4	11.1	11.3	8.5	7.7	5.8	6.0	4.8	5.1	4.4	4.9
	6.4	7.4	4.0	4.5	7.7	5.7	4.5	3.0	5.5	4.1	3.4	3.5	3.1	2.9	2.7	2.1	2.8	1.4	2.1	1.8	2.3
alpha 1	9.0	10.8	11.1	11.8	12.5	12.2	10.2	8.9	10.4	9.9	9.3	8.8	8.7	6.8	7.8	4.7	5.5	3.8	5.1	4.7	5.2
	3.6	6.4	5.0	4.6	5.1	6.5	4.0	3.1	5.2	6.4	4.0	5.3	4.7	4.1	4.5	2.5	4.2	1.6	4.0	2.8	3.9
alpha 2	5.7	6.0	6.7	6.5	8.8	8.4	6.9	7.7	7.8	7.9	7.3	9.2	9.2	6.7	9.0	4.6	4.6	4.6	5.3	5.0	5.3
	2.4	2.2	2.7	1.9	4.1	4.2	2.8	3.1	3.6	4.5	3.4	5.3	5.3	4.0	5.5	2.8	2.6	2.2	3.4	3.0	2.8
beta 1	7.9	6.5	5.8	5.8	6.2	5.8	5.6	5.4	6.2	5.1	5.3	5.8	5.5	4.4	4.7	3.7	3.7	3.5	3.3	3.4	3.7
	4.7	2.5	1.4	1.6	2.9	2.0	1.9	1.4	2.5	2.2	1.3	2.7	2.4	1.9	1.9	1.2	1.4	1.8	1.2	1.8	1.9
beta 2	8.8	5.8	5.5	5.2	6.1	4.9	4.8	4.7	5.4	4.9	4.7	5.3	6.1	4.4	4.9	3.8	3.8	3.8	3.8	4.3	4.5
	5.8	1.8	2.0	1.6	4.8	1.4	2.1	1.4	24.	2.7	1.4	2.3	4.4	2.0	2.1	1.2	1.2	1.5	1.4	2.6	1.7

T3-T5 and T4-T6																					
delta	23.7	22.9	18.7	20.0	19.8	16.9	16.1	16.1	15.5	14.8	13.9	12.7	14.1	9.8	8.5	7.2	7.6	6.9	6.5	6.3	5.9
	6.8	5.3	4.7	6.5	6.7	4.6	6.0	7.2	5.7	5.6	4.7	4.4	11.7	3.8	3.0	2.4	3.4	2.0	2.2	2.0	2.3
theta	17.6	17.8	17.9	21.1	19.4	16.8	15.1	13.9	15.2	14.0	13.3	10.4	10.5	9.8	7.8	5.4	6.4	5.4	5.4	5.5	5.0
	7.7	5.0	7.7	8.1	7.0	7.5	5.6	6.3	6.1	6.1	6.5	4.2	3.7	5.0	3.6	2.2	4.6	1.8	2.7	2.8	2.7
alpha 1	6.7	7.4	8.2	10.5	13.4	11.1	11.1	12.5	15.0	13.2	12.4	11.6	10.1	9.3	9.6	7.2	8.9	7.9	7.6	9.0	8.0
	2.0	2.1	4.1	3.6	5.4	5.8	4.2	7.0	10.8	8.2	8.5	7.4	6.0	6.7	6.5	6.8	8.6	5.4	5.8	7.4	7.3
alpha 2	5.0	5.1	5.1	6.5	7.6	7.5	8.3	10.1	9.8	10.3	9.5	11.4	10.7	8.9	11.0	6.7	8.1	8.8	7.7	9.4	8.3
	1.5	1.4	1.5	2.6	3.0	3.4	4.7	5.8	4.2	8.0	5.5	7.4	5.1	4.5	5.5	4.8	5.0	5.3	5.9	5.0	4.8
beta 1	8.4	7.4	6.6	7.3	7.7	6.8	6.6	6.9	8.3	6.8	6.4	6.3	6.5	5.7	6.0	4.7	4.9	5.5	4.6	5.1	4.5
	2.8	2.7	2.0	2.7	3.5	2.5	1.6	3.1	2.3	2.8	2.0	2.7	2.9	2.2	1.8	1.7	1.9	3.2	1.3	2.4	1.6
beta 2	12.5	8.3	7.0	7.5	7.8	5.7	5.7	5.7	6.6	6.0	4.9	4.9	5.7	4.8	5.2	4.6	4.7	5.7	4.5	5.1	4.5
	6.7	3.4	2.4	3.7	4.3	1.5	1.7	2.7	2.2	2.7	1.5	2.0	3.6	2.1	1.5	2.0	1.8	3.2	1.5	1.8	1.6
P3-01 and P4-02																					
delta	28.3	28.3	20.1	21.7	23.4	18.0	19.2	16.7	16.8	16.8	16.4	13.9	13.4	11.1	9.3	7.5	7.4	7.0	6.3	6.9	6.1
	9.3	8.0	3.8	5.3	9.6	5.6	7.2	4.0	6.0	7.7	6.8	5.6	5.2	4.3	3.6	2.8	3.6	1.9	2.2	3.9	2.1
theta	19.0	22.5	21.1	22.4	25.6	21.7	17.7	16.5	18.0	16.7	17.0	12.6	12.6	10.1	9.6	6.6	7.0	6.5	6.2	6.3	5.8
	7.0	7.7	7.3	6.6	11.2	8.7	6.2	6.1	7.9	8.0	8.2	4.9	5.0	4.5	3.8	3.0	4.0	2.3	3.1	3.5	3.0
alpha 1	8.7	9.4	12.0	15.3	23.4	18.0	17.0	18.2	20.5	19.2	20.0	16.6	17.0	12.5	12.8	10.7	11.7	8.9	9.5	10.1	10.6
	3.3	3.3	6.1	5.3	12.2	6.7	7.0	7.3	10.3	9.6	11.7	9.2	9.0	8.3	7.4	10.0	9.2	5.8	6.5	7.6	8.5
alpha 2	5.7	6.0	6.8	7.9	11.0	11.1	11.6	14.5	14.0	14.4	15.1	17.8	18.8	13.5	16.6	12.4	12.8	11.8	11.4	13.1	13.1
	2.0	1.8	2.0	2.1	5.1	4.3	6.6	6.5	5.7	11.0	7.4	10.5	10.2	8.4	8.3	9.0	9.0	6.6	8.8	7.1	7.6
beta 1	7.3	6.0	7.1	7.3	9.0	8.0	7.5	7.7	8.9	7.7	8.5	8.4	8.0	6.6	7.2	5.8	5.3	6.1	5.1	5.4	4.8
	3.8	1.5	1.6	2.1	4.5	2.6	2.0	2.1	3.2	3.4	2.6	4.0	3.0	2.7	2.5	2.2	2.3	4.6	1.7	2.0	1.9
beta 2	9.1	5.9	5.7	5.5	6.2	5.2	5.3	5.5	5.9	5.6	6.2	7.0	6.8	5.2	6.5	4.9	4.5	4.9	4.6	5.5	5.1
	5.5	1.3	1.7	2.5	2.9	1.3	1.9	1.7	1.9	2.8	2.4	3.3	2.9	2.0	2.6	1.7	1.9	2.0	2.1	2.1	2.4

[a]From Matoušek and Petersén (1973).

utility of a variety of different parameters for calculation of clinically useful ADQs was explored and is discussed in detail in their original paper (Matoušek & Petersén, 1973a). While definite variations were found between the predictive or diagnostic precision of different versions of the ADQ, the amount of slow activity was most important. The advantage of the basic principle was obvious no matter which variation of the theme was examined.

EEG spectra vary in a characteristic manner with age and head region. Interpretation of any individual spectrum demands that it be compared with the spectrum expected from that corresponding head region in a healthy person the same age as the patient. By constructing the ADQ, the ratio of the average normal value to the observed individual value, the data from the patient can be transformed to a metric relative to a normal reference point. If the individual value is approximately the same as the average normal value, the value of the quotient normal/individual will be about unity. If low frequencies contribute greater energy than normal to the spectra from particular head regions, the ADQ will become less than unity. In general, the more abnormal the activity of a local region, the lower the ADQ.

Matoušek and Petersén defined .8 as a suspiciously low ADQ. By examining the distribution of ADQ value for different head regions, it was possible to construct gradients of abnormality related to the localization of pathological process. This will be discussed further in Chapter 5. Figure 3.7 illustrates a computer-produced topographic display of the results of spectral analysis transformed into ADQ values. Note that, in contrast with the methods described earlier in this chapter, *this method provides an explicit numerical evaluation of abnormality and locates that abnormality topographically*. This was perhaps the first example of truly neurometric evaluation of the EEG.

In our opinion, this is a unique advantage of this procedure. Normalization of electrophysiological data with respect to chronological age and anatomic locus will be an essential feature of successful computer methods for diagnostic evaluation of electrophysiological measures, providing a metric for relative abnormality of different brain regions. This metric should be invaluable for inferences about the nature of the disorder, its probable functional consequences, prognoses of recovery, and surveillance of the progress of abnormality and the relative efficacy of various modes of treatment.

B. Symmetry Analysis

The methods for quantitative analysis of the EEG considered thus far emphasize the frequency composition of the EEG signal. Examination of the standard deviations in Table 3.1 shows that there is substantial variation around the mean value of energy in any frequency band within samples of EEG from groups of normal subjects classified according to age. This variability is sufficient so that fairly large deviations from these norms of band-specific activity must occur in a

```
EEG FREQUENCY ANALYSIS      PATIENT: 31. 9.15. - EEG EXAM.: 72. 2.24.
----------------------

EAQ 20 (MEAN VALUE  57):

             *
            * *
       *************
      *             *              SYMMETRY:
     *               *
     *  49       62  *                0.79
     *               *
     *     29   52   *                0.57
     *               *
     *  26       48  *                0.54
     *               *
     *    103   97   *                0.94
      *             *
       *************           VALUES < 80 (SYM. <0.80) ARE PATHOLOGICAL

EVALUATION OF BACKGROUND ACTIVITY:

*  SEVERE
*  ABNORMALITY
*  IN THE
*  CENTRAL
*  AND
*  TEMPORAL
*  REGION
*  ON THE LEFT SIDE

SOME DISTORTION BY ARTIFACTS CANNOT BE EXCLUDED.
COMPARE THE RESULTS WITH EAQ 8 (NEGLECTING DELTA AND BETA 2 ):

             *
            * *
       *************
      *             *              SYMMETRY:
     *               *
     *  65       68  *                0.96
     *               *
     *     51   57   *                0.88
     *               *
     *  56       64  *                0.86
     *               *
     *     60   59   *                0.98
      *             *
       *************           VALUES < 80 (SYM. <0.80) ARE PATHOLOGICAL

EVALUATION OF BACKGROUND ACTIVITY:

*  MODERATE
*  ABNORMALITY
*  OCCURING DIFFUSELY
*  WITH UNCERTAIN MAXIMUM IN THE
*  CENTRAL
*  REGION
*  ON THE LEFT SIDE

(LESS INFLUENCED BY ARTIFACTS?)
```

FIG. 3.7 Top: Detection of artifacts during automatic processing. In a patient (male, 41) with a meningioma in the left frontocentral region, a discrepancy was found between ADQs obtained with and without delta and beta 2 activity. The computer prints two alternatives, leaving the final decision to the electroencephalographer. However, making a distinction between artifacts and EEG slow activity is even difficult in visual assessment of the record. Note the less certain result as it appears in the second alternative (bottom). (From Matoušek and Petersén, 1973a)

particular individual in order to be considered significant (see Tables 6.1–6.3 in Chapter 6). Most electroencephalographers consider bilateral symmetry of waveform and amplitude as an important feature of the EEG, sensitive to neuropathology. Symmetry measures place lower importance on the absolute frequency composition of the EEG signal, and rely more on the use of the EEG from one hemisphere as the "control" relative to the other. Strategically, this seems well advised since the variability in frequency spectra within a normal group can be quite substantial, while the difference between the two hemispheres of an individual is usually quite small. Matoušek and Petersén (1973b) computed bilateral symmetry between different frequency bands of the EEG from various homologous derivations. They found that the symmetry values for the different frequency bands were usually comparable and corresponded well to the symmetry of the overall EEG activity. The symmetry ratio (low value over high value) for percentage of theta activity, averaged across all age groups and derivations, was found to be .91, with a standard deviation of about .06. In temporal derivations, a significant increase of asymmetry was found with age. Symmetry is higher when the EEG activity in the band is expressed as a percentage, while greater asymmetry is observed if absolute amplitude values are compared.

Some years ago, John and Laupheimer devised a symmetry analyzer, intended to provide rapid quantitative measurements of the symmetry of waveform and of amplitude between two EEG signals (1972). The computation of waveform symmetry was based upon utilization of polarity coincidence correlation methods. *Polarity coincidence correlation* consists of making a large number of comparisons of the *polarity* (positivity or negativity) of two simultaneous electrical signals. If A stands for the number of instances in which the two signals were of the same sign, B stands for the number of instances when the signals were of opposite sign, and M stands for the total number of measurements, then the polarity coincidence correlation coefficient, or PCC, is equal to $(A - B)/M$. It can be seen that if the two signals are identical and in phase, then $A = M$, $B = 0$, and PCC $= (M - 0)/M = 1$. If the two signals are identical but 180° out of phase, then $A = 0$, $B = M$, and PCC $= (0 - M)/M = -1$. If the relation between the signals is random, then on the average $A = B = M/2$ and PCC $= (M/2 - M/2)/M = 0$. Ruchkin (1971) has shown that the PCC is equal to the arc sine of the familiar Pearson product moment correlation coefficient.

We selected the PCC in preference to the Pearson correlation because of the convenience of implementing polarity coincidence correlation in a small special-purpose computer and because of the physiological significance of polarity of the EEG signal (see Volume 1, Chapter 2). Signals were sampled at a very high rate (10 kHz) so that the frequency response of the correlator far exceeded the bandwidth of EEG signals. A digital delay circuit permitted either signal to be delayed up to 49 msec with respect to the other, thus permitting quantification of phase shifts or "lead-lag" analyses, sometimes useful in ascertaining latency shifts or propagation times.

This symmetry analyzer also measured amplitude symmetry by rectifying and integrating each input signal and computing the ratio of the integrated absolute amplitudes, referred to as the *signal ratio*, or SR. The signal ratio is a number defining the ratio of the larger to the smaller signal. SR values greater than 1.5 are rarely encountered in normal subjects, and the great majority of values are less than 1.2.

Using an instrument based upon this design (Neurodata Symmetry Analyzer SA2200), Harmony and her co-workers (1973) gathered normative data from frontal, temporal, central, occipital, centrooccipital, temporooccipital, centrotemporal, frontocentral, and frontotemporal derivations in a large group of normal subjects (Harmony *et al.*, 1973b). Those data showed good test-retest replicability and moderate dispersion. The average values and standard deviations from groups of normal subjects and the changes in symmetry found in groups of patients with tumors, cerebrovascular accidents, and epilepsy, can be found in Chapter 5. At this point in our discussion, it suffices to mention that neurometric evaluation of bilateral symmetry of waveform and amplitude of the spontaneous EEG was somewhat more accurate than conventional visual examination of the same data by a group of electroencephalographers for the identification of abnormalities related to tumors and cerebrovascular accidents, but was of lesser utility in detecting abnormality in records from epileptic patients.

The results just described relate to symmetry of the total EEG signal, from .5 to 40 Hz. Although Matoušek and Persén (1973b) found that in normal subjects amplitude symmetry was quite constant in the various frequency bands, it may well be that in neuropathology, both amplitude and waveform asymmetry may be restricted to a particular band (especially delta). Note the close relationship between the band-limited cross correlation just described and the coherence spectrum discussed by Dumermuth and Keller (1973) and others.

Thus, symmetry analysis provides another method for neurometric representation of important features of the spontaneous EEG. Amplitude asymmetry, asynchrony of similar processes, and waveform differences can be readily quantified. Since symmetry measures use each hemisphere as the control for the other, the problem of interindividual variability in spectral analysis is circumvented. Computation is far more simple, rapid, and economic than spectral analysis. Whenever possible, the quantitative analysis of amplitude and waveform symmetry as well as the frequency composition of the EEG would seem desirable.

II. AVERAGE EVOKED RESPONSE AND VARIANCE COMPUTATIONS: GENERAL CONSIDERATIONS

The second major class of electrophysiological measures of brain activity which is available is sensory evoked responses. The physiology of the evoked response, the basic principles of average response computation, and some of the evidence showing that sensory, perceptual, and cognitive processes of the brain are reflected in evoked response measurements have been discussed in Chapter 2, 9, and 10 of Volume 1 and Chapter 2 of this volume. In contrast to the spontaneous EEG, which reflects ongoing electrical transactions between and within various anatomic regions of the brain related to its intrinsic organization, the evoked response provides insight into the reactivity of various functional systems of the brain to afferent input and tells us something about how the system processes different kinds of information.

A. The Average Evoked Response

Suppose we take a series of n samples of the EEG fluctuations that occur as a specified stimulus is repeatedly presented. Each sample begins at the instant, t_0, when the stimulus occurs and is T msec in duration. This period T is often referred to as the *analysis epoch*. Each sample will contain two kinds of activity: (1) the transient oscillation, which constitutes the evoked response phase-locked to the time of stimulus onset, t_0 to which we refer as the "signal," $S(t)$. $S(t)$ will have a reproducible waveshape reflecting the responsivity and transmission latencies of the anatomic pathways whose activity contributes to the response to the stimulus; (2) the ongoing activity reflecting neuronal processes unrelated to the effects of the stimulus, to which we refer as the "noise," $N(t)$). $N(t)$ will display random variations in amplitude and polarity, since the ongoing activity is in no reproducible phase relationship to the arbitrary time of presentation of the stimulus and changes constantly in waveshape. Any single sample of the evoked response can thus be considered as a composite voltage oscillation, $V(t)$, such that

$$V(t) = S(t) + N(t).$$

The accuracy with which the evoked response, $S(t)$, can be defined by a single sample obviously depends upon the relative contributions of these two components to $V(t)$, or the ratio of signal to noise (S/N) in the sample. In practice, especially in unanesthetized, behaving subjects, N is at least as large as S and is often significantly larger. In order to obtain a reasonably accurate estimate of $S(t)$, therefore, it is necessary to devise a procedure to increase the signal-to-noise ratio. This can be accomplished by averaging a set of samples of the EEG fluctuations, $V(t)$, time-locked to the repetitive stimuli. After n presentations of the stimulus, the contribution of the signal to the average will be proportional to the number of samples, or $nS(t)/n$, since the signal is in reproducible phase relation to the time of stimulus onset, t_0. However, since the "noise" is in random phase relation to t_0, the contribution of the noise to the average will be proportional to the *square root* of the number of samples, or $\sqrt{n}\, N(t)/n$. Thus, after n stimulus presentations, the averaged evoked responses can be described as

$$\frac{\Sigma V(t)}{n} = \frac{nS(t)}{n} + \frac{\sqrt{n}\,N(t)}{n}$$
$$= S(t) + \frac{N(t)}{\sqrt{n}}.$$

It can easily be seen that whereas the ratio of the contributions of signal and noise to a single sample of the evoked response was S/N, the relative contributions of signal and noise to the average of n evoked responses will be $\sqrt{n}S/N$. The averaging process therefore improves the signal-to-noise ratio, or the accuracy of definition of the evoked response, by an amount proportional to the square root of the number of samples obtained.

The above considerations hold true for an invariant signal imbedded in Gaussian noise. In neurophysiology, neither condition may exist. First, the evoked response is not an invariant signal, but can assume a variety of waveshapes, depending upon momentary changes in the state of the brain. As mentioned in previous chapters, this fact has been amply documented by us (John, 1973a; John *et al.*, 1973), using methods developed by Ruchkin (1971). Second, the ongoing activity is often not at all Gaussian, but may have very stable features, for example, high amplitude rhythms in the alpha frequency band. The interested reader is advised to consider these fuller expositions to avoid the pitfalls of assuming that averaged responses necessarily provide an insight into the "prototypic" response of the brain to a stimulus. We will return to this question briefly in our treatment of pattern recognition methods later in this chapter.

Average response computation can be implemented in small special-purpose average response computers, computers of average transients (CAT), or in general-purpose digital computers. In either case, the principle is the same. The analog EEG signal is led to an analog to digital (A/D) converter, where it is represented by making the number of pulses per unit time proportional to the analog voltage. When the stimulus occurs, the computer begins to "sweep," storing the pulses occurring in each successive unit of time, Δt_i, into a set of M registers. The number of registers multiplied by the time Δt per register (or "dwell"), $M\,\Delta t$, equals the *analysis epoch*, the time period T throughout which the conversion and storage process took place. As this process is repeated, the number of pulses stored in the ith register becomes proportional to the average voltage of the evoked response at a latency Δt_i after the stimulus. In this fashion, the M registers construct a digital representation of the average evoked response at each of M time points across the analysis epoch as a sequence of numbers.

B. Variance of the Average Evoked Response

Let us define the average evoked response as a series of numerical values of voltages at sequential time points. We have computed the value of the average voltage at each time point, t_i, which can be represented as $[\Sigma^n V(t_i)]/n$. Suppose that while we compute this quantity, we also compute the mean squared value of the voltage, $[\Sigma^n V^2(t_i)]/n$. The *variance* is defined as the mean square minus the squared mean; therefore, we can compute the variance of the averaged evoked response at each latency by squaring the value of the AER and subtracting it from the mean squared voltage. Thus, we see that the variance, $\sigma^2(t_i)$, can be described as

$$\sigma^2(t_i) = \frac{\Sigma^n V^2(t_i)}{n} - \frac{\Sigma^n V(t_i)}{n}^{\;2}$$

We now have an estimate of the variability of each component of the averaged evoked response. If the variance is due to the presence of random noise superimposed upon a stable signal, the curve of variance versus time will be *flat* across the analysis epoch. This indicates *homogeneity* of variance of the signal. The amplitude of the variance will be proportional to the percentage of noise in the channel measured. On the other hand, if certain components of the evoked response are differentially variable due to diverse origins, then fluctuations in different components of the signal will make disparate contributions to the variance. As a result, peaks will appear in the variance curve at those latencies in the analysis epoch where variations occur in the evoked response. Peaks in the variance curve indicate inhomogeneity of variance, and reflect the fact that different classes of evoked response are being elicited by the stimulus and erroneously averaged together. If such evidence of inhomogeneity is obtained, the investigator is well advised to identify the factors responsible and not to consider the resulting average response as an accurate estimate of some "typical" evoked response morphology. Mere reproducibility of the shape of an AER does not provide any reassurance, because different evoked response types or *modes* may occur in reproducible proportions under the test conditions. This reproducible mix of different response types will display a misleading replicability of AER shape.

C. *t* Test for the Significance of Differences between AERs

Once the variance becomes available, two AERs (AER_1 and AER_2) obtained simultaneously from bilaterally symmetric derivations or sequentially from the same derivation under two different conditions can be compared quantitatively, by means of the familiar *t* test. *t*, at each point t_i along the analysis epoch, can be defined as

$$t(t_i) = \frac{AER_1\,(t_i) - AER_2\,(t_i)}{[\sigma_1{}^2(t_i) + \sigma_2{}^2(t_i)]^{1/2}}$$

The significance of the difference between two AERs at any point t_i can be evaluated by computing the *t* test across the epoch and consulting a conventional statistical table giving the probability of obtaining any given value of *t* between two samples by chance. The possibility of obtaining significant values of *t* by chance because so many *t* tests are being computed can be handled by calculating the sequential probability of obtaining a number of successive significant *t* values in a row.

III. OBJECTIVE COMPARISONS BETWEEN AERS

A. Symmetry of the AER

Particularly in view of the inhomogeneities which have been observed within samples of evoked responses during the course of many years of working with averaged evoked responses, as well as the evidence cited in previous chapters that the waveshape of the AER is sensitive to endogenous influences in addition to the characteristics of the afferent input, it is difficult to decide whether a particular AER waveshape displays normal morphology. As with the spontaneous EEG, one strategy to circumvent this issue consists of using one hemisphere as the control for the other. Evidence cited in Chapter 2 indicates that AERs from homologous placements in normal subjects display excellent symmetry.

Two methods for quantitative evaluation of AER symmetry exist: first, the cross correlation between AERs derived simultaneously from bilaterally symmetric derivations can be computed; second, the significance of differences between bilateral pairs of AERs can be computed using the *t* test. Normative values for AER symmetry in a variety of head regions of a large population of normal subjects have been published (Harmony *et al.*, 1973b) and have established that symmetry values are normally quite high. These data are summarized in Chapter 5.

B. Assessment of Effects of Altered Conditions

AERs obtained under two different conditions can also be compared quantitatively using cross correlation and *t*-test indices. The basic design of such comparisons is as follows: first, a sample of AER under the first condition is obtained, to which we will refer as AER_{11} ; second, a replication of this sample is obtained under identical conditions, referred to as AER_{12} ; third, the relevant situational parameter is changed and the first sample under the new condition is computed, referred to as AER_{21} ; finally, a replication of the measurement under the new condition is obtained, referred to as AER_{22}.

If AER_{11} is significantly different from AER_{12} , the base-line condition is unstable. The source of instability may be too small a sample size or variation in a variable significantly altering state, such as attention or habituation. Until stabilization is achieved, proceeding to change conditions can only produce ambiguous results. However, let us assume that AER_{11} is not significantly different from AER_{12}. Acceptable replicability has been achieved. Now compare AER_{12} with AER_{21}. If a significant difference is found, the change may be due to the new condition or to emergence of instability in the system. The latter alternative can be excluded with reasonable confidence if AER_{22} is *not* significantly different from AER_{21}; that is, if the new response can be replicated. The significance of interaction effects might be estimated by reversing the sequence of conditions and comparing the responses to the same condition in the different sequences.

The procedure described above, which might be termed the "successive *t*-test strategy," lends itself to a wide variety of applications where it is desirable to use AER measures to identify significant changes in brain response but where the precise nature of the response is uncertain or difficult to define in advance morphologically. Such applications might consist of ascertaining drug dosages required to produce a significant change in brain state, identification of thresholds for sensory acuity, or changes brought about by experience, as examples. In Chapter 6, neurometric studies using the successive *t*-test strategy to measure visual acuity, color, and shape perception will be described. Those studies were carried out using a small special-purpose computer designed to perform successive *t* tests on line (Neurodata Statistical Analyzer VT-100).

IV. PATTERN RECOGNITION METHODS[1]

The methods described in Section III permit quantitative comparison of AERs recorded simultaneously from symmetrically located electrode pairs or sequentially from the same electrodes under two different conditions. While those neurometric methods circumvent the necessity to deal with the particular morphology of the AER, they also evade the issue of how morphology can be described quantitatively. There exists a substantial body of evidence that brain damage or dysfunction or change in state causes characteristic changes in morphology of EPs as well as EEGs (see Chapters 2, 5, and 6). Unless we are willing to accept the necessity that such changes be visually evaluated, which would constitute a regression to the subjective methods based upon personal

[1] We wish to acknowledge our indebtedness to several of our colleagues, without whom we would still be strangers to pattern recognition procedures. Dr. Daniel S. Ruchkin collaborated with us in our original explorations of factor analysis. Dr. Paul Easton substantially extended the power and utility of the method by devising powerful computational innovations and by introducing Varimax techniques into our laboratory. Dr. Eric Schwartz introduced us to other statistical pattern recognition methods. Most of the sections on cluster analysis and multidimensional scaling were taken from a survey of these methods which he was kind enough to make available.

judgment and experience already judged as unacceptable for EEG analysis, we must come to grips with the necessity to devise neurometric or quantitative descriptions of AER as well as EEG waveshapes. The remainder of this chapter will describe techniques intended to cope with this problem.

A. Template Methods

One class of pattern recognition techniques can be described as "template" methods. By this we mean that the data analysis procedure involves assumptions about the features of some particular event being sought or the general nature of the structure of the solution to some problem. A variety of template methods are described in the electrophysiological literature and are discussed briefly below.

1. Amplitude Sorting

We referred earlier to the hazards of assuming that sets of EPs are homogeneous and discussed the necessity for decomposing samples of EPs into homogeneous subsets, or modes, before averaging or classification attempts. In our discussion of variance computation, we pointed out that peaks in the variance curve will be located at those latencies in the analysis epoch where nonstationarities exist in the EP waveshape. If a particular component can vary with respect to amplitude or latency in a graded fashion, or can be present or absent, while other portions of the EP behave differently or are stable, a peak will appear in the various curves. Ruchkin (1965) devised a method for classifying, or "sorting," single EPs based upon this phenomenon. His method first examines the distribution of EP amplitudes in the set, at the latency where the variance peak occurs, and tests that amplitude distribution to establish whether it deviates from Gaussian. If so, amplitude values are specified that truncate the observed distribution into significantly different subgroups. The individual EPs are then classified one by one by comparing their amplitudes at the critical latency with the selected voltage reference levels. This process is continued until the average of the EPs classified similarly displays a variance curve which is essentially flat. For examples of this procedure, see John (1973a) and John, Bartlett, Shimokochi, and Kleinman (1973).

This method has produced striking correlations between EP waveshape and behavior (see Chapters 9 and 10, Volume 1). The relevant templates are defined by examination of the variance curve, and are based upon the assumption that heterogeneity of variance of EP waveshape may well reflect the different functional significances of the various members of a set of EPs.

"Light" or "sliding" averages. The attempt to use amplitude sorting, or any other method of classifying single EPs, may founder upon an unacceptably poor signal-to-noise ratio in the data. We have found that a remarkably effective

stratagem which often resolves this problem consists of constructing an average response "window" only a few EPs wide, that "slides" along the data in time; viz.

$$\frac{EP_1 + EP_2 + EP_3}{3}, \quad \frac{EP_2 + EP_3 + EP_4}{3}, \quad \frac{EP_3 + EP_4 + EP_5}{3}, \quad \text{etc.,}$$

comprise a light average of three EPs sliding along a series at the rate of 1 EP shift per light average. The size of the light average which sufficiently enhances the signal-to-noise ratio to permit further analysis without obliterating short-term state changes must be empirically determined. This method rests upon the assumption that most changes of brain state will occur relatively slowly compared with the intervals between successive EPs.

2. Cross-Correlation Methods

Whether applied to the spontaneous EEG, the AEP, or the individual EP, template methods can be viewed as a process of scanning a long train of samples of electrical waveshapes to identify recurrences of the template. Computation of the cross-correlation coefficient between the template and the train of samples seems one straightforward approach to the problem. A major shortcoming of this method is that cross correlation is energy-sensitive; fast events, even if quite large, are readily overwhelmed by low-frequency activity. Weinberg and Cooper (1972) were relatively successful in overcoming this difficulty with human electrophysiological data by separating the high- and low-frequency activity in the template and carrying out separate computations. This procedure, although yielding interesting results in the study cited, is computationally time consuming. Template cross-correlational methods have been applied to the detection of epileptiform spikes in the EEG by Zetterberg (1973) who detected spikelike pulses by passing a received EEG signal through a suitably shaped digital filter. Using a similar strategy, Kaiser, Petersén, and Magnusson (1973) digitized each EEG full wave as a 14-bit data word describing critical wave features and compared this word to those representing stored reference waves. Both of these procedures essentially rely for identification of the sought for feature upon an acceptable match (sufficiently high cross correlation) between the template shape specified by the digital filter and the EEG signal.

In a somewhat more sophisticated attempt to evade the problem of how to define the template, Viglione and Martin (1973) have utilized a two-phase procedure. The first step forms frequency characteristics of different sleep stages from patterns selected by human analysts. Next, a learning algorithm weights various features of the EEG differently to modify the decision logic classifying the spectral patterns.

3. *Cross-Spectral Analysis*

It is not necessary to calculate the cross-correlation function by shifting the template along the sequence of ongoing activity and actually computing the cross correlation between the template and each segment. Instead, one can make use of the shifting theorem and the cross-spectral theorem (Duda & Hart, 1973) of Fourier analysis. These theorems show that one can obtain the cross-correlation function between a given template and a given segment of ongoing spontaneous activity by performing the Fourier transform, the inverse Fourier transform, and multiplication (Bendat & Piersol, (1971). Since the Fourier transform and its inverse may be calculated very efficiently by use of the fast Fourier transform (FFT) algorithm (Cooley & Tukey, 1965), the amount of time required to scan data for such an analysis can be reduced by a factor of about 100 (Bendat & Piersol, 1971). If huge volumes of data, such as are typical in clinical applications, are to be examined for the appearance of certain features, computational time becomes a critical consideration.

4. *Adaptive Filtering*

The adaptive filter method (Woody, 1967) can be viewed as an extension of the cross-spectral technique in which the template is iteratively redefined. Each successive pass through the EEG sample locates those points where the cross correlation is maximal. Those segments are then averaged together to define a new template and the process is repeated. This iterative process has been shown to converge rapidly for a variety of electrophysiological signals (Woody, 1967). Adaptive filtering, combined with cross-spectral analysis based upon the FFT, provides an extremely powerful tool for the efficient and accurate location of recurrences of a criterion or template process in a large volume of electrophysiological data.

B. Cluster Analysis

Clearly, all of the template methods discussed in Section IV.A above suffer from the constrain that someone must specify the basic characteristics of the process or event for which a search is to be conducted. Even in the case of adaptive filtering or learning algorithms, not only is the starting point for the search defined by an a priori template specification, but the success of the search depends upon establishing the existence of a reliable relationship between some electrophysiological feature and some independently judged function, a behavior or an event. Suppose one wishes not to make any assumption about the nature of the solution, both because a priori assumptions often bias solutions along certain pathways and because one well may not have any compelling intuition or insight into the nature of the structure of a solution to the problem of concern.

Cluster analysis is the name given to a class of analytic methods that describe the details characterizing the structure of a body of data. Cluster analysis methods have been reviewed by Ball (1965) and many considerations relevant to the use of cluster analysis have been thoughtfully discussed by Larsen, Ruspini, McNew, Walter, and Adey (1973), Meisel (1973), Chen (1973), Duda and Hart (1973), and Fukunaga (1972).

Basically, cluster analysis offers the enormous advantage that it permits the structure of a body of data to be analyzed in the absence of any prior assumptions about the nature of this structure. In many of the problems confronted as one attempts to devise analytic neurometric solutions to electrophysiological problems, there exists little or no theoretical guide to the significant structures that may exist in the data; in fact, it is just this information which is being sought. For example, one cannot assume that a set of evoked potentials follows a multivariate normal distribution. Such statements, if true, must be established by the analysis itself.

Nonetheless, the investigator must provide some guidance, which is limited to four kinds of decisions:

1. He must specify the features of the data to be considered. These features constitute a vector, an array of numerical attributes that most efficiently characterize the data. The choice of what these features may be is very much dependent on the intuition and experience of the experimenter; in addition, the identification of the most significant features is itself an important goal of the analysis. Specifically, the feature vector for an evoked potential might be the amplitude sampled at regular time intervals, (Ruchkin, 1971) or its derivatives (Hjorth, 1970). It might be the phase and amplitude spectrum of the individual EP, or any measure of the data that seems significantly useful in abstracting the information carried by the data. The correct choice of the feature vector is crucial to the success of the following cluster analysis. If the feature vector fails to include the significant features of the data, it must be obvious that any subsequent analysis is doomed to failure. An old dictum about computation states: "garbage in, garbage out!" Computational procedures are enormously and increasingly powerful, but they have not yet and probably never will relieve the investigator of the requirement that he be able to think sensibly about his problem.

2. Some measure(s) of similarity must be defined. Such measures can essentially be considered as a distance function in the multivariate data space, and will be the basis for construction of quantitative statements about the degree to which a given EP or EEG sample is "like" another. Similarity measures that might be used are the correlation coefficient between two EPs (Easton, unpublished results 1971) which is equivalent to the Euclidean distance metric, the generalized Minkowski r metric (Kruskal, 1964), the divergence distance of information theory (Kullback, 1959; Ryan, 1968), or a wide variety of other

distance functions that have been suggested in the literature for various purposes.

3. Criteria must be defined such that two EEG or EP samples, related through their *feature vectors* by some *measure of similarity*, will be included in the same or different clusters. A variety of approaches to this problem has been suggested. Perhaps the most general is the "similarity matrix" approach (Chen, 1973). An alternative procedure is to present the data structure graphically and to use the pattern recognition capabilities of the experimenter. This method is risky, depending upon the occurrence of unequivocal structure in the data to avoid the problems of subjectivity and bias on the part of the experimenter. This shorter method will be referred to as *multidimensional scaling* (Kruskal, 1964; Sammon, 1969) to distinguish it from cluster analysis, which presupposes that some decision criterion has been provided. An interesting approach is that proposed by Ruspini (1969, 1970) to handle the problem of "fuzzy" sets, that is, sets to which a degree of belongingness is allowed. Rather than lying within or outside a given set, a particular point can be inside to a greater or lesser degree A mean square distance measure of belongingness is minimized, but conditional probabilities provide estimates that a point belongs to any of the possible sets.

C. Discriminant Analysis

An early and interesting application of discriminant analysis to electrophysiological data was provided by Donchin, Callaway, and Jones (1970). Discriminant analysis can be considered as another analytic method for separating data into two or more groups, or clusters, for deciding which of several possible clusters most probably contains a particular data point.

Discriminant analysis can be conceptualized as a mapping of each feature vector into a multivariate space. Let us suppose that it is desired to ascertain the features characterizing or discriminating two groups of data. After projecting the feature vectors into the hyperspace, the centroids of the two groups are then computed and a line or discriminant vector is drawn in the space. The individual points representing the feature vectors are then projected upon this line. A criterion point is defined, usually the mean of the two projected centroids, and the two groups are then classified with respect to where the points representing the members of each group fall with respect to that criterion. The success of the discrimination is defined by the amount of *overlap* between the resulting distributions. A measure for the amount of overlap is the ratio of the distance between the means to the spread within the distribution. Discriminant analysis is a technique for finding an optimal vector, which minimizes the spread within distributions and maximizes the separation between the centroids or the two sets of points projected upon that vector, thus providing the best possible discrimination between the two bodies of data.

In a sense, discriminant analysis is the converse of cluster analysis. Discriminant analysis ascertains, given a set of features characterizing two or several groups of data, which features best discriminate between the two groups and what are the clusters of feature vectors corresponding to that discrimination. Given a set of feature vectors, cluster analysis ascertains how many groups are contained within that structure and which elements most probably belong to each group.

D. Multidimensional Scaling

Multidimensional scaling is a graphic approach to cluster analysis (Kruskal, 1964; Sammon, 1969). One computes a distance between all pairs of feature vectors in the multidimensional data space. This distance is a single number for each pair of feature vectors, $N(N) - 1)/2$ in all for N feature vectors. Then a search is made in a lower dimensional space for a set of vectors for which the $N(N-1)/2$ distances computed in the original feature vector space is the same. In particular, if the final space is two or three dimensional, it is possible to make a scatter plot of the final configuration of points. Visual inspection of this scatter plot may directly reveal the desired information about the structure of the data, making a formal cluster analysis unnecessary. In the event that a formal cluster analysis is performed, multidimensional scaling may be still used for a graphic presentation of the data; in this case, boundaries in the data that separate clusters would be determined by the cluster analysis; the multidimensional scaling would be merely for data presentation.

To summarize: multidimensional scaling is a (nonlinear) mapping of distances between data points. The original distances are in the (high-dimensional) space of the feature vector, and the distance function is the similarity measure; the final space is a two- (or three-) dimensional space of points with the property that the distances between these points are equal to the original distances. The distance function used in the final space is not necessarily the same as the similarity measure; a possible choice is the Minkowski L^r metric (Kruskal, 1964) for some r. A number of questions that are raised by this procedure are as follows:

1. *Feasibility*. It is not clear that there exists a two-dimensional representation of data points scaled down from an arbitrary and large-dimensional feature space that satisfies the description given above. In fact, it is necessary to define a quantity, the *stress*, which is a sort of least squares measure of the fit of the final set of distances to the original set. Practically, the stress is specified in a given multidimensional scaling run that is satisfactory; the computer iteratively varies the final configuration of points, and stops when the specified stress is achieved. Monte Carlo studies of the statistical properties of the stress have been performed, and curves are available that provide a confidence level for the reliability of a given scaling with a given final stress value (Klahr, 1969).

2. *Creating nonexistent structure.* It is, of course, undesirable to obtain structure in the final space that does not exist in the original feature space! A closely related problem is that of losing structure in the final space that actually existed in the original. Data indicate that, for more than 10–15 feature vectors in a given run, there is essentially zero probability to obtain artifactual structure in the final space (Klahr, 1969).

As far as we know, we are the first group to apply multidimensional scaling techniques to the problem of classification of evoked potentials. This method has been applied with extremely promising results to the analysis of single evoked potentials recorded from electrodes chronically implanted in cats in a differential generalization paradigm. The resulting clusters achieved an impressive separation of EP waveshapes elicited by the same indifferent stimulus in behavioral trials leading to performance of two different conditioned responses (Ramos, Schwartz, & John, 1974). In a different application, illustrated in Chapter 5, it was possible to achieve excellent separation of AERs to visual stimuli computed from groups of normal subjects and patients with brain tumors (Schwartz & John, unpublished observations, 1974).

V. MULTIVARIATE FACTOR ANALYSIS

A. Limitations

Factor analysis can be considered as a particular example of cluster analysis. Factor analysis attempts to infer the structure of a set of data from correlations between all pairs of elements in the set. Cluster analysis treats the multivariate structure of the data directly. Factor analysis can be effective if an adequate description of the structure is provided by the second-order statistics of the data. If this is not the case, factor analysis will be inadequate. Distributions of data for which second-order statistics fail to distinguish essential features of structure can readily be devised. If the data are multimodal and this is not explicitly recognized, the covariance matrix will be critically sensitive to the distance between the means of the two modes. The covariance matrix itself, derived by computation of the correlation coefficients between vectors representing pairs of waveshapes, is inherently insensitive to high-frequency, low-energy components; this same weakness was pointed out in our discussion of the use of cross correlation as a similarity measure in pattern recognition methods utilizing the template approach. Major influences in the data overwhelm those that occur more rarely. Weak effects or rare events are likely to be obscured. Atypical data points with extreme values can have an inordinate effect on the covariance matrix, even though they occur rarely. Factor analysis permits an infinite set of solutions for the orientations of the axes that span the data space. Recourse to arbitrary definitions must be used to select a particular rotation of the basis vectors, or axes. No physiological rationale exists for selecting any particular orientation on theoretical grounds. Most of the mathematical methods for selecting a rotation are unduly influenced by the locus of the largest cluster of data points. Finally, and most serious, the mutual orthogonality of the set of basis vectors or axes provided by *any* factor analytic procedure is equivalent to the assumption that the essential multivariate features imbedded in the set of data are independent, and that any feature vector can be accurately described as a linear combination of a limited subset of those independent features. There is no justification for the assumption, for example, that the neuroanatomic systems responsible for the electrogenesis of the disparate features of the sensory evoked potential operate independently from one another. In contrast, cluster analysis techniques are not limited by most of these constraints. They take advantage of the actual structure of the data set and utilize its multivariate dependence to restrict permissible descriptions of inherent structures. For all these reasons, it seems likely that although factor analysis methods have been effective tools for the analysis of electrophysiological data, as will be discussed in Chapter 5, the development of effective methods of cluster analysis and multidimensional scaling potentially offers a more powerful approach to establishing the real structure and dimensionality of electrophysiological data.

The long list of shortcomings and reservations about inherent properties of the factor analytic approach, enumerated above, may seem to the reader equivalent to an indisputable demonstration that factor analysis is highly unlikely to be of any significant utility for analysis of electrophysiological data. Actually, factor analysis has proved to be a remarkably effective technique for the quantitative and parsimonious representation of orderly relations within large quantities of electrophysiological data. This may indicate that the reservations above do not apply in a strong form to data representing the electrical activity of the brain, or may merely mean that thus far the effects studied have been sufficiently strong to emerge in spite of those shortcomings. The remainder of this chapter will outline the methods used and describe some of the results that have established the utility of factor analysis for the concise and meaningful description of a variety of types of electrophysiological measures. In Chapter 5, evidence will be presented showing that factor analysis provides as powerful a method as any other yet studied, with the possible exception of multidimensional scaling, for the automatic computer detection of neuropathology. The fact that an analytic method with all the shortcomings detailed in the preceding discussion can be so effective gives us every reason to believe that, when the more powerful methods now on the horizon reach the point of routine applicability, the achievements of neurometric evaluation of electrophysiological data will not only be quite impressive, but will assuredly enable a far higher magnitude of resolution of fundamental brain mechanism affected by pathology and involved in sensory, perceptual, and cognitive processes than has been available with the qualitative methods still overwhelmingly in daily use.

B. Factor Analysis Methods

After conversion to digital form, electrophysiological data can be described quantitatively in a number of ways. The quantitative data must not only offer a precise description of the original activity, but ideally should also provide a way to extract from that activity parsimonious descriptions of the fundamental processes underlying electrophysiological activity; differences between those processes in different brain regions or in different brain states should also be expressed.

We have made substantial progress toward these goals using *factor analysis* of sets of averaged evoked potentials. Some of our results will be used to illustrate what such methods can accomplish in evaluating the effects of drugs and various types of neuropathology on the AER. Since the focus of this chapter is on concepts rather than specific technical details, we will not attempt to describe the mathematical procedures that have been presented elsewhere (John *et al.*, 1964, 1972, 1973). A general description of the method of factor analysis is provided, however, to facilitate understanding of the remainder of this chapter.

Figure 3.8 depicts a factor analysis of a number of averaged evoked potentials, which appear in the top row. The first three factors are shown in the column at the left. The row of waves to the right of the first factor shows the contributions

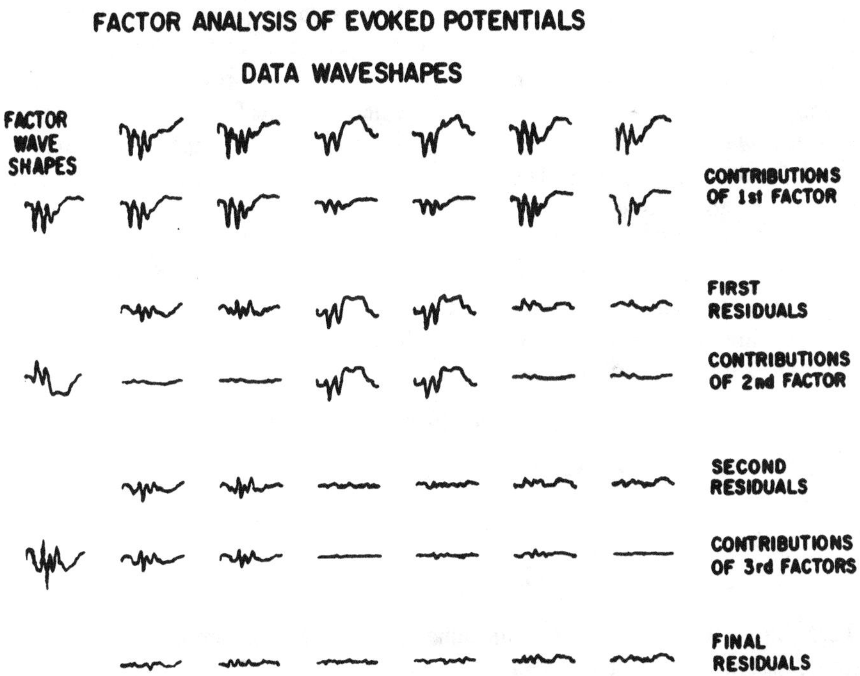

FIG. 3.8 Factor analysis of a set of evoked potentials, in graphic representation.

of this factor to each of the data waveshapes. Each of these is simply the factor multiplied by the appropriate weighting constant. The weighting constant may be negative, which has the effect of inverting the factor shape. The weighting constant is called the *factor loading* of the data waveshape, and is in fact the correlation coefficient between the original data waveshape and the factor.

The next row of waves, labeled "first residuals," shows what is left after the contribution of the first factor has been subtracted from the original data waveshape. In the following rows this process is repeated for the second and third factors; the "final residuals," which are unaccounted for by the three factors, are displayed at the bottom. In this way, one can represent a great many diverse waveshapes as different combinations of only a few basic factors. Each data wave can be described by the set of its factor loadings, leading to great compression of the data. If the final residual for a particular wave were zero, the sum of the squares of its factor loadings would equal one, and we could say that 100% of the energy of the wave was accounted for by the factors. As the residual becomes larger, the percentage of the energy accounted for by the factors becomes proportionately smaller.

The collection of evoked potentials that can be described by a given set of factors is called the *signal space* defined by those factors. The *dimensionality* of the signal space is the number of factors required to account for a predetermined percentage of the energy of the original set of waveshapes.

The most common method of factor analysis is called *principal component analysis* (Harmon, 1960). Principal component analysis accounts for the energy in the set of signals in the most parsimonious way. Various rotations of this set of factors, or *axes*, can be defined, each of which accounts for the energy in the space or *spans* the space. Analogously, the quadrants of the compass might have been defined in various ways. The method we have found to yield orientations of axes that correspond best to physiological processes is called the *Varimax* rotation (Kaiser, 1958).

Whatever method is utilized, the general feature of factor analysis is that a set of evoked potentials, W_i, representing the responses of different derivations to the same stimulus or of the same derivation to different stimuli, can be described in terms of a common set of factors, or axes, as shown below:

$$W_1 = a_{11}F_1 + a_{12}F_2 + \ldots + a_{1K}F_K,$$

$$W_2 = a_{21}F_1 + a_{22}F_2 + \ldots + a_{2K}F_K,$$

$$\vdots$$

$$W_N = a_{N1}F_1 + a_{N2}F_2 + \ldots + a_{NK}F_K.$$

Each wave, W_i, is described as an equation in which the amount a_{i1} of the process described by factor 1, $a_{i1}F_1$, is added to the amount a_{i2} of the process described by factor 2, $a_{i2}F_2$, and so on, until a predetermined amount of the energy in the wave has been described with specified accuracy.

If the separate waves in the set are completely independent or individually determined, the number of dimensions, K, needed to describe a set of N waves will be equal to N. If the N waves are generated by the interaction of a smaller number of underlying processes, K will be smaller than N. The important fact for electrophysiology is that the number of factors needed to account for a set of responses from different derivations is usually far smaller than the number of derivations (John *et al.*, 1964, 1972, 1973). It has been shown that as the number of waveshapes recorded from electrodes permanently implanted in the cat increases, the number of dimensions necessary to span the signal space reaches an asymptote and stabilizes (John *et al.*, 1964). The sum of the squared coefficients for the terms in each equation must account for 100% of the variance. Thus, each coefficient defines the percentage contributed by that factor to the whole process, or the square of the projection of the process upon the corresponding dimension.

This analysis, then, meets the desired criterion of allowing us to describe many different waveshapes with great precision, in terms of a small number of common processes which are combined in differing amounts to account for the observed electrophysiological phenomena. Changes in state should either change the relative amount of these processes in particular regions of the brain, or should cause new dimensions to appear. This method particularly focuses upon the subtle details of response waveshape, but replaces the skilled eye of the expert by an automatic "pattern recognition" procedure.

C. Classification of Drugs

This neurometric method has been successfully applied to diverse electrophysiological problems (Bennett *et al.*, 1971; Donchin, 1966; Elmgren & Lowenbard, 1969; Halas & Beardsley, 1969; Naitoh *et al.*, 1971; Suter, 1968, 1970). Recently, we used it to evaluate the action of several different doses of each of four drugs upon evoked potentials from arrays of electrodes permanently implanted into a variety of brain structures in three cats (John *et al.*, 1972, 1973). Drug doses were at least a week apart. Averaged evoked responses to two different stimuli were obtained from each of 12 derivations under each of 24 conditions; 11 control conditions before drug administration, and 13 experimental conditions after administration of various drugs. These waveshapes were arranged in a matrix of 24 columns (12 derivations multiplied by 2 stimuli) by 24 rows (11 control and 13 experimental conditions). Each column from this matrix, consisting of the 24 averaged evoked potentials recorded from a single derivation in response to one type of conditioned stimulus under 11 control and 13 experimental conditions, was subjected to a separate factor analysis. We called such treatment *column* analysis. The raw data that formed the input to a typical column analysis and the regression equations that described the waveshapes as a result of factor analysis are shown in Fig. 3.9.

Note that the control and postsaline waveshapes received most of their energy from factor 1, the waveshapes after chlorpromazine (CPZ) injections from factor 2, waveshapes after MJ injection [MJ is the tranquilizer Buspirone (Wu *et al.*, 1972)] from factor 3, the waveshapes after phenobarbital (PHENO) injection from factor 2, and the waveshapes after methamphetamine (METH) from factor 5. These results indicated that the set of control waveshapes coexisted in a space well described by factor 1. The various drugs added new "dimensions" to the brain signal space defined by factors 2, 3, and 5.

These new dimensions are shown in Fig. 3.10, which depicts the vectors describing the average state of three different cats provided by column analysis after three doses of CPZ, three doses of MJ, three doses of PHENO, and two doses of METH. Notice that each drug moves the state vector in a characteristic direction in this 2-3-5 factor hyperspace. Normally, the signal vectors of the brain have no energy in this space at all. While some drugs act only in one dimension, merely increasing their loading in that direction as the dose increases, other drugs change the dimension of their action at different dosages.

D. Factor Analysis of AER in Humans

These methods, when applied to data from humans, may have much clinical value. Using a cluster analysis method described by Wishart (1969), implemented on a CID201A, Valdes (1973) at CENIC[2] analyzed the averaged evoked responses obtained from left occipital monopolar derivations in 30 normal subjects. He obtained five clusters of waveshapes. The waveshapes (a–e) correspond to the five cluster centers shown in Fig. 3.11 (top). These results were confirmed in subsequent work (Valdes *et al.*, 1975) and correspond to those previously described by Arnal *et al.* (1972) using correspondance factor analysis. Valdes then submitted the five cluster center waveshapes to principal component factor analysis. It was possible to account for better than 85% of the energy in that system with 2 factors, C_1 and C_2, which are illustrated in Fig. 3.11 (bottom).

Valdes went on to analyze the averaged evoked responses recorded from 10 homologous derivations. The waveshapes from all these derivations within a single subject could be described by three factors. The original data waveshapes, factor waveshapes, and the coefficients of the regression equations reconstructing the data waves as linear combinations of the factors are shown in Fig. 3.12.

Valdes, Harmony, and John then carried out factor analyses with Varimax rotations on averaged evoked potentials recorded from bilateral occipital, tem-

[2]National Scientific Research Center for Cuba (Centro Nacional de Investigaciones Científicas). (Personal communication.)

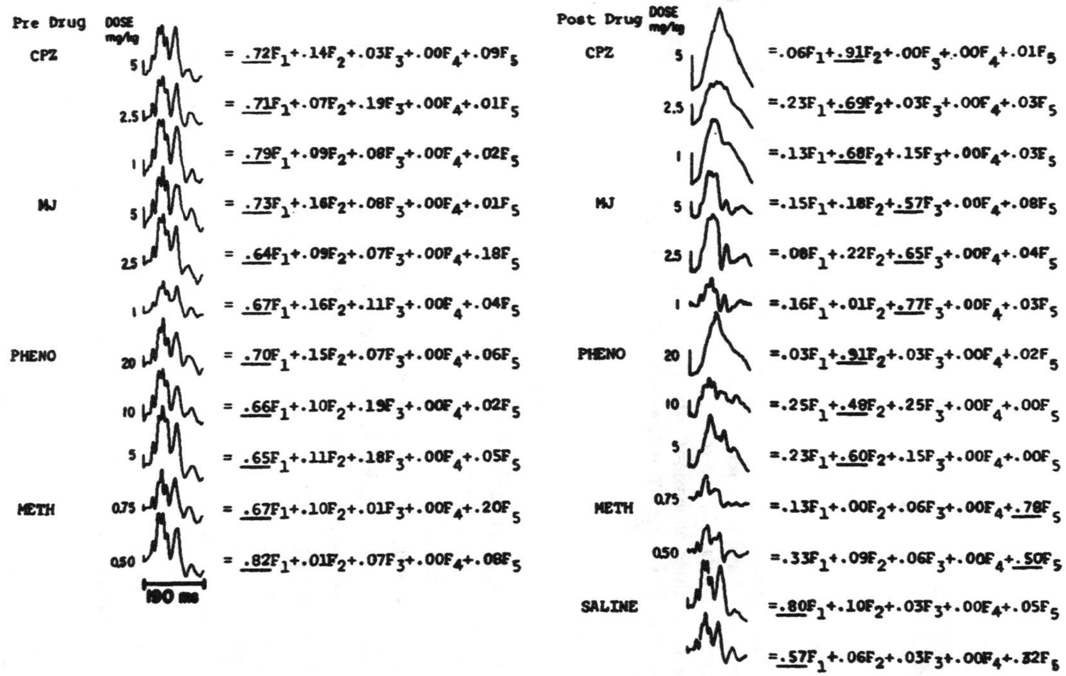

FIG. 3.9 Average response waveshapes elicited from the visual cortex of a cat, bipolar derivation, in response to 200 presentations of an approach CS (5 HZ/sec) when one of four drugs [chlorpromazine (CPZ), buspirone-(MJ), phenobarbital (PHENO), or methamphetamine (METH)] had been injected, during 11 control sessions (left) and during 13 experimental postdrug sessions (right). The regression equation to the right of each waveshape accounts for the energy of the waveshape as a linear combination of the Varimax factors, with each coefficient defining the percentage of the total energy of the waveshape contributed by the corresponding factor. Analysis epoch, 190 msec. (From John *et al.*, 1973.)

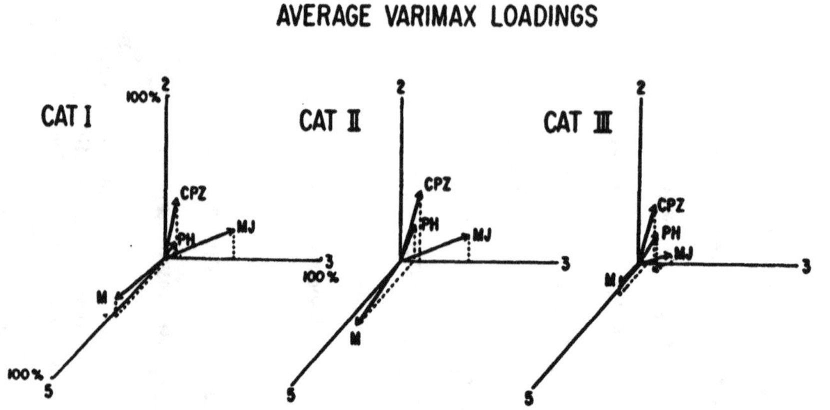

FIG. 3.10 The effects of four different drugs on the evoked responses of three cats, described in three-dimensional subspace, representing Varimax factors 2, 3, and 5. The length of each vector defines percentage of the total signal energy in the subspace, while the component along each of the three dimensions corresponds to the percentage of the energy distributed to the vector by that factor. The diagram for each cat is based upon the analysis of 576 average response waveforms. (From John *et al.*, 1973.)

poral, and central monopolar and bipolar derivations for 10 normal subjects. Data were amplified by an Alvar EEG, averaged with a CAT 400C, and plotted on an $X \leftrightarrow Y$ plotter for manual digitizing. Factor analyses were performed using a CID201A. Three factors spanned the space occupied by the full set of 120 AERs (6 derivations X 2 sides X 10 subjects). For each subject, as few as two and no more than three factors were needed to account for an average of 94% of the energy in the full set of data.

These results showed that, for a given derivation, a limited variety of AERs existed in a group of subjects. These interindividual varieties could themselves be accounted for as combinations of a few factors, with different loadings. Further, the intraindividual waveshapes from various areas in a given subject could also be accounted for by a few factors.

These findings indicated that different linear combinations of a relatively small number of basic processes might account for the waveshape of the visual evoked potential from any scalp area in any subject. Could the observed interindividual and intraindividual factors be reconciled into a combined set capable of describing the signal space of "normal subjects"? To answer this question, we took the full set of 10 evoked potentials from each of 48 normal subjects and factor analyzed them (John, Ahn, & Easton, unpublished, 1974). We found that 8 factors could account for 93% of the energy of a total of 480 evoked potentials, obtained from bilateral occipital, temporal, central, centrooccipital, and occip-

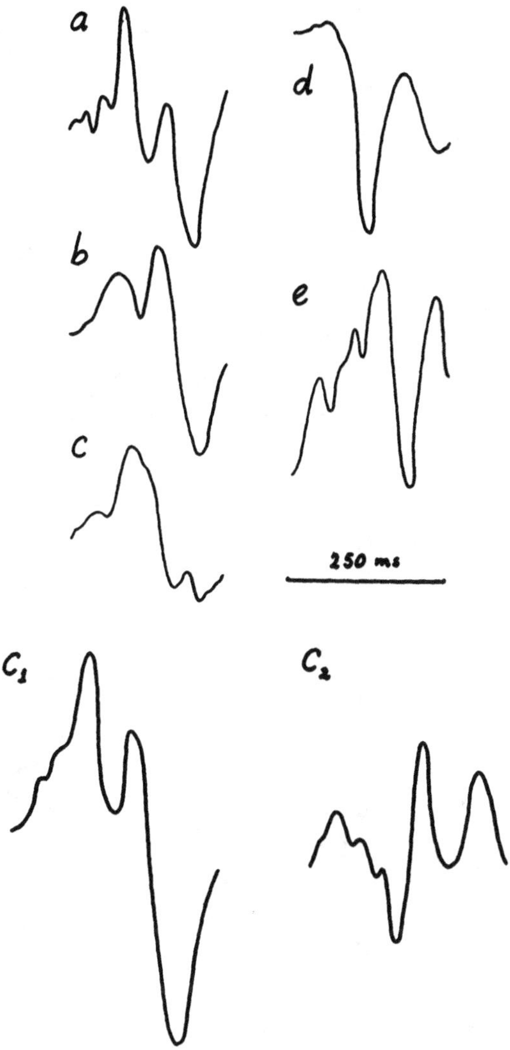

FIG. 3.11 Top (a – e): Cluster centers describing the 5 different types of visual evoked responses obtained from monopolar occipital derivations in 30 normal subjects by use of cluster analysis. Bottom (C_1 and C_2): Factors accounting for 85% of the energy. (From Valdes, unpublished, 1974.)

itotemporal derivations from each of 48 normal subjects. These ideas are illustrated in Fig. 3.13.

The enclosed volume is the hyperspace containing evoked responses from 10 derivations from a group of 48 normal subjects. A small group of factors was capable of describing every derivation in normal subjects, with a small overall

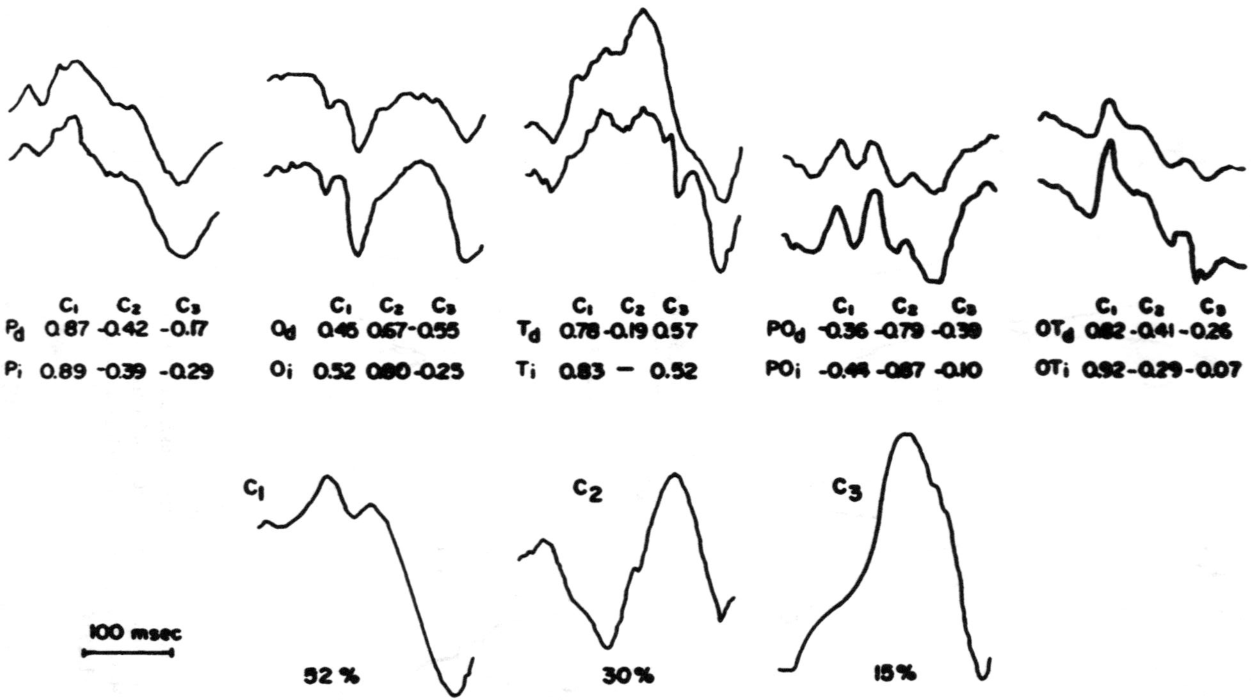

FIG. 3.12 (A) Visual evoked potentials bilaterally recorded from a normal subject with five derivations. In each pair, the top waveshape came from the right side and the bottom from the homologous region on the left. (B) C_1, C_2, and C_3 are the weighting coefficients describing the contribution of factors C_1, C_2, and C_3 to each data waveshape. The three-factor waveshapes are shown together with the percentage of the energy in the original set of 10 data waves contributed by each of the factors. Derivations were parietal (P), occipital (O), temporal (T), parietooccipital (PO), and occipitotemporal (OT). d = right side (derecho); i = left side (izquierda). (From Valdes, unpublished, 1974.)

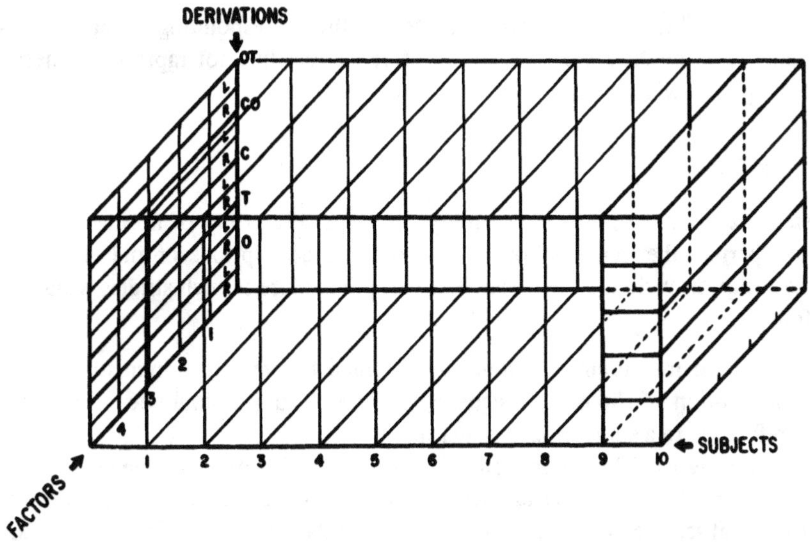

FIG. 3.13 Diagrammatica representation of the factor space of normal subjects. For a group of normal subjects, all averaged evoked potentials recorded from a set of derivations can be described by a limited number of factors. Regression equations can then be written which account for most of the energy of any averaged evoked potential as a linear combination. Ninety-three percent of the energy in 480 average responses, recorded bilaterally from five derivations in 48 normal subjects, could be accounted for by 8 factors. (For clarity, the "subject" axis only extends to 10 subjects.) For details, see text.

error. As a first approximation, based on a reasonably large sample, the visual evoked potential signal space of normal subjects can be described by about 8 factors. Sufficient data on which to base an adequate analysis should yield a set of factors capable of describing any visual evoked potential recorded from any derivation in normal humans.

E. "Normal Spaces" and Screening for Pathology

For effective utilization of factor analysis in screening for neuropathology, it will be necessary to construct a comprehensive set of tests of various aspects of brain function and to obtain neurometric indices of the responses to these tests from a large population of normal subjects. Presumably, such a set of tests would include indices of gross neuropathology and of sensory, perceptual, and cognitive processes. Factor analysis of the data so obtained should provide the definition of the normal signal space reflecting each of these processes. Grouping waveshapes with similar loadings on the different factors should yield clusters corresponding to groups of subjects with electrophysiological and perhaps functional similarities. In principle, individuals whose responses, for some items of the tests, did not lie within the normal signal space could be tentatively identified as "not normal," with respect to the corresponding functions. Potentially, this method would seem to offer a procedure for rapid neurometric screening for a wide variety of brain dysfunctions.

F. Drug Subspaces

Systematic administration of a variety of drugs to a group of normal subjects would permit the construction of neurometric descriptions of the effects of those drugs on various brain areas. The basic procedure in such studies would be as follows:

1. Standard measurements would be obtained before drug administration, so that description of the normal signal space of each subject could be constructed with reference to a standardized signal space.
2. The drug would be administered, and another set of measurements would be obtained. These new measurements would be regressed on the original normal factors; that is, one would determine the percentage of energy in the responses after drug administrations that could be accounted for by the normal factors. The drug might (a) cause a shift in the factor loadings of that individual to a position markedly different from his usual position, but still within the normal space; or (b) cause a shift which carries some or all of the energy of the response out of the normal space, into a new subspace created by the action of the drug. Shifts within the normal space would be revealed as a change in the coefficients defining the contribution of each normal factor to the postdrug state. Shifts out of the normal space would be revealed by the inability of the normal factors to account for the AER after drug administration. This new subspace could be defined by subtracting the portion of the vector inside the normal space from the vector after drug administration and constructing the residual vector, which lies completely outside the normal space, in a new drug space. This procedure, which we call regression factor analysis, is diagrammatically represented in Fig. 3.14.

We have used these methods to analyze the effects of a large variety of centrally acting drugs. Determination of the extent to which different drugs alter the electrical responses of a brain region so that they can no longer be described by normal factors provides an absolute metric that permits comparison of the *amount* of effect exerted by the different drugs. Factor analysis of the set of residual vectors obtained after administration of a number of different drugs provides a relative metric by which to evaluate the *quality* of the effects of the individual drugs. Since the residual waves thus obtained represent effects solely due to drugs, they define the "space of drug effects." Factor analysis of this space provides the dimensionality of drug action on the brain. We are now determining whether drugs with similar behavioral or clinical effects share the dimensionality of their residual waves, defined by this method. Figure 3.15

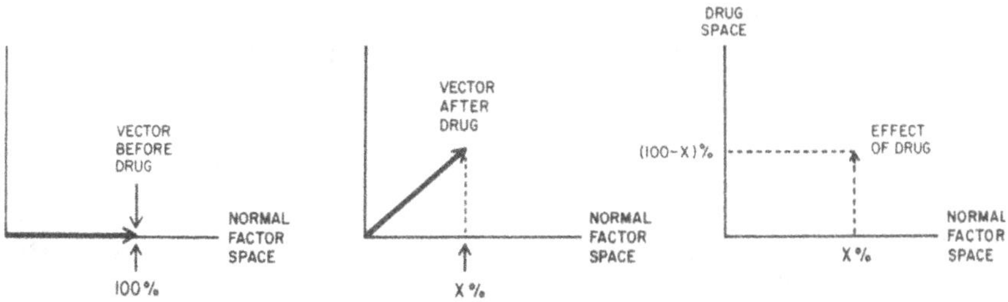

FIG. 3.14 Diagrammatic representation of the definition of a residual subspace created by a particular drug, derived from sequential factor analysis. For further details, see text.

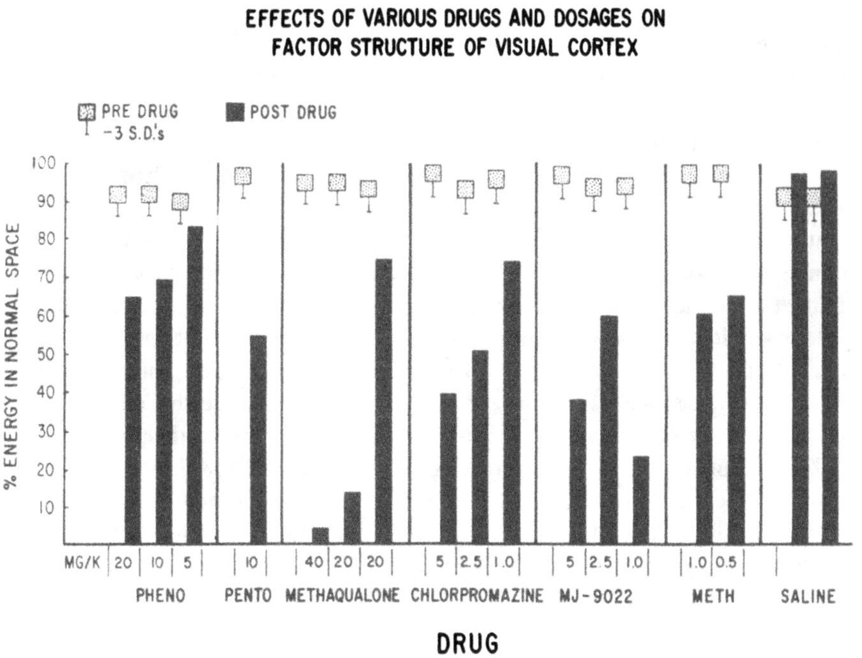

FIG. 3.15 Amount of energy in the normal factor space of the visual cortex of a cat after different doses of various drugs. The stippled squares represent the percentage of the energy of the averaged visual evoked potential before drug administration, and the bracket extends three standard deviations below that value. The black bars represent the amount of energy in the response after drug administration which can be accounted for by the first set of factors. Drug doses were administered intramuscularly at one-week intervals according to a latin square design.

illustrates the comparative amount of alteration in response of the visual cortex caused by different doses of various drugs, as ascertained by the procedure described above (John & Auerbach, unpublished observations, 1974).

G. Pathological Subspaces

By using the same methods, it should be possible to conduct a systematic survey of examples of different sorts of neuropathology, and to construct a quantitative description of the pathological subspace. Surveys could be made for children at different developmental stages, children with various behavioral disorders, mental retardates, psychiatric patients, geriatric patients, and narcotic addicts. The amount of pathological deviation would be defined as the percentage of the energy of the measurements of the individual patient which could not be accounted for by the normal factors. The cerebral region most affected might well be that deviating most extremely from the normal space. The portion of the energy of the measurements lying within the normal space would be subtracted from the overall measurements, so that the residual vector for each patient would be constructed. The region of the pathological subspace occupied by the residuals from an individual patient would probably aid in diagnosis of the pathology. Results of applying this neurometric method to diverse types of neuropathology are presented in Chapter 5.

In other words, a set of residual measurements would be obtained from groups of subjects representing different developmental stages as well as different types of pathology. The dimensionality of this total pathological subspace would be determined by factor analysis of the residuals. The regression equations describing the residual vectors in terms of the dimensions of the pathological subspace would constitute a classification of pathologies, assisting in diagnosis. Examination of the factor structure of pathology in an individual, the details of the factor loadings, and the change in such indices with successive determinations might prove particularly useful in providing a prognosis of the progress of the disorder in a given patient and an objective criterion for the efficacy of various therapeutic procedures. The flow chart of regression factor analysis just described is illustrated in Fig. 3.16, and the procedure is diagrammatically represented in Fig. 3.17.

VI. CONCLUSIONS

In this chapter, we have provided a review of the major presently existing methods for neurometric analysis of the electrophysiological activity of the brain. It should be clear to the reader that enormously powerful tools for precise quantification and systematic description of electrophysiological activity have already been developed. In conjunction with the review of the insights provided

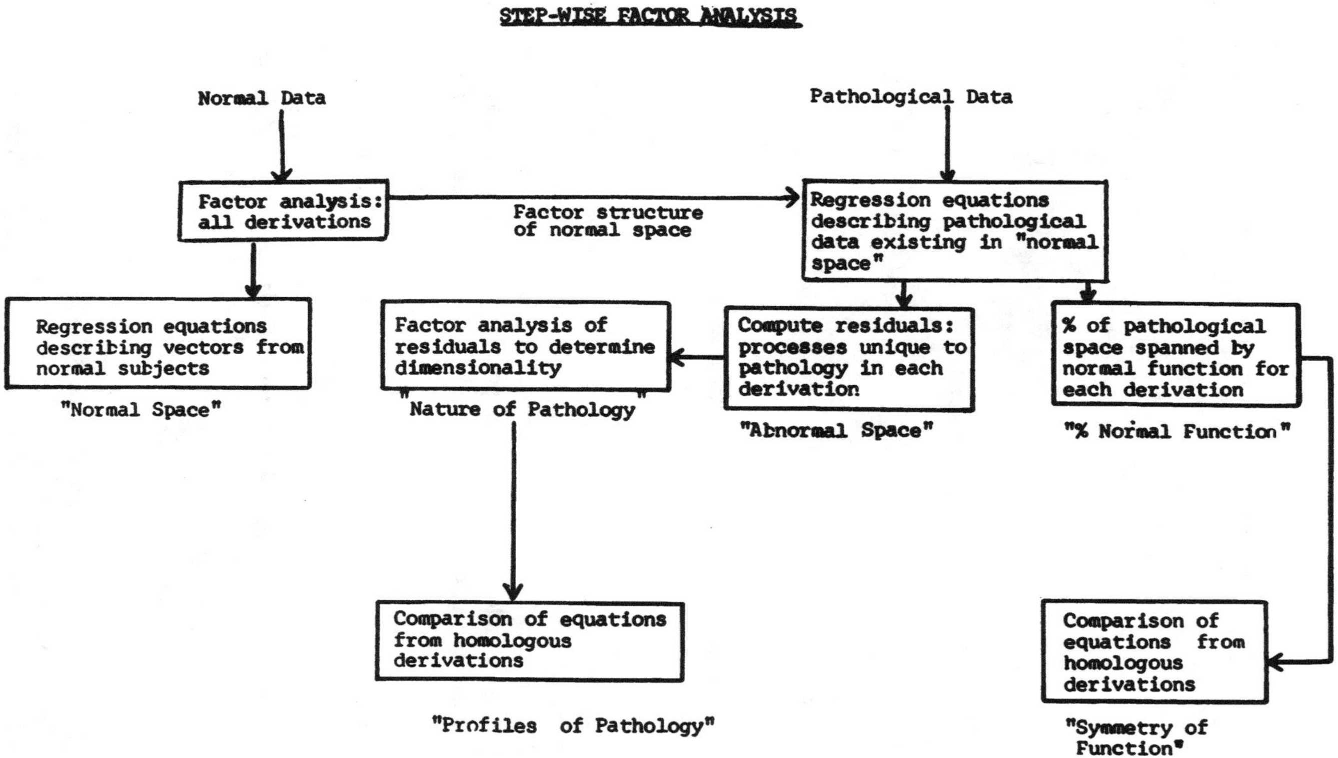
STEP-WISE FACTOR ANALYSIS
Normal Data
Pathological Data
Factor analysis: all derivations
Factor structure of normal space
Regression equations describing pathological data existing in "normal space"
Regression equations describing vectors from normal subjects
"Normal Space"
Factor analysis of residuals to determine dimensionality
"Nature of Pathology"
Compute residuals: processes unique to pathology in each derivation
"Abnormal Space"
% of pathological space spanned by normal function for each derivation
"% Normal Function"
Comparison of equations from homologous derivations
"Profiles of Pathology"
Comparison of equations from homologous derivations
"Symmetry of Function"

FIG. 3.16 Flow charts for the procedure of regression factor analysis.

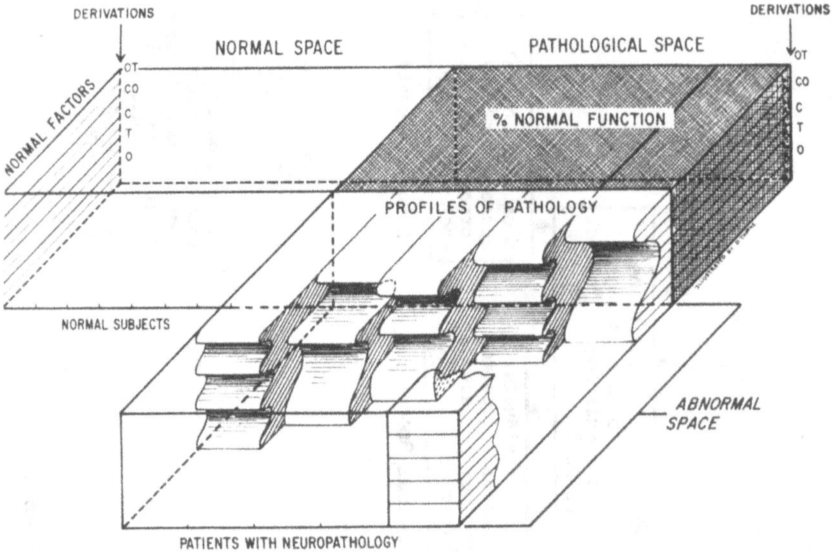

FIG. 3.17 Diagrammatic representation of the procedure of regression factor analysis.

by electrophysiological data into functional integrity of the brain and into brain processes related to sensation, perception, and cognition, already substantially reviewed in Chapter 2 and further presented in Chapter 6, it is self-evident that within the foreseeable future it will be possible to construct automatic, computer-administered and computer-evaluated neurometric assessment procedures, which will provide insights into subtle aspects of normal and abnormal brain function hitherto unavailable. Whether the endeavor of defining normal and abnormal spaces for various brain functions, as described in the previous section, is carried out using the techniques of factor analysis or the more powerful techniques already becoming available in the computer armamentarium, the same strategy outlined for regression factor analysis can be utilized. It does not matter what mathematical tools we use to define the relevant spaces. These developments are inevitable.

Three essential conditions can be identified that must coexist before this goal can be realized:

1. Digital EEG data acquisition and analysis systems must be constructed that provide the necessary neurometric indices in a fashion requiring the minimum expenditure of time, effort, and money, and in a format permitting interpretation by physicians who are not trained explicitly in neurology or even by

paramedical personnel. Such a digital EEG system has already been constructed in our laboratories and is described in detail in Chapter 4.

2. A mass of data comprising a normative data base for neurometric indices of functional integrity and brain functions related to sensory, perceptual, and cognitive processes must be gathered, together with an adequate sample of comparable data from large groups of patients afflicted by each of the many disorders disrupting normal brain function. Data must also be gathered on the changes in brain state which are brought about by the most commonly used centrally active drugs. Normal brain spaces, pathological brain spaces, and drug brain spaces must be defined for each class of electrophysiological activity. Such spaces will necessarily vary as a function of developmental stage, and must therefore be constructed for separate populations, from infancy to old age.

This undertaking is no less than the creation of a new field of medical science, with enormous implications for the identification, treatment, and hopefully prevention of a host of diseases and disorders. This new field, *clinical neurometrics*, will eventually make major contributions to pediatrics, psychiatry, psychology, neurology, neurosurgery, ophthalmology, otolaryngology, obstetrics, and internal medicine. It can be expected to bring about major changes and improvements in rehabilitation and remedial procedures. However, in order for all this to come about, a vast effort will be required. The scope of the requisite data not only exceeds the capability of any single laboratory or scientific discipline, it exceeds the resources of any single nation. The goal which we envisage will require a well-coordinated and extensive international collaboration. In our opinion, such a collaboration will produce far richer fruits in terms of meeting human needs than are likely to result from the vast treasure expended upon space exploration.

3. Yet, for all this to occur, the *third* essential condition must be fulfilled. Highly trained and motivated neurometricians, who understand the complex scientific bases for these developments and who master the difficult technology involved, must come into existence in large numbers. It will be they who construct the necessary data bases, learn how to interpret the results of the quantitative computer assessments, and develop the new methods required to translate those results into effective interventions. It is our hope that this volume will constitute the initial primer used for the formation of this new breed of neuroscientist.

4

Automatic Acquisition and Analysis of Electrophysiological Indices of Brain Functions[1]

In order to make electrophysiological measurements more readily available and more precise, and to extend their utility into new areas of application, several steps are required:

1. Optimal instrumentation must be devised that will gather data in a standardized fashion and will maximize the efficiency and accuracy of data processing. These goals, in our opinion, can best be accomplished by integrating specially constructed amplifiers, programmed stimulators, and a minicomputer with certain kinds of peripheral equipment into a computer-controlled digital electrophysiological data acquisition and analysis system (DEDAAS). We have

[1] The digital electrophysiological data analysis and acquisition system (DEDAAS) described in this chapter was designed and built by a team of collaborators. The system requirements were specified by E. Roy John, the A/D and computer hardware specifications were devised by Paul Easton, the A/D system was designed by Arthur Rose, the EEG amplifiers and stimulator were designed by Howard Bailin and built by Matthew Avitable, Waymon Samuels, and Jerry Cohen. The system of computer programs and the actual routines for data acquisition, construction of gain, artifact and stimulation protocols, and analysis of all data were written by Paul Easton and Daniel Brown. It is this program system which is the heart of DEDAAS.

The quantitative electrophysiological or neurometric test battery (NB) described in this chapter, as well as methods for quantification of the results of each challenge, were devised by E. Roy John to comprise a maximum number of meaningful challenges of sensory, perceptual, and cognitive processes in a minimum of time, while also providing measures of intrinsic reactivity and integrity. All necessary computer programs were written by Paul Easton.

The original work described in this chapter was supported by Grant No. GI 34946 from the RANN Program of the National Science Foundation.

designed and constructed such a system, which has been in operation about 2 years. The first portion of this chapter will describe the basic organization and operating principles of the DEDAAS system.

2. Once such a system is constructed, a set of challenges must be devised that can be expected to reflect important aspects of brain function in particular electrophysiological measurements. There is no reason to expect that any stimulus delivered to the central nervous system will provide useful insights into the manner or adequacy with which some particular function is performed by the brain, unless the context of the situation in which the stimulus is imbedded, as well as the stimulus itself, is devised to exercise that particular capability. Mere presentation of blank flashes, for example, need not necessarily reveal anything about how visual spatial information is processed (Regan, 1972). We refer to test situations which seem to be plausible probes of particular functions as "challenges" rather than stimuli to emphasize this point. The review of EEG and EP literature in Chapter 2 described many findings suggesting challenges that might be included in such a test battery. Using that literature as well as our own experiences as a guide, we have constructed a systematic set of challenges which we refer to as the Neurometric Test Battery, or NB. The second part of this chapter will describe and discuss the items that constitute the NB.

3. The essential features of each of these measures must be quantified, so that they can be represented numerically and manipulated statistically and mathematically. The last part of this chapter will discuss some of the basic considerations relevant to the data analysis procedures used to evaluate the NB.

I. AN AUTOMATIC DIGITAL ELECTROPHYSIOLOGICAL DATA ACQUISITION AND ANALYSIS SYSTEM (DEDAAS)

The essential considerations for reliable automatic acquisition and efficient quantitative analysis of EEG and EP data are as follows:

1. Amplifiers. Amplifiers must be of precise, fixed gain, with great stability and freedom from drift. They should have a very high common-mode rejection ratio plus a sharp 60-Hz notch filter to eliminate any need for a shielded recording room, and they should have a low noise level. The higher the input impedance, the less critical will be the impedance of the scalp electrode. The preamplifiers should ideally be photically isolated to comply with recent safety recommendations, particularly if the system is to be used in a newborn nursery or upon patients being sustained by life-support systems or monitored in intensive care units, because of the danger of circuit interactions which might result in electrical shock to the patient. The amplifiers used in DEDAAS have an input impedance of 10 MΩ, a fixed gain of 10,000, frequency response from .3

to 70 Hz (Roll Off 6 dB/octave), common-mode rejection ratio of about 106 dB, a sharp 60-cycle notch filter (greater than 40 dB with a Q = 30, adjustable), and a noise level of less than 2 μV, with a photically isolated, battery powered preamplifier (Bailin & Avitable, 1973, unpublished results). The amplifier is built upon one printed circuit board, with components costing about $50 at this time (1975). The circuit is shown in Fig. 4.1.

2. 24-channel amplifying system. Twenty-four of the amplifiers described above are mounted in a single rack. The full 10/20 system is recorded monopolar, using linked earlobes as the references (see montage computation, Section 9, below), occupying 19 channels. The remaining channels are used for a transorbital electrode pair to detect eye movement, an accelerometer to detect head movement, and EKG or other measures as desired.

3. Display system. In order to permit monitoring of the EEG, since no ink record is available in the recording chamber, 19 light-emitting diodes (LED) are arranged so that an LED is monitored at every electrode position corresponding to the international 10/20 system on a display of a head. Each LED monitors the output of the amplifier connected to the electrode at the corresponding position of the patient's head. If the output voltage swings *positive*, the LED turns *red.* If the output is *negative,* the LED turns *green.* Thus, looking at the head display, the technician sees a display of lights which flicker from red to green at a rate synchronized with the zero crossings determined by the spontaneous EEG rhythms. Not only does this display inform the technician about blocking due to patient movement, loose electrodes, and other types of artifacts, but it provides a "toposcope" display that reveals many interesting features of synchronization between derivations, propagation of voltage patterns, and asymmetries. This information is not utilized in DEDAAS, where the intent of the display is solely for monitoring. A selector switch permits the output of any channels, to which the technician's attention is drawn by the LED display, to be examined on a multichannel oscilloscope.

4. Impedance testing. The amplifiers perform excellently with electrode impedances as high as 50,000 Ω. A properly applied scalp electrode should have an impedance well below 10,000 Ω, and impedances between 2,000 and 4,000 Ω are routinely obtained. A circuit included in the amplifier itself permits the computer to interrogate each electrode for impedance. If such impedance testing, performed periodically, reveals unacceptably high impedance at any electrode, the LED at that location on the display *turns and remains* red until the undesirable condition is corrected by the technician. If desirable, this circuit could sound a buzzer or even interrupt data acquisition until appropriate steps are taken, but these features have not been programmed into DEDAAS. The impedance testing circuit is shown in Fig. 4.2.

5. Variable gain analog-to-digital conversion. The output of the amplifying system goes to an analog-to-digital (A/D) conversion system (Rose, unpublished, 1973). Each of the 24 amplifier channels is digitized in sequence, at a rate of

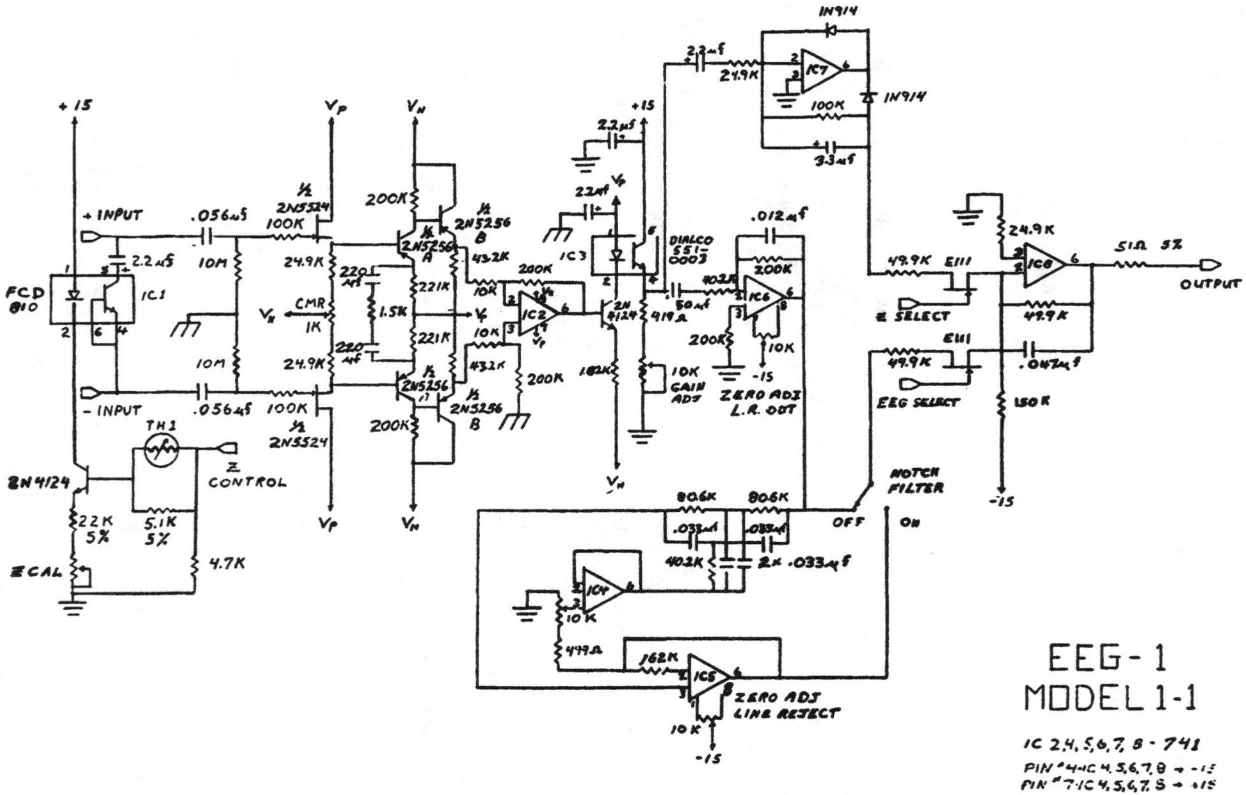

FIG. 4.1 Circuit diagram of EEG amplifiers.

200 samples/channel/sec, and the resulting 10-bit words are multiplexed as input to a PDP 11/45. The peak-to-peak excursion of voltage within each channel is constantly monitored by the computer throughout data acquisition. The A/D converter for each input channel is of variable gain, up to a factor of 256, and the computer can adjust this gain at regular intervals for optimal resolution of the incoming signal.[2] A *gain protocol* is constructed for every EEG channel. (See Section 8, below.)

6. Automatic artifact control. Portions of the data acquisition program monitor the eye movement and accelerometer channels, as well as all 19 EEG channels. A threshold value can be set independently for the maximum permitted actual voltage (*dV*/*dt* could also be used as an adjunct to artifact identification but we have not found this necessary). If this voltage is exceeded, two options are available. Option 1 accepts the data but enters a protocol note commenting that the data in a particular channel is contaminated by artifact. An *artifact protocol* is constructed for every channel. Subsequent analysis of those channels free of artifact can be accomplished by specific disabling of a program routine that prevents analysis of any channel if data from any other channel is not artifact-free during that interval. Plotting such data (see below) will cause an artifact indicator to appear on the baseline and the artifactual signal will be underlined in any channels in which it occurred. Option 2 refuses to accept data while any EEG channel displays artifact, or if eye movement, eye blink, or head movement has occurred within the last two seconds. In many circumstances, such as when attempting to record from infants, hyperkinetic children, or patients incapable of cooperating, it is simply impossible to obtain reliable EEG or EP data without this capability for automatic exclusion of artifact periods while brief interspersed intervals of usable data are accumulated. For recording under special conditions, unacceptable data can be eliminated by providing manually controlled or transducer-generated input to an artifact-detecting channel.

Figure 4.3 shows plotter output of EEG data.

7. Station multiplexing. The present configuration of DEDAAS permits data to be gathered simulatneously from up to four installations or "stations," each with the features just described. Data can be gathered with complete independence, with no requirement that similar portions of the QB be administered at the different stations simultaneously. Any station can begin, interrupt, or end data acquisition independently. With minor hardware augmentation (principally core expansion), DEDAAS could handle up to 10 stations independently at the same time.

8. Digital recording, encoded protocols, and automatic analysis. In approximately 15 years of processing EEG and EP data initially recorded on analog

[2] Thus far, it has not been found necessary to utilize this feature. Data acquisition with a fixed gain of 10,000 has provided satisfactory data for our analysis programs. Gain changes have been limited to output or display only.

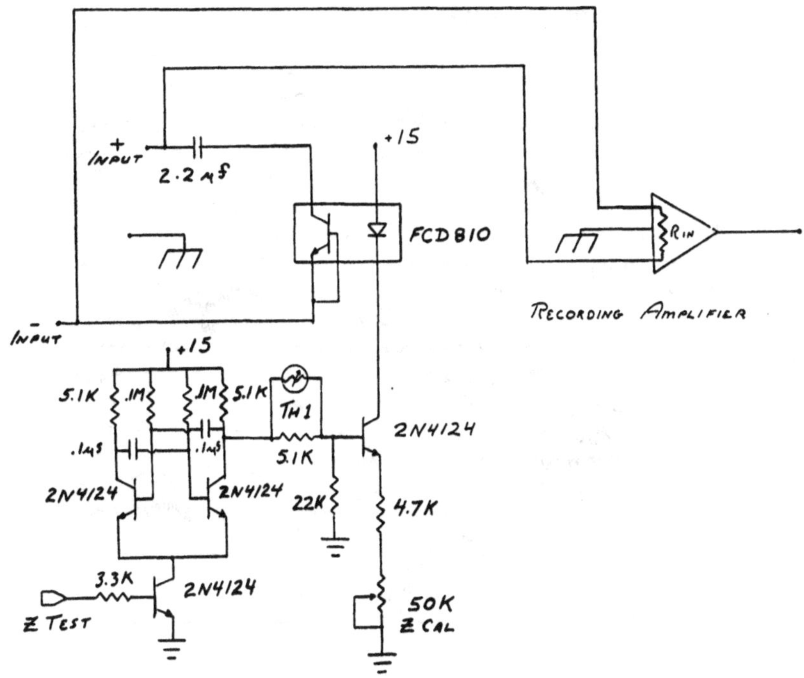

FIG. 4.2 Diagram of impedance monitoring circuit.

tapes using a written recording protocol, we learned that location of desired data on the analog tape, editing of unacceptable portions, and analog-to-digital conversion required an interval about 10 times as long as that required for the original recording. Once digital conversion is accomplished, selection of the analytic routines appropriate for processing different portions of the data, and corrections for gain changes in the original data, can occupy a substantial further amount of operator as well as computer time.

In addition to the unique advantage which the capability for automatic selection of artifact-free intervals affords when recording from hyperactive children (the group for which DEDAAS was explicitly devised), another major advantage provided by this procedure is on-line transformation of all data into digital representation, using a format which permits subsequent automatic computer analysis without the need for operator intervention or consultation of a written protocol made during the recording session.

DEDAAS accomplishes this by incorporating a gain protocol and an artifact protocol directly on the digital recording constructed for each EEG channel during the data acquisition session. Further, a *stimulation* protocol (see Computer-controlled stimulator, Section 10, below) is also encoded on the tape whenever any instructions are sent to the stimulator by the computer. As a consequence of the presence of all of these protocols encoded directly upon the

COMPUTER GENERATED EEG-ARTEFACTS SHOWN

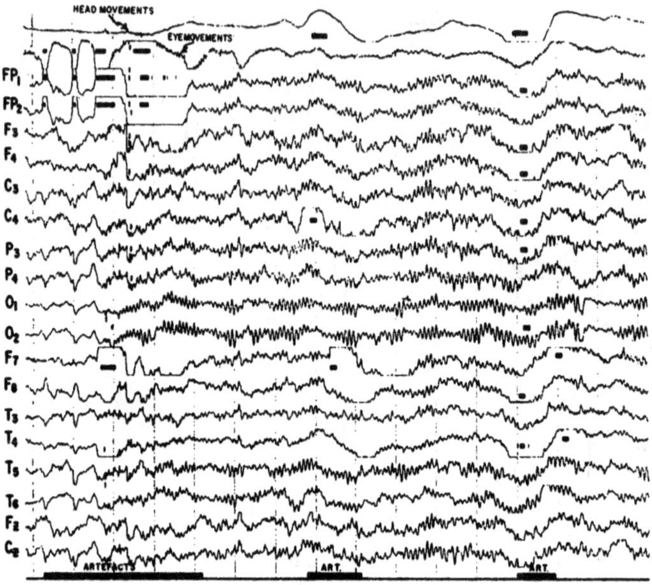

COMPUTER GENERATED EEG - ARTEFACT SUPPRESSED

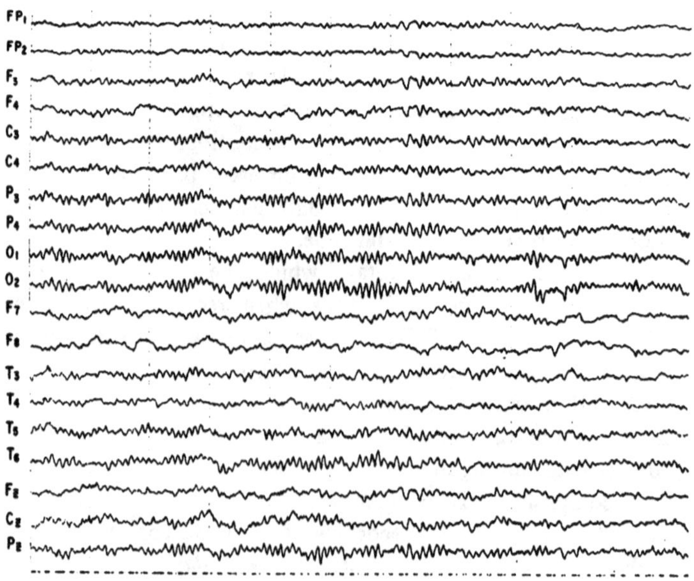

FIG. 4.3

data tape itself, the computer can be self-instructed how to process the data on any segment of tape by decoding these protocols. In this way, DEDAAS eliminates the need for operator intervention to locate any desired type of data, to monitor or edit during A/D conversion, or to make scaling corrections in order to compensate for changes in amplifier gain. Analytic programs automatically select appropriate portions of data for analysis, using the protocols encoded on the digital tape.

9. Computation of all bipolar montages. In conventional procedures, a bipolar derivation is defined by the output of a differential amplifier with inputs from two adjacent electrodes. The differential amplifier rejects portions of the signal common to both electrodes and outputs only their difference. This process is accomplished completely by computation in DEDAAS. During data acquisition, the full 10/20 system is recorded referentially. For data analysis, all data from electrodes that are adjacent along anterior–posterior or left–right lines of the array are subtracted one from the other. Thus, all conventional bipolar derivations, or any specially defined derivations, can be constructed in the computer by simple arithmetic operations performed upon data from pairs or appropriate groupings of monopolar channels. Currently, we routinely compute three montages, equivalent to simultaneously recording 57 derivations (19 monopolar, 19 coronal bipolar pairs, 19 sagittal bipolar pairs), for EPs as well as the EEG. Another major advantage of DEDAAS, therefore, is that access is gained to a very large effective channel capacity without increasing the actual number of EEG channels, fatiguing the subject, or preempting technician and equipment time with iterated runs for diverse montages. Additional montages can be specified at any time by typing in the desired electrode combinations for computation.

10. Computer-controlled stimulator. In order to ensure that stimulation would be delivered in a reproducible fashion, especially for some of the complex challenges in the NB which require precise timing, a computer-controlled stimulator was constructed. This compound stimulator contains a flash photo stimulator, a source of sinusoidally oscillating light, an automatic slide projector with a strobe light source, circuits which produce clicks or pure tones (250, 500, 1,000 and 3,000 Hz) of specified intensity from a high-fidelity loudspeaker, and a tactile stimulator which delivers a tap of constant intensity. Appropriate se-

FIG. 4.3 Examples of EEG which were digitized by DEDAAS and then reconstructed by the matrix printer. *Top:* example of recording obtained from a child, without using the program option to suppress artifacts. Under this option, artifact-contaminated data is accepted for recording, but is underlined on the reconstructed record. First channel shows artifacts from head movements, detected by a small accelerometer, while the second channel shows artifacts produced by blinking and eye movements; *Bottom:* example of data recorded from same child using option to suppress artifacts. Under this option, only segments of artifact free data at least 2.5 seconds in duration are accepted for recording and are joined to construct a "continuous" record.

quences of these stimuli permit construction of all of the challenges in the NB. Each of the challenges is defined by a subroutine in DEDAAS. The subroutines can be requested individually by the operator or can be presented in an automatic sequence during the data acquisition session. In either case, a stimulation protocol is encoded onto the data tape, specifying the challenge being presented and representing every element of the challenge by an entry identifying the nature of the element and its time of occurrence. These entries serve as trigger information for subsequent evoked potential analyses. Each type of element in each different challenge is separately processed during data analysis. The stimulator and the console containing the EEG amplifiers and LED display are shown in Fig. 4.4.

11. Computer system. DEDAAS has been programmed for our PDP 11/45[3] which has 16K of memory core, a 1.2 million word disk, a digital display oscilloscope, a teletype, and an industry-compatible digital tape drive operable at either 800 or 1,600 bits per inch.

12. Plotter. All hard copy data output is provided from a Gould electrostatic matrix printer. This device outputs alphanumeric or graphic data at the rate of as much as 7 linear inches of full-width (8 inches) data per second. Data are represented by spots with a density of up to 100/inch. The EEG example shown in Fig. 4.3 were produced by this plotter. The computer system and plotter are shown in Fig. 4.5.

13. Block diagram of DEDAAS system. A block diagram of DEDAAS is presented in Fig. 4.6.

14. Economic advantages of DEDAAS. Depending upon the number of stations it is eventually intended to service the hardware for a DEDAAS system with three terminals costs about $70,000–80,000. This is roughly the cost of six conventional 16-channel EEG machines. Because of the huge effective channel capacity, DEDAAS requires one-third the data acquisition time of a conventional system, so three times as many patients can be examined per unit time in each installation. If an institution has three data-acquisition installations or terminals, the hardware cost per patient is already lower for DEDAAS than for conventional EEG.[4] Digital tape for a standard EEG examination (spontaneous EEG eyes open and eyes closed, photic stimulation, hyperventilation, and post-ventilation recovery) costs about $5 per patient, less than EEG paper, and yields a compact record which can be dynamically reproduced for comparisons later on. Conventional EEG records in familiar montages can be obtained from the matrix printer; either the whole session or selected portions found to be

[3] For data acquisition only, DEDAAS can be implanted on a much smaller PDP11/10 configuration. We use these smaller systems as data acquisition terminals, which are relatively mobile.

[4] Note that these installations need not be in the immediate vicinity of the computer. Interactions between the computer and a remote terminal can be transmitted by cables, since no low-level or high-frequency signals are involved. Telephone or telemetered interactions are also possible, but have not yet been developed.

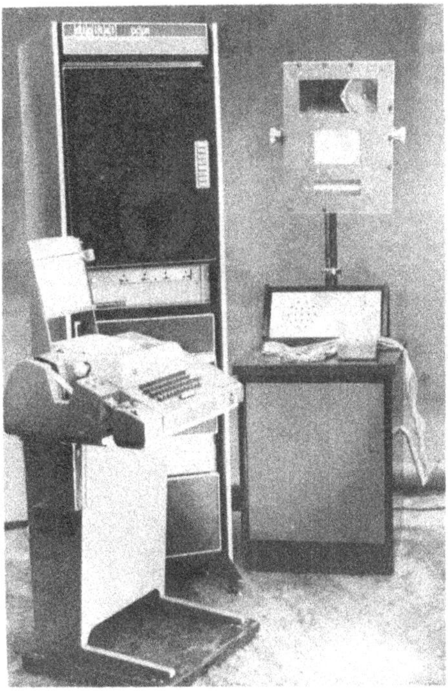

FIG. 4.4 Picture of portable acquisition system including a teletype, stimulus console, magnetic tape drive, and a PDP 11/10 computer.

abnormal by the criteria of the analytic programs can be reproduced. Final paper costs depend, of course, upon how much of the raw data need be reproduced. In general, it seems fair to say that the costs of a DEDAAS installation *for data acquisition* would be less than conventional EEG systems, provided that the installation is intended to service a very large number of patients.

The striking economic fact is that once the data acquisition volume reaches the capacity of two to three installations, the data analysis cost of DEDAAS becomes essentially zero. Expenses will be incurred for supplies and for the salary of the operator, but the hardware costs can be considered as defrayed by the costs of conventional EEG machines capable of producing paper records for an equivalent volume of patients.

Translating this into practical terms, any large institution such as a municipal health service, a hospital, or a university (or a network of clinics) which contemplates the purchase of five or six conventional EEG machines can get free quantitative analysis of EEG as well as a substantial number of other advantages by building a DEDAAS system instead. As DEDAAS systems are adopted, we expect a great increase in the ease of access of patients to EEG examinations, an

FIG. 4.5 Picture of teletype, PDP 11/45, magnetic tape drive and printer/plotter.

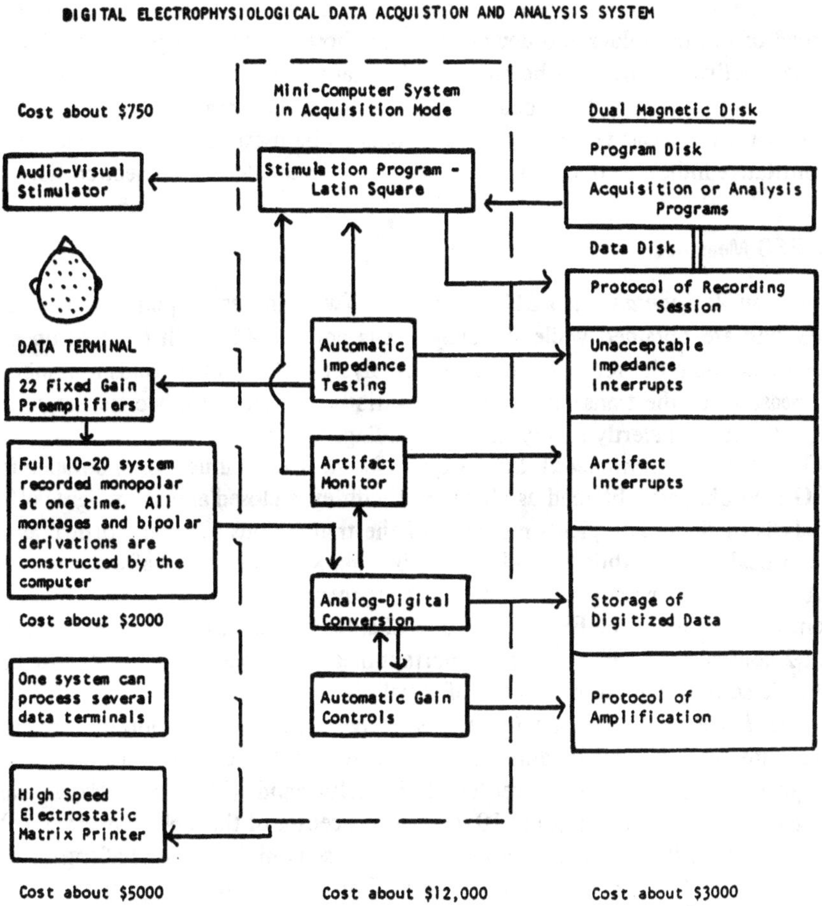

FIG. 4.6 Block diagram of DEDAAS.

increase in the accuracy and sensitivity of the examination, a decreased dependence of the medical practitioner upon the electroencephalographic specialist for preliminary decisions about CNS involvement, the initiation of mass screening procedures, and sharp decreases in the cost of the individual EEG examination.

II. A QUANTITATIVE ELECTROPHYSIOLOGICAL TEST BATTERY (NB)

The items comprising the NB are enumerated in this section. Although their intent will be discussed, no attempt is here made to document the plausibility of each challenge. Their rationale is based upon findings presented in earlier chapters (particularly Chapter 2 and in Chapter 6), and in many cases, hopefully, will already have intuitive face validity for the reader. Studies are in progress to ascertain the functional significance and utility of these measures, analyzing the spread of normal values and deviations from those values in various conditions. Some positive findings will be presented in Chapters 5 thru 8. It is probably safe to assume that many of the challenges in the NB will eventually be displaced by more sensitive measures, but some are going to be useful for a long time. The quantitative indices extracted from these challenges are defined in Section III.

A. EEG Measures

Condition 2: Spontaneous EEG – eyes open. Two minutes of spontaneous EEG recording are obtained while the subject sits comfortably with his eyes open, free to examine the recording chamber. Analysis of these data yields a number of measures of the transactions between various cortical and subcortical regions while the subject alertly surveys his surroundings.

Condition 2: Spontaneous EEG – eyes closed. Two minutes of spontaneous EEG recording are obtained as above, but with eyes closed and room lights off. Analysis of these data yields measures of the transactions between cortical and subcortical regions while the subject receives markedly decreased afferent input.

Aside from providing sets of measures of intrinsic value, as already known from years of use of the EEG reviewed in Chapter 2, comparison of Conditions 1 and 2 constitutes Challenge 1, an important measure of reactivity: the change in intrinsic activity caused by removal of visual afferent input.

Condition 3: Stimulation with sinusoidal lights. A total of 6 min of stimulation with sinusoidally modulated light, divided into four periods: (a) at a frequency of 2.0 Hz, in the center of the delta band; (b) at 5.25 Hz, in the center of the theta band; (c) at 10.0 Hz, in the center of the alpha band; and (d) at 19.0 Hz, in the center of the beta band. These stimuli define Challenges 2A, 2B, 2C, and 2D, which constitute additional measures of reactivity: the changes in intrinsic rhythms that can be induced by photic driving at the center of each of the different spectral bands.

Condition 4: Habituation to sinusoidal stimulation. A stimulation of 2 min with sinusoidal flicker at 1 Hz. This condition defines Challenge 3, which is the rate of habituation to a sustained series of monotonous and inconsequential events, reflected in the EEG rhythms.

B. Evoked Potential Measures

Condition 5: Phasic habituation to simultaneous visual, auditory, and somatosensory stimuli. A compound stimulus (FCT), consisting of a simultaneous flash (F), click (C), and tap (T), is delivered five times at regular 1-sec intervals. After a 4-sec pause, this *set* of five-compound stimuli is repeated. Twenty-five such sets are presented. This condition defines Challenge 4, which is the rate of habituation to a series of monotonous events, reflected in the features of the AER.

Condition 6: Dishabituation. As soon as Condition 5 ends, 50 flashes (F), 50 clicks (C), and 50 taps (T) in *random* sequence are presented at a rate of 1 stimulus per second. This condition serves several purposes: first, it serves to interrupt the phasic habituation procedure, for reasons which will shortly become obvious; second, it provides the basis for an estimate of the AER to random, unpredictable flashes, clicks, and taps, which is used as part of Challenge 8.

Condition 7: Rehabituation to compound visual, auditory, and somatosensory stimuli. As soon as Condition 6 ends, 25 sets of simultaneous FCT are again presented, following exactly the same procedure as Condition 5. This condition contributes to the definition of several challenges, that is, Challenge 5, did the interruption of the initial phasic habituation sequence cause "dishabituation"?; Challenge 6, does rehabituation occur more rapidly the second time than initially?; Challenge 7, does tonic or "long-term" habituation occur?

Condition 8: Regular alternation of visual, auditory, and somatosensory stimuli. A series of 150 stimuli consisting of 50 flashes (F), clicks (C), and 50 taps (T) is presented with a regular interval of 1 sec between stimuli, in the regular order F, C, T, F, C, T, and so on. This condition provides the data for portions of Challenges 8, 9, 10, and 11. Challenge 8 measures the ability of the subject to distinguish between random, unpredictable sequences of environmental events and regular predictable sequences. It gives an index of the ability of the subject to discern orderly relations between events in time. Challenge 9 measures the responses to simultaneous presentations of stimuli in several modalities and to separate presentations of stimuli in individual modalities and gives an index of sensory-sensory interaction. Challenges 10 and 11 give a measure of cross- and ipsimodal sensory inhibition, an index of the ability to structure figure–ground relations.

Condition 9: Presentation of regular alternative visual, auditory, and somatosensory stimuli during meaningful visual stimulation. A sequence of 150 stimuli consisting of flash (F), click (C), and tap (T) is presented as in Condition 8, while the subject views a color TV screen showing only the picture with the sound turned off. This condition gives a measure of internal inhibition, the suppression of meaningless afferent input both in the visual modality (ipsimodal) and in the auditory and somatosensory modalities (cross-modal) while attending to meaningful visual experience. These measures, systematically compared with those obtained during Condition 8, are used to construct the measures of Challenge 10.

Condition 10: Presentation of regular alternating visual, auditory, and somatosensory stimuli during meaningful auditory stimulation. This condition is identical with Condition 9, except that a tape recording of music with a definite strong rhythm replaces the video picture. This condition gives the same measure as Condition 9, but for meaningful *auditory* experience and is used similarly to evaluate the results of Challenge 11.

Condition 11: Sensory–sensory conditioning. This condition is used to obtain an electrophysiological estimate of conditionability (Challenge 12). It also provides an estimate of visual and auditory recovery cycles (Challenge 13). The procedure is as follows: (a) 150 stimuli consisting of 50 flashes (F_1), clicks (C_1), and taps (T_1) are presented at regular intervals of 1/sec in regular order F, C, T, F, C, T, and so on; (b) 200 pairs of flashes (F = CS) and clicks (C = US) are presented, with flash always preceding click by 250 msec; (c) 150 stimuli consisting of 50 flashes (F_2), clicks (C_2), and taps (T_2) are presented, as in (a); (d) 200 pairs of clicks (C = CS) and flashes (F = US) are presented, with click always preceding flash by 250 msec; (e) 150 stimuli consisting of 50 flashes (F_3), clicks (C_3), and taps (T_3) are presented, as in (a). (Note: In the notation above, CS stands for conditioned stimulus and US for unconditioned stimulus.)

Condition 12: Sensory acuity and encoding. This condition is used to obtain an estimate of auditory and visual acuity, perception of visual shapes, and perceptual invariance. Fifty stimuli are presented in some or all of the following classes, at a rate of 1/sec. Each class of stimulus is presented as a block: (a) "gray" flash consisting of 85 lines/inch as a 50% transmission grid; (b) checkerboard of 27 lines/inch (50% transmission); (c) checkerboard of 7 lines/inch (50% transmission). These stimuli are used for Challenge 14, estimation of visual acuity; (d) pure tone at 45 dB, 250 Hz; (e) pure tone at 45 dB, 500 Hz; (f) pure tone at 45 dB, 1,000 Hz; (g) pure tone at 45 dB, 3,000 Hz. These stimuli are used for Challenge 15, estimation of auditory acuity; (h) large square, 16 square inch area; (i) small square, 4 square inch area; (j) large diamond, 16 square inch area; (k) small diamond, 4 square inch area. These stimuli are used for Challenge 16, discrimination between different geometric forms of equal area, and Challenge 17, perceptual invariance to similar forms of different area; (l) the letters "b," "d," "p," and "q." These stimuli are used for Challenge 18, discrimination between forms causing reversal problems in reading letters.

Condition 13: Spontaneous activity, eyes open. This is a repeat of Condition 1, intended to serve as a measure of the reliability of the spectral estimates of Condition 1 and changes related to fatigue caused by administration of the NB. Challenge 19, stability of spontaneous EEG with eyes open and effects of fatigue, compares Condition 13 with Condition 1.

Condition 14: Spontaneous activity, eyes closed. This is a repeat of Condition 2, with the same rationale as Condition 13, and is used to construct the measures of Challenge 20.

III. QUANTITATIVE NEUROMETRIC INDICES EXTRACTED FROM NB CHALLENGES

Features considered to reflect important properties of the activity of the brain and the response to each challenge must be extracted from the raw data. In general, all items referred to under "EEG Measures" in Section IIA are preprocessed by wideband spectral analysis and the absolute energy and relative (%) energy in the delta, theta, alpha, and beta bands are computed for every derivation. All items referred to under "AER Measures" in Section IIB are preprocessed by computation of the averaged evoked responses for each type of stimulus in the challenge, and the variance of each type of AER is also computed. The remainder of this section specifies the details of this preprocessing and then deals with further operations that are carried out upon the preprocessed data to ascertain the effect of the challenges.

A. Neurometric Indices Extracted from EEG Measures

The following basic neurometric indices are computed for each derivation under each condition defining an EEG measure in the NB:

1. *Distribution of energy in different frequency bands.* The absolute and relative (%) amounts of energy are measured in the delta 1 (0.5–1.5 Hz), delta 2 (1.5–3.5 Hz), theta (3.5–7 Hz), alpha (7–13 Hz), beta 1 (13–19 Hz), beta 2 (19–25 Hz), and gamma (25–40 Hz) bands for every derivation.
2. *Age-dependent quotient.* For all measures where normative data exist, the age-dependent quotient is calculated separately for every derivation, and is defined as the ratio of the average measure for a group of normal individuals of a specified age to the value obtained from a subject of the same age.
3. *Energy ratios.* The ratios of energy between various frequency bands and combinations of bands is calculated separately for each derivation, including the ratio of delta to alpha, theta to alpha, delta plus theta to alpha, and delta plus theta to alpha plus beta.
4. *Energy symmetry.* For all possible pairs of symmetric (homologous) derivations, the ratio of the total energy and the energy within each of the frequency bands is calculated between the two hemispheres.
5. *Waveform symmetry.* For all possible pairs of symmetric derivations, the complex cross correlation between the left and right side is computed for the total EEG and for the signal within each of the defined frequency bands.

Challenge 1. The effect of removal of visual input upon intrinsic activity, a type of reactivity, is specified by subtracting the value of each index computed under Condition 2 (C_2) from the comparable value observed under Condition 1 (C_1). The results are stated in absolute terms, that is, $(C_{2i}-C_{1i})/(C_{1i})$.

In these expressions, C stands for any test condition, the subscript number identifies the particular conditions compared in a challenge, and the subscript refers to any of the i indices which can be computed under those conditions and which, therefore, provide different ways to assess the effect of a challenge.

Challenge 2. a. Reactivity to light sinusoidally modulated in the delta frequency band is computed for every index by subtracting the index obtained under Condition 1 from the value of the comparable index under Condition 3, that is, value during driving with sinusoidal 2-Hz flicker, minus value during spontaneous activity. This type of reactivity may be viewed as a way to estimate what engineers call the "transfer function." Results are expressed in absolute terms, that is, increase of energy in delta band expressed in microvolts squared, and in relative terms, that is, percent increase of energy in the delta band. b. Reactivity to flicker in the theta frequency range, evaluated as Challenge 2a. c. Reactivity to flicker in the alpha range, evaluated as Challenge 2a. d. Reactivity to flicker in the beta range, evaluated as Challenge 2a.

Challenge 3. Rate of habituation to sinusoidal driving at 1 Hz can be considered as an index of a rudimentary adaptive process, the inhibition of response to meaningless, slightly noxious afferent input. For each of the indices defined for Condition 1, two separate measurements are obtained during Condition 4, one during the first 30 sec (early) and the second during the last 30 sec (late). By subtracting the "late" from the "early" measures, any diminution in effectiveness of driving becomes discernible. Such changes are expressed in absolute and relative terms for each index.

B. Neurometric Indices Extracted from EP Measures

The following basic neurometric indices are computed for each derivation under every condition that defines an evoked potential measure in the NB: (1) averaged evoked response (AER). A separate AER is computed for each different type of stimulus or during each indicated time interval; (2) the variance is similarly computed; (3) the amount of AER energy is measured in the whole analysis epoch and in various sub-epochs; (4) symmetry of energy is calculated by taking the ratio of energies between symmetric derivations in each of these time domains; (5) symmetry of waveshape is calculated by computing the cross correlation; (6) the significance of left–right asymmetries is calculated by computing the t test between AERs from homologous leads; (7) the latency of successive major peaks and troughs is measured; (8) latency differences are calculated for corresponding processes; (9) factor analysis is used to describe the morphology of the waveshape; (10) multidimensional scaling is used to place the individual AER waveshape within the class of waves to which it "belongs"; (11) if variance peaks indicate inhomogeneity of process, split-half reliability is computed to see whether the measurement is reliable. If not, it is not further processed. (Factor analysis and multidimensional sorting are carried out only on selected bodies of data. It is expected that discriminant functions will be generated in the future that will permit an AER waveshape to be classified rapidly by taking advantage of previous factorial and scaling analyses.)

Challenge 4: Phasic habituation to compound stimuli. Five AERs are computed: the first is the average of the initial stimulus in each series of five, averaged across the 25 sets (AER_1); the second is comprised of the second stimulus in each series of five, averaged across the 25 sets (AER_2); similarly, averages AER_3, AER_4, and AER_5 are computed. Phasic, that is, "short-term" habituation, can be defined as the systematic decrement observed as one goes from AER_1 to AER_5, examining the change in response as a function of the position of the stimulus within the "fine structure" of the set. Although all of the AER indices defined above are computed and compared, the major signs of phasic habituation are diminutions in amplitude and increases in latency of components when AER_5 is compared with AER_1. The *t* tests are computed between AER_1 and AER_5 to test the significance of any changes observed. Changes are stated in absolute, that is, $(AER_1 - AER_5)$, and relative, that is, $(AER_1 - AER_5)/AER_1$ terms. The percentage of phasic habituation (% HP) is equal to $(AER_1 - AER_5)/AER_5$. When clear corresponding peaks are found across the five AERs, the rate of decrement between adjacent points in the set is calculated for each peak.

Challenge 5: Dishabituation. If the subject is continuing to monitor his environment effectively even though specific afferent input is dynamically inhibited, a change in the stimulus features will cause dishabituation. The extent of dishabituation is determined, after random F and C and T interrupt monotonous sets of FCT, when the FCT sets are resumed. Comparing an AER computed across the *last* five sets of Condition 5 (the average of FCT presentations 100 through 125) with an AER computed across the first five sets of Condition 7 (the average of FCT presentations 125 through 150) indicates dishabituation if the AER from Condition 7 is larger than the AER from Condition 5.

Challenge 6: Rehabituation. If the subject has short-term memory for the previous habituation condition, habituation in Condition 7 will be more rapid than in Condition 5. In order to estimate the extent of savings upon rehabituation, a measure of short-term memory, five averages are again computed across the second series of 25 sets of five stimuli. AER 6 is the average of the 25 initial stimuli in each set. AER_7 is the average of the 25 second stimuli, and AER_8, AER_9, and AER_{10} are similarly computed. A quantitative estimate of savings upon rehabituation is defined by $AER_1 - AER_6$, $AER_2 - AER_7$, and so on. Differences between the rates of decrement in initial habituation and rehabituation are computed. These differences are stated in absolute and relative terms.

Challenge 7: Tonic habituation. A distinction was recognized long ago between the mechanisms mediating phasic and tonic habituation (Sharpless & Jasper, 1956). Phasic habituation involves the intralaminar nuclei of the thalamus while tonic habituation depends more upon influences of the mesencephalic reticular formation. Tonic or long-term habituation is quantified by subtracting the AER computed across the last five sets of Condition 7 (AER_L = the average of FCT presentations 225 through 250) from the AER computed across the first five sets of Condition 5 (AER_F = the average of FCT presentations 1–25). The percentage of tonic habituation (%HT) is equal to $(AER_F - AER_L)/AER_F$. A substantial time (10 min) elapses between these two data samples, which is long relative to the five 1-sec intervals assessed in the phasic habituation measure. The tonic and phasic measures are compared by computing HP/(HP + HT), which describes the percentage of total habituation arising from the phasic contribution.

Challenge 8: Perception of orderly temporal relations. As described in Chapter 2, late components appear to reflect uncertainty about stimuli. This is especially true for a late positive wave usually found at a latency of 300 msec (P300). AERs are separately computed for the predictable flashes (F_p), predictable clicks (C_p), and predictable taps (T_p) presented in Condition 8. AERs are also computed for the random flashes (F_r), clicks (C_r), and taps (T_r) which were presented in Condition 6 (dishabituation). By subtracting indices computed to regular stimuli from those computed to random stimuli, the AER processes related to uncertainty are unmasked. Specifically, the absolute effects of uncertainty in the visual, auditory, and somatosensory modality are separately computed: $AER(F_r)–AER(F_p)$, $AER(C_r)–AER(C_p)$, $AER(T_r)–AER(T_p)$. The morphology and anatomic distribution of the processes related to uncertainty are computed by factor analysis of the difference waves obtained at every derivation. The area under the difference wave defines the quantitative magnitude of the effect. The factor structure and loading coefficients define the morphology of the effect in each derivation. These distributions and magnitudes are compared for the three different sensory modalities, to establish whether the physiological mechanism underlying this prediction process is sensory-specific or nonsensory specific. Similarly, the relative (%) effects of uncertainty are computed by normalization, that is, by dividing the energy of the difference wave by the energy of the AER to the unpredictable stimulus.

Challenge 9: Sensory–sensory interaction. It has long been known (Buser & Borenstein, 1956) that when visual, auditory, and somatosensory stimuli are presented simultaneously, the anatomic distribution and morphology of EPs on the cortex is altered. It has been shown that such alteration arises via interactions mediated by the rostral midline thalamus. By subtracting the sum of AERs separately computed to stimuli within several single modalities from the AER computed when the same stimuli are presented simultaneously, one can construct the difference wave revealing this sensory-sensory interaction. Accordingly, we compute the sum of the first 10 sets of stimuli presented during Condition 5 [$AER(FCT_r)$]. This new AER represents the average of the responses to 50 compound stimuli (FCT) presented at regular intervals of 1/sec. Next, we sum the values of the three independent AERs obtained in Condition 8, viz., $AER(F_r)$, $AER(C_r)$, $AER(T_r)$. We next compute the difference:

$$AER\,(FCT_r) - [AER(F_r) + AER(C_r) + AER(T_r)]\,.$$

The resulting difference wave provides an index of sensory–sensory interaction. This interaction may be either facilitatory or inhibitory. The absolute and relative amount of sensory-sensory interaction at each site is determined and its anatomic distribution is examined. This index gives an estimate of the reactivity of the midline thalamus and establishes whether the diffuse projection system interacts differentially with various cortical regions.

Challenge 10: Cross- and ipsimodal inhibition by meaningful visual stimuli (figure–ground relations). It has been believed for some time (Birch, 1964) that the ability to structure figure–ground relationships is deficient in many children suffering from learning disabilities. These children seem to lack the ability to "pull" information from their surroundings by identifying significant events. This challenge provides an estimate of this capacity. AERs to random flashes [AER(F_r)], clicks [AER(C_r)], and taps [AER(T_r)] are computed from the data obtained in Condition 8 (note that these AERs were previously utilized in Challenges 8 and 9). AERs are similarly computed to these stimuli when presented while the subject watches a color TV program with the sound turned off, as defined in Condition 9. These averages, presumably altered by the inhibitory processes arising while attention is focused upon the video picture, are denoted as AER (F_i), AER(C_i), and AER(T_i), where the subscript "i" stands for "inhibited." By computing the difference waves [AER(F_r)–AER(F_i)], [AER(C_r)–AER(C_i)], and [AER(T_r)–AER(T_i)], the absolute effects of ipsimodal (visual–visual) and cross-modal (visual–auditory and visual–somatosensory) inhibition are obtained for a meaningful visual stimulus. The absolute and relative magnitudes of these effects and their anatomic distribution are computed in the usual manner.

Challenge 11: Cross- and Ipsimodal inhibition by meaningful auditory stimuli (figure–ground relative). In a fashion completely analogous to the computations used in Challenge 10, but utilizing the data from Condition 10, the inhibitory effect of meaningful auditory experience upon visual (cross-modal), auditory (ipsimodal) and somatosensory (cross-modal) stimuli can be quantified and its anatomic distribution ascertained.

Comparisons of the results of Challenges 10 and 11 may yield insights into the sensory modality likely to be most effective for imparting information to an individual child with learning disability. The defect in structuring figure–ground relations may be greater with one or another input modality or may be nonsensory-specific.

Challenge 12: Sensory–sensory conditioning. When a stimulus in one sensory modality (CS) regularly precedes a stimulus in a different sensory modality (US), a change takes place in the response evoked by the CS, which comes to elicit EPs like those usually produced by the US (for reviews of the extensive literature on sensory–sensory conditioning see John, 1961, 1967a; Morrell, 1961a). The procedures used to quantify this effect, the only challenge in the NB intended to provide a direct estimate of incidental learning, are probably already obvious to the reader. The necessary data are obtained from Condition 11. Comparison of the responses to flash before and after F–C pairing shows the effect of conditioning on the ability of the CS to elicit US-like responses. Comparisons of the responses to click before and after F–C pairing show sensitization effects on the

US. Comparison of the responses to tap before and after F–C pairing show nonspecific changes unrelated to pairing per se such as changes in arousal or habituation. Analogous measures are available after C–F pairing. In general, one expects to find US-like responses being elicited by the CS on cortical regions specifically related to the modality of the US. These changes are quantified in the usual manner: $(F_2-F_1)/F_1$ = % conditioned visual response; $(C_2-C_1)/C_1$ = % sensitization; $(T_2-T_1)/T_1$ = % nonspecific change; $(C_3-C_2)/C_2$ = % conditioned auditory response; $(F_3-F_2)/F_2$ = % sensitization; $(T_3-T_2)/T_2$ = % nonspecific change (notation for stimuli defined in Condition 11).

Challenge 13: Recovery cycle. A substantial literature, referred to earlier, indicates that in some conditions (particularly psychosis) the recovery cycle may be altered. By recovery cycle is meant the time required for the brain to return to a baseline level of excitability after a sensory stimulus. Recovery cycles are usually measured by comparing the response elicited by the second of two identical stimuli to that elicited by the first member of the pair, as the interval between the two members of the pair is systematically decreased. Because of the time required by the usual tedious procedure, systematic mapping of the recovery cycle was considered impractical. Further, we felt that cross-modal recovery measures were inherently more interesting than the usual ipsimodal procedures, because of evidence that in certain conditions (particularly psychosis) a lag becomes apparent when it is necessary to shift attention from one sensory modality to another. Access to an estimate of the amount of recovery from cross-modal stimulation is available from the data of Condition 11. Specifically, the response to C during F–C pairing (C_{US}) is compared to the response to C alone before F–C pairing (C_1). Differences are quantified in the usual manner and reflect the diminution in response to C_{US} because of incomplete recovery from F_{CS} and because of the lag in cross-sensory shift. Similarly, the response to F_{CS} during C–F pairing is compared to the response to F alone after F–C pairing (F_2). Differences are quantified as usual and reflect the diminution in response to F_{US} because of incomplete recovery to C_{CS} and the lag in cross-sensory shift. These measures can be conceptualized as an index of a sort of perceptual moment of inertia, reflecting the ease with which attention can shift from modality to modality.

Challenge 14: Estimation of visual acuity.

Challenge 15: Estimation of auditory acuity.

Challenge 16: Discrimination between different geometric forms equated for area.

Challenge 17: Perceptual invariance to similar forms of different size.

Challenge 18: Discrimination between language symbols causing reversal problems. The necessary data for computation of the results of these challenges come from Condition 12. Details of analysis of Challenges 14–18, in case they are not already obvious to the reader, are discussed in Chapter 6.

Challenge 19: Estimate of stability of spectral analyses and effects of fatigue, spontaneous EEG, eyes open. The measures computed for Condition 1 are again computed, and the patterns of activity at the end of the NB session are compared with those obtained at the beginning.

Challenge 20: Estimate of stability of spectral analysis and effects of fatigue, spontaneous EEG, eyes closed. This is analyzed as Challenge 19, but compared with Condition 2.

IV. CONCLUDING COMMENTS

In this chapter, we have presented a description of DEDAAS, which we consider to be a radical innovation in electrophysiological data acquisition and analysis technology. We have also described the details of a quantitative electrophysiological, or neurometric, test battery (NB) and discussed the nature of the descriptions of functional interactions and integrity and the challenges to sensory, perceptual, and cognitive functions it was designed to impose upon the subject. In a very real sense, this chapter concludes the most significant portion of the task which we have undertaken; that is, a detailed examination of the present state of our knowledge about the basic mechanisms which mediate the most significant integrative functions of the brain, a review of the present and potential diagnostic utility of neurometric measures for the assessment of those functions, and description of a new technology and a set of measures which we believe chart the first systematic course toward the goal of optimal utilization of these powerful methods for a wide variety of practical clinical applications.

At this point in the development of functional neuroscience, it is unrealistic to expect definitive demonstrations of the advantages and subtleties of these new methods. DEDAAS and NB are still infants. Nonetheless, some of the measures they provide have already been demonstrated to be effective for a variety of purposes, whether in the specific form outlined in this chapter or in a form that is congruent in its essential details. These early indications of the utility of this technology will be presented in the remainder of this volume. Before we turn to those demonstrations, there are several general considerations to be discussed:

1. DEDAAS and NB, no matter how powerful they may ultimately become, are useless until normative data are acquired, and until systematic correlations are established between the indices of brain function which they provide and functional measures obtained by the conventional techniques of neurology and psychology. We are in the closing phases of a major endeavor to establish such correlations for one cross section of the population. A large group of 9-year-old children who perform well in school and on a conventional psychological test battery have been compared with a group of their age mates who have learning

disabilities and psychometric and/or neurological profiles suggesting an organic basis for the learning impairment. This large-scale testing has already demonstrated that DEDAAS enables the rapid, efficient acquisition and precise analysis of huge amounts of electrophysiological data. While the multivariate analysis of those data is still in process, results of univariate analysis show that many NB measures reveal strong differences between the normal and learning disabled groups. The results of this study will be made available in a subsequent volume. These data, as pointed out in the final section of Chapter 3, are the rudimentary beginning of the normative data base and the profiles of abnormal sub-groups which must be constructed tediously before these methods can truly come to fruition.

2. The reader must be aware that DEDAAS and NB generate an appalling mountain of facts about any individual. Where previously we suffered from a lack of quantitative information, now the peril is that we will be overwhelmed by an avalanche of numbers. Techniques must be devised to segregate the meaningful numbers (signal) from the meaningless numbers (noise). We are keenly conscious of this danger and actively engaged in devising ways to combat it. A salutory first step will be to subject all data to Z transformations and to disregard all data except those which reach a sufficiently high level of significance. A necessary second step, posing far more rigorous difficulties, is to devise a way to encompass the vast body of data produced for a single individual by a discrete set of descriptors. These descriptors must achieve a great deal of data compression, yielding profiles of deviation from expected or probable values for various functions which permit classes of individuals with similar etiologies to be identified. It is possible that some variant of multidimensional scaling, applied to feature vectors constructed of only the significant deviations from the mean on all 20 challenges, will provide a way to gain an overview of the total set of facts about each individual. In our opinion, this problem and the associated problem of devising ways to display data which compel attention to those features that are most important constitute the major obstacles now to be surmounted. We hope to deal with these problems in the subsequent volume, already in its preliminary phase.

The quantitative measures we presently extract from the AER waveshapes obtained in each NB condition and challenge are defined in Appendix 4.1.

APPENDIX 4.1: QUANTITATIVE FEATURES EXTRACTED FROM AER MEASURES

Below are definitions of measures derived from averaged evoked responses, from MERGE, a computer program by P. Easton.

Let X_{ist} be a digitized single evoked potential sweep where X is voltage in digitizer counts (0–1024), or some linear transform of it, and time $t = 1, 2, \ldots,$

T (usually 200); in 5-msec steps so that $t = T = 200$ is usually 1 sec after $t = 0$; s is an index for the successive sweeps, $s = 1, 2, \ldots, S$, and i is an index for individual subjects, $i = 1, 2, \ldots, I$.

Let $\bar{X}_{i \cdot t}$ be the averaged evoked potential for subject i as a function of t, that is, averaged at each point in the epoch over the sweeps that went into a particular $\bar{X}_{i \cdot t}$.

Let

$$\sigma \bar{X}_{i \cdot t} = \frac{1}{S} \sum_{s=1}^{S} X_{ist}^2 - \frac{1}{S^2} \left(\sum_{s=1}^{S} X_{ist} \right)^2$$

be the variance for each i as a function of t, that is, at each point in the epoch, over the sweeps that went into a particular $\bar{X}_{i \cdot t}$.

Let

$$\sigma \bar{X}_{i \ldots} = \frac{1}{T} \sum_{t=1}^{T} \bar{X}_{i \cdot t}^2 - \frac{1}{T^2} \left(\sum_{t=1}^{T} \bar{X}_{i \cdot t} \right)^2$$

be the variance over the epoch of the point in an averaged evoked potential.

Then, S or *signal energy* is

$$S_i = \frac{1}{T} \sum_{t=1}^{T} \bar{X}_{i \cdot t}^2$$

and equals $\sigma^2_{\bar{X}_{i \ldots}}$ only if $\bar{X}_{i \ldots} = 0$, that is, if there is no DC component in the $\bar{X}_{i \cdot t}$.

N or *noise* is defined as

$$N_{i \ldots} = \frac{1}{T} \sum_{t=1}^{T} \bar{X}_{i \cdot t}$$

and is the average of the variance over the epoch of the points in $\bar{X}_{i \cdot t}$.

SRAT or the *signal-to-noise ratio* is defined as

$$\text{SRAT} = \frac{S_i}{S_i + N_i}.$$

MSN or the *normalized signal energy* is defined as

$$\text{MSN} = \frac{1}{T} \sum_{t=1}^{T} \frac{\bar{X}_{i \cdot t}}{\sigma X_{i \cdot t}}$$

that is, is a $S_{i \cdot t}/N_{i \cdot t}$ at each point in the epoch and then averaged over the epoch.

FDF or the *first difference* is defined as

$$\text{FDF} = \frac{1}{T-1} \sum_{t=2}^{T} (\bar{X}_{i \cdot t} - \bar{X}_{i \cdot t-1})^2$$

that is, it is proportional to the mean square frequency times the signal and so varies directly with both signal amplitude and frequency content, and any changes in it should be compared to those in S, to determine if frequency content changes are important. The $\bar{X}_{i \cdot t}$ curves should be examined directly as well if anything interesting appears. Evoked potential difference measures between pairs of $\bar{X}_{i \cdot t}$ are preceded by a P if the pair are from bilaterally homotopic brain areas and by a D if they are from the same brain area under 2 different experimental conditions.

DS is the *difference in signal energy:*

$$\mathrm{DS} = \frac{1}{T} \sum_{t=1}^{T} (\bar{X}_{i \cdot t} - \bar{Y}_{i \cdot t})^2$$

where $\bar{Y}_{i \cdot t}$ is the other $\bar{X}_{i \cdot t}$ when there is a pair of these measures.

NDS is the *normalized DS:*

$$\mathrm{NDS} = \frac{\frac{1}{T} \sum_{t=1}^{T} (\bar{X}_{i \cdot t} - \bar{Y}_{i \cdot t})^2}{2}$$

$$= \frac{\mathrm{DS}}{2\, \sigma_{\bar{X}_{i \cdots}}\, \sigma_{\bar{Y}_{i \cdots}}}.$$

EASY is the normalized energy difference due to amplitude (energy amplitude asymmetry):

$$\mathrm{EASY} = \frac{\frac{1}{T} \sum_{t=1}^{T} \bar{X}_{i \cdot t}^2 + \frac{1}{T} \sum_{t=1}^{T} \bar{Y}_{i \cdot t}^2 - \frac{2}{T} \sum_{t=1}^{T} \bar{X}_{i \cdot t} \quad \bar{Y}_{i \cdot t}}{2 \sigma_{\bar{X}_{i \cdots}}\, \sigma_{\bar{Y}_{i \cdots}}} - 1.$$

SASY is the normalized energy difference due to shape difference (shape asymmetry):

$$\mathrm{SASY} = 1 - \frac{\frac{1}{T} \sum_{t=1}^{T} \bar{X}_{i \cdot t} \bar{Y}_{i \cdot t} - \frac{1}{T} \sum_{t=1}^{T} \bar{X}_{i \cdot t} \quad \bar{Y}_{i \cdot t}}{\sigma_{\bar{X}_{i \cdots}}\, \sigma_{\bar{Y}_{i \cdots}}}$$

$$= 1 - \gamma_{\bar{X}_{i \cdot t}\, \bar{Y}_{i \cdot t}}$$

and is larger the more dissimilar the $\bar{X}_{i \cdot t}$ and $\bar{Y}_{i \cdot t}$.

Note that NDS = EASY + SASY. That is, the difference in signal energy is the sum of an amplitude and a shape component.

5

Neurometric Assessment of Brain Dysfunction in Patients with Neuropathology

In Chapter 2, we discussed the major features of the EEG and AER which are the basis for clinical evaluations, and described some of the abnormalities which are often encountered in patients suffering from the neurological disorders of greatest incidence. The reader may have noticed that amplitude decreases, frequency slowing, and asymmetry were abnormal signs found in a variety of different disorders. In Chapter 3, the salient methods of computer evaluation of EEG and AER activity were discussed and a number of examples were provided in which such methods were applied to the assessment of neuropathology. Most of these examples were characterized by the common feature that, although the brain electrical activity was evaluated by precise frequency analysis or average response computation, the results of the computer analysis were then qualitatively assessed.

Displays of spectral analyses, whether compressed or not, pictures of the distribution of activity in different EEG bands at different electrode positions, pictures of AERs computed under various conditions or from homologous electrodes all require subjective judgment for their evaluation. Yet, such computations can be further reduced to scaled numerical values or mathematical expressions. It is paradoxical to go to great efforts to describe various features of the electrophysiological phenomenon produced by the brain in a precise, quantitative manner and then to evaluate those quantitative descriptors qualitatively. Unless the extracted electrophysiological feature is reduced to a numerical equivalent, which is evaluated relative to a criterion value representing the "normal" or average value of that feature in an appropriately defined healthy population, the potential quantitative power of the computed descriptor has not been fully utilized. This step, achievement of a numerical representation of a neurophysiological process, is an indispensable prerequisite for neurometric analysis. This chapter is limited to examples of computer assessment of electrical measurements from patients with neuropathology in which the resulting descriptors were evaluated by comparison with an objective criterion to yield a neurometric index. Although such examples are still few in number, remarkably good results have been achieved in every instance. Since the methods have been rather different, this suggests that *any* treatment of EEG or AER activity which generates sensible quantitative descriptors and refers them to some objectively defined normal criterion offers substantial discriminative power for identification of abnormal electrical features associated with many neurological disorders.

I. NEUROMETRIC INDICES EXTRACTED FROM SPONTANEOUS EEG

A. Indices Derived from Frequency Analysis

The utility of descriptors derived from the spectral analysis of the EEG has been clearly demonstrated in two recent papers that used different methods.

1. Abnormality Index

Gotman and his colleagues (1973) showed that it was possible to accomplish impressive separation of normal subjects from patients with brain tumors by use of an "abnormality" index derived from spectral analysis of the EEG. The abnormality index was defined in terms of weighted ratios of activities in various frequency bands within each derivation. These ratios were multiplied by a factor proportional to the bilateral symmetry between homologous derivations. These features were then summed across the head to yield a single abnormality index, as discussed in Chapter 3.

The results obtained using the abnormality index were quite good, as was shown in Fig. 3.5. However, the general applicability of the method awaits further study because these patients represented a special class: the abnormal EEG records were selected for computer analysis on the basis of prior visual inspection which revealed gross and consistent abnormal rhythms. Furthermore, all records came from patients with tumors relatively near to the surface of the brain.

2. Age-Dependent Quotients

Matoušek and Petersén (1973b) showed that excellent *localization* of tumors could be achieved by constructing a display of the topographic distribution of deviant spectral profiles measured in bilateral frontotemporal (F_7T_3/F_8T_4), central (C_3C_0/C_4C_0), temporal (T_3T_5/T_4T_6), and parietooccipital (P_3O_1/P_4O_2) derivations. For each of these derivations, they computed what they called the "age-dependent quotient," or ADQ. Actually, a variety of versions of such ADQs were defined and their diagnostic utility was explored by studying correlations between each version and such variables as the extent and severity of abnormality judged by visual examination of the EEG, side and site of lesion as established by independent methods, and so on. The common feature of all versions of the ADQ was that a ratio was constructed between two values of a quantitative function, heavily weighted by the amount of slow activity in the spectral analysis: the numerator of this ratio represented the average value of this function found in a particular head region in a group of normal individuals of comparable age as the patient; the denominator represented the actual value of the function obtained from the corresponding head region in the individual patient. Thus, ADQ (each derivation)

$$= \frac{\text{normal value (derivation, age)}}{\text{patient value (derivation, age)}}.$$

If the spectral features of the patient's EEG in a given derivation correspond well to the average values found in a group of normal individuals of comparable age, the value of the ADQ will be approximately unity. As slow wave features, usually associated with abnormality, begin to predominate in the spectral analysis, the denominator will increase and the value of the ratio will drop below 1.0. Matoušek and Petersén considered values below .8 as suggestive of the presence of pathology. By mapping the ADQ obtained from the eight derivations onto a diagram of a head, and by considering the gradients of excess slow wave activity thus revealed, it was possible to determine the region that was most abnormal. In this fashion, Matoušek and Petersén were able to correctly localize 45 of 46 tumors in the population they studied.

They also constructed a directory of language statements corresponding to comments about a variety of possible patterns of EEG activity and defined a set of logical decisions that would permit a computer to select those comments which were appropriate to the EEG patterns detected in the individual patient. The computer was programmed to type these comments next to the head diagram describing the ADQ topography. Two examples of the results achieved by this procedure were illustrated in Fig. 3.7. As yet, we cannot evaluate the precise discriminative power of the ADQ, for various reasons:

1. The tumor population studied by Matoušek and Petersén were previously identified cases. No results have yet been provided about the utility of this method for discrimination between normal subjects and neurological patients.
2. The criterion value of .8 was selected because it represented one standard deviation from the mean ADQ.

If the ADQ is normally distributed, this criterion will by definition yield 35% false positives in a normal population. This is about three times higher than the 12% incidence of false positives with conventional procedures (Gibbs & Gibbs, 1964). Clearly, the incidence of false positives would have to be decreased before this method could be used as a routine mass screening procedure. One obvious way to accomplish this would be to lower the critical value, say to .6, which would yield about 5% false positives if the ADQ were normally distributed (.6 corresponds to two standard deviations from the mean). This might increase the incidence of false negatives to an unacceptable level.

Parametric studies of the effects of changing the criterion value on the incidence of false positives and false negatives, and studies of the increased sensitivity which might be achieved by combining the ADQ with some of the other measures described below, would be extremely valuable. While such further studies are awaited, the cited results strongly suggest that this method will constitute a very powerful and sophisticated diagnostic tool.

B. Indices Derived from Symmetry Analysis

Some studies of spontaneous EEG activity have utilized a different approach to the problem of mass screening for neuropathology. Of particular interest is a method based upon the assumption that most neurological diseases, but tumors, strokes, and epilepsy in particular, involve a unilateral disruption of the structure and/or function of the brain and might be expected to cause a concomitant decrease in electrical symmetry. A simple special-purpose computer, called a symmetry analyzer, has been used (1) to quantify the symmetry of the waveforms simultaneously recorded from pairs of symmetrically placed electrodes by computing the polarity coincidence correlation coefficient (PCC), and (2) to compare the amplitudes of those signals by computing the signal ratio (SR). These measures were discussed in Chapter 3.

Results of normative studies with the symmetry analyzer were published by Harmony and her co-workers (1973b), based upon 9 derivations recorded in each of 144 normal subjects. Similar measurements were obtained for 36 patients with brain tumors, 65 patients with cerebral vascular lesions (CVA), 61 patients with epilepsy, 21 patients with Parkinsonism and 38 neurological patients with "miscellaneous other diseases." Standard clinical EEGs were also taken from each patient and were evaluated by at least one, usually two, and sometimes three electroencephalographers. Diagnosis of a patient was based upon clinical evaluation, pneumoencephalograms, and angiography, often confirmed by subsequent surgery (Otero *et al.*, unpublished results 1973, 1975a, b).

1. Discriminant Functions for Identification of Neuropathology Using Measures of EEG Symmetry

Tables 5.1 and 5.2 show that the mean values of the PCC and SR for the tumor and the CVA patients were significantly different from the normal subjects for many derivations. The individual values of PCC and SR obtained from all derivations in 144 normal and 35 tumor patients were then used to calculate a discriminant function, as described in Chapter 3.

TABLE 5.1
Values of PCC and SR obtained from Normal Subjects and Tumor Patients at Rest and During Photic Driving

Derivation	Rest					Flicker				
	T[a]		N[b]			T		N		
	X	*SD*	*X*	*SD*	P[c]	*X*	*SD*	*X*	*SD*	*P*
PCC										
F	.55	.16	.53	.18	NS[d]	.53	.14	.52	.18	NS
C	.53	.15	.55	.18	NS	.56	.14	.57	.16	NS
O	.49	.13	.52	.18	NS	.49	.11	.53	.18	.01
T	.36	.13	.36	.13	NS	.36	.16	.39	.14	NS
FC	.50	.16	.57	.15	.0005	.51	.19	.57	.14	.0005
CO	.51	.11	.59	.15	.0005	.50	.13	.59	.15	.0005
OT	.39	.15	.48	.15	.0005	.31	.15	.49	.18	.0005
TF	.32	.19	.45	.15	.0005	.31	.15	.49	.18	.0005
CT	.24	.10	.50	.12	.0005	.25	.13	.40	.13	.0005
SR										
F	1.28	.17	1.23	.12	.05	1.28	.22	1.24	.13	NS
C	1.31	.28	1.21	.11	NS	1.30	.20	1.22	.06	NS
O	1.31	.17	1.22	.09	.0005	1.34	.21	1.23	.10	.05
T	1.31	.22	1.25	.12	NS	1.35	.25	1.26	.11	NS
FC	1.29	.16	1.23	.10	.025	1.31	.20	1.22	.10	NS
CO	1.31	.20	1.21	.11	.05	1.34	.10	1.22	.11	.025
OT	1.33	.18	1.27	.11	NS	1.39	.21	1.25	.11	.01
TF	1.38	.23	1.24	.10	.05	1.34	.22	1.24	.13	.05
CT	1.39	.30	1.26	.11	NS	1.45	.33	1.26	.12	.025

[a]T = Average values obtained from 35 patients with brain tumors.
[b]N = Average values obtained from 144 normal subjects, taken from Harmony *et al.* (1973a). *X* = mean value; *SD* = standard deviation.
[c]P = Significance level of the difference between the means of the two groups as assessed by *t* test.
[d]NS = Not significant.

Application of discriminant analysis correctly identified 25 out of 36 tumor patients as abnormal (71%), while 17 of the 144 normals were misclassified (12%), giving an overall separation accuracy of 87%. The tumor group included 11 patients with brain stem or pituitary tumors, 7 of which were detected by this method. Discriminant analysis correctly identified 56 of 65 patients with cerebrovascular problems (83%), but misclassified 34 of the 144 normals (24%) with an overall separation accuracy of 80% (Otero *et al.*, 1975a, b).

These results can be better appreciated when contrasted with the outcome of conventional EEG evaluation by which 10 of the tumor cases and 26 of the CVA cases were erroneously classified as normal; this constituted a tumor detection

TABLE 5.2
Comparison of 144 Normal (N) Subjects[a] and 65 Cerebrovascular Accident (CVA) Patients[b]

Deviation	Rest					Flicker				
	CVA		N			CVA		N		
	X[c]	*SD*	*X*	*SD*	*P*	*X*	*SD*	*X*	*SD*	*P*
PCC										
F	.51	.18	.53	.18	NS	.52	.12	.52	.18	NS
C	.51	.14	.55	.18	.025	.53	.16	.57	.16	.01
O	.47	.17	.52	.18	.01	.48	.15	.53	.18	.005
T	.36	.14	.36	.13	NS	.35	.13	.39	.14	.01
FC	.48	.14	.57	.15	.0005	.49	.15	.57	.14	.0005
CO	.51	.13	.59	.15	.0005	.52	.14	.59	.15	.0005
OT	.40	.13	.48	.15	.0005	.42	.12	.51	.16	.0005
TF	.39	.11	.45	.15	.005	.41	.18	.49	.18	.0005
SR										
F	1.27	.17	1.23	.12	.05	1.28	.17	1.24	.13	.05
C	1.27	.26	1.21	.11	NS	1.24	.12	1.22	.06	NS
O	1.29	.17	1.22	.09	.0005	1.28	.14	1.23	.10	.0005
T	1.31	.17	1.25	.12	.0005	1.33	.18	1.26	.11	.05
FC	1.27	.17	1.23	.10	.05	1.27	.16	1.22	.10	.01
CO	1.28	.20	1.21	.11	.0005	1.26	.15	1.22	.11	.025
OT	1.32	.16	1.27	.11	.025	1.30	.17	1.25	.11	.025
TF	1.28	.16	1.24	.10	.025	1.28	.14	1.24	.13	.025
CT	1.31	.17	1.26	.11	.01	1.34	.17	1.26	.12	.0005

[a]From Harmony *et al*. (1973a).
[b]From Otero *et al.* (1973).
[c]See Table 5.1 for explanation of symbols.

accuracy of 73% and a CVA detection accuracy of 62%, a level usually considered acceptable (Cobb, 1963b; Gibbs & Gibbs, 1964). Eight of the 10 tumor cases and 22 of the 26 CVA cases with apparently normal EEGs were correctly classified by the quantitative analysis of symmetry. This method was found to have no utility for detection of patients with epilepsy or Parkinsonism. Nonetheless, this relatively simple method achieved somewhat higher accuracy than conventional EEG evaluation for detection of tumors and strokes.

2. Numerical Taxonomy of Neuropathology: Neurometric Discrimination between Types of Neurological Diseases

Because of the differences observed between the centroids of subgroups of patients with different neurological diseases, Otero and her colleagues used multiple discriminant analysis to test whether such subgroups could be significantly distinguished from each other as well as from normals (Otero *et al.*, 1975a,b).

A total of 36 separate measurements from each patient were available for this multiple discriminant analysis: PCC and SR values during flicker driving and at rest for each of 9 derivations. First, the PCC values were subjected to a z transformation. Next, in order to reduce the number of variables, a stepwise discriminant function was computed between the group of normal subjects and the five groups of neurological patients. Six of the 36 measures were found to provide the most significant contributions to the stepwise discriminant analysis. These 6 variables, in their order of importance, were the values of PCC in derivation CO, SR in CT, and SR in OT (all during flicker), PCC in C at rest, PCC in OT during flicker, and SR in OT at rest. The mean values of these 6 most discriminating variables were calculated for each of the 6 groups of patients and normal controls and are given in Table 5.3.

Using these 6 variables, multiple discriminant analysis yielded a discriminant equation with three canonical variables. The centroid for each of the 6 groups of patients and the normal control group was then described in terms of these canonical variables. Finally, every subject in the sample was classified as a member of the group for which the Mahalanobis distance between that subject and the corresponding group centroid was minimal. *This classification procedure constitutes the first example of the use of true numerical taxonomic methods for differential diagnosis of neurological diseases*. (The actual values contributed by each variable to the 3 canonical variables, the canonical variable values corresponding to each group centroid, and the definitions of operations used to calculate the Mahalanobis distance used in this neurometric classification, are provided in Footnote 1.)[1]

The results of this neurometric classification are shown in Table 5.4A. Examination of Table 5.4A indicates that discriminable differences were found between these various groups of patients with different neurological diseases. It is noteworthy that the tumor group displayed relatively consistent characteristics

[1] The values of the three canonic variables obtained from the multiple discriminant analysis were as follows:

	1st canonic variable	2nd canonic variable	3rd canonic variable
PCC C	−.849	−3.205	3.315
PCC CO[a]	3.313	−3.415	−2.050
PCC OT[a]	1.534	1.751	−3.826
SR OT	2.169	4.223	3.635
SR OT[a]	−4.906	−1.490	−6.076
SR CT[a]	−2.522	−3.117	.434

[a] During flicker stimulation.

continued

which were markedly different from the largest subset of the stroke patients. This distinction reflects the fact that the asymmetry due to tumors usually affects a more widespread anatomical region than that caused by a stroke, as if not only the afflicted region were altered but also adjacent regions to which it sends aberrant output. Tumors seem to cause a more widespread change in neuronal transactions within the involved hemisphere than strokes.

Of the 144 normal subjects, only 73 were correctly classified. Twenty-three of the 65 patients with cerebrovascular disease were correctly classified, and 33 were assigned to other categories of neuropathology, while 9 were erroneously identified as normals. Twenty of the 36 patients with brain tumors were correctly classified, and 11 assigned to other categories of neuropathology, while 5 were misidentified as normal. Differential diagnostic accuracy was less adequate in the other groups. The borderline was not well defined between all groups and the normal domain. Thus, when all types of pathology were combined, the discrimination between normal and abnormal became more difficult. Other methods were more successful, as described later in this chapter. This

Footnote contd.

The centroids for each of the 6 groups were as follows:

	1st canonic variable	2nd canonic variable	3rd canonic variable
Normals	.646	.154	–.332
C. vasc.	–.171	.496	.163
Tumors	–.951	.024	–9.175
Miscellaneous	.082	–.505	–.049
Epilepsies	.356	.178	.192
Parkinson	.038	–.347	.102

The Mahalanobis distance d from any subject i to each centroid is equal to

$$d = \left|\sum_{j=1}^{3} (X_{ij} - \bar{X}_j)^2\right|^{1/2}, \tag{5.1}$$

where j is canonical variable = 1, 2 or 3; k is EEG symmetry variable = 1, 2, . . . , 6; i, is subject = 1, 2, . . . , N; p is neuropathology group = 1, 2, . . . , 6;

$$X_{ij} = \sum_{k=1}^{6} Y_{ik}E_{kj};$$

$$\bar{X}_j = \sum_{k=1}^{6} \bar{Y}_{pk}E_{kj};$$

$$Y_{ik} = Z_{ik} - \tfrac{1}{6}\sum_{p=1}^{6} \bar{Y}_{pk};$$

where Z_{ik} is the value of variable k for subject i, $\bar{Y}_{pk}$ is the mean value of variable k for group p, and $^1/_6 \sum_{p=1} Y_{pk}$ is the average of averages, where E_{jkj} is the discriminant equation (a $j \times k$ matrix).

TABLE 5.3

Mean Values of Most Critical PCC[a] and SR[b] Values

Measure	C. Vasc.	Tumors	Miscellaneous	Epilepsy	Parkinson's	Normals	Grand average
PCC C rest	0.56	0.59	0.66	0.65	0.67	0.61	0.62
PCC CO flicker	0.56	0.55	0.67	0.64	0.65	0.69	0.63
PCC OT flicker	0.45	0.43	0.49	0.50	0.50	0.56	0.49
SR OT rest	1.32	1.33	1.26	1.30	1.28	1.27	1.29
SR OT flicker	1.29	1.39	1.27	1.26	1.27	1.25	1.29
SR CT flicker	1.34	1.43	1.37	1.28	1.37	1.26	1.34

[a] PCC values were subjected to Fisher's z transformation.
[b] Absolute values of the SR were used.

TABLE 5.4A
Discriminant Analysis

Group	Number of subjects classified as						
	Cerebrovascular	Tumors	Miscellaneous	Epilepsy	Parkinson's	Abnormal	Normal
Cerebrovascular	23	15	6	8	4	56	9
Tumors	5	20	4	2	0	31	5
Miscellaneous	5	4	13	3	4	29	9
Epilepsy	10	8	11	12	5	46	15
Parkinson	2	3	8	3	2	18	3
Normals	15	9	17	22	8	71	73

discriminant equation was evaluated further in a blind study on 43 neurological patients. In this replication, all patients suffering from cerebrovascular disease were correctly identified, and the overall level of identification of neuropathology remained at the high level of the previous study, as shown in Table 5.4B.

II. NEUROMETRIC INDICES EXTRACTED FROM AER

A. Indices Derived from AER Symmetry Analysis

In Chapter 2, we discussed the major features of the AER thus far found useful in diagnosis of neuropathology. Asymmetry of the amplitude or waveshape of AERs from homologous derivations and latency discrepancies between corresponding components were generally agreed to reflect abnormality.

Normative values for a number of measures of AER symmetry have been ascertained for a large group of healthy subjects and have recently been published by Harmony *et al.* (1973b). The features for which norms were provided were *waveshape similarity*, as measured by the Pearson correlation coefficient (r), *signal ratio* (SR), *amplitude differences* (percent), and *maximum latency lag*.

The normative data for waveshape similarity are presented in Table 5.5. A total of 87% of all lags in latency were less than or equal to 5 msec. This simultaneity of congruent peaks was found in all five scalp derivations and in each of the latency segments for which the analysis was conducted. Only 5% of all peaks were unilateral. A total of 90% of all AERs showed less than a 40% difference in peak amplitudes. Similarly, SR values were very close to unity, indicating AERs from homologous derivations were approximately equal in energy.

Shortly after these norms appeared, they were used in a systematic analysis of AER features in various neurological disorders. Significant differences in the mean value of the correlation coefficients (r) and signal ratio (SR) between AERs recorded from homologous derivations were found between 64 normal subjects and four pathological groups consisting of 34 tumor, 50 CVA, 55 epileptic patients, and 38 patients suffering from a variety of less common disorders (Harmony, Otero, Ricardo, & Valdes & Fernandez, 1975). Using these values of r and SR, a discriminant function was computed between the group of normal subjects and the combined group of 177 patients with different neurological diseases.

1. Discriminant Function Separating Normal Subjects from Neurological Patients on the Basis of AER Symmetry

The discriminant function obtained in this study by Harmony *et al.* (1975) is reproduced below, since it can be applied to AER data obtained from the

TABLE 5.4B
Replication of Discriminant Analysis: Blind Sample

	Number of subjects classified as						
Group	Cerebrovascular	Tumors	Miscellaneous	Epilepsy	Parkinson's	Normal	Total pathologic
Cerebrovascular	5	0	0	0	0	0	5
Tumors	1	2	0	0	0	0	3
Miscellaneous	4	0	3	7	1	5	15
Epilepsy	6	1	2	1	2	3	12

TABLE 5.5
Mean and Standard Deviation of Correlation Coefficients for VEPs by Derivations and Subdivisions of Analysis Time

Derivation	Total	0–62.5 (msec)	65–125 (msec)	127.5–187.5 (msec)	190–250 (msec)
Central	.927±.079	.714±.239	.911±.162	.905±.173	.861±.182
Occipital	.908±.156	.717±.294	.822±.306	.861±.223	.883±.202
Temporal	.874±.140	.504±.410	.888±.160	.892±.210	.785±.205
Centro-occipital	.871±.142	.817±.216	.791±.382	.865±.205	.811±.266
Occipito-temporal	.895±.113	.703±.324	.882±.215	.843±.299	.827±.216
Derivation	**Total**	**0–125 (msec)**	**127.5–250 (msec)**	**252.5–375 (msec)**	**377.5–500 (msec)**
Central	.935±.040	.919±.128	.944±.154	.893±.241	.806±.269
Occipital	.899±.073	.863±.127	.838±.272	.855±.168	.756±.243
Temporal	.876±.087	.751±.239	.895±.125	.814±.235	.756±.243
Centro-occipital	.900±.089	.866∓.172	.883±.131	.861±.186	.860±.143
Occipito-temporal	.876±.128	.860±.150	.890±.146	.841±.173	.870±.101

Analysis time: 500 msec (n = 30)

Analysis time: 250 msec (n = 50)

corresponding derivations on any individual for which Pearson product moment correlation coefficients (r) and signal ratios have been computed:

$$\begin{aligned} DF = \; & .156\, C_r - .184\, C_{\mathrm{SR}} + .466\, O_r \\ & + .074\, O_{\mathrm{SR}} + .384\, T_r + .207\, T_{\mathrm{SR}} \\ & + .142\, CO_r + .182\, CO_{\mathrm{SR}} \\ & + .414\, OT_r + .118\, OT_{\mathrm{SR}}. \end{aligned}$$

In this equation, each term is obtained by multiplying the value of r or the SR from the indicated derivation by the weighting coefficient. C refers to values obtained from electrodes C_3 versus C_4 in the 10/20 system, O to values from O_1 versus O_2, T to values from T_5 versus T_6, CO to C_3O_1 versus C_4O_2, and OT to O_1T_5 versus O_2T_6.

The centroid for the normal group of 64 subjects was 5.98. The centroid for the combined group of 177 neurological cases consisting of 34 brain tumors, 50 cerebrovascular accidents, 55 epileptics, and 38 cases of miscellaneous disease entities was 4.66. The midpoint between the two centroids was 5.32. Individuals with discriminant values below 5.32 were classified as abnormal, while individuals with values above 5.32 were classified as normal (Harmony *et al*., 1975). The results obtained using this discriminant function are presented in Table 5.6.

It is apparent from inspection of Table 5.6, that the accuracy of detection of neuropathology by neurometric analysis of the AER was high in these cases. The results achieved in this study are particularly impressive if we realize that these were unselected cases with a wide variety of etiologies. There were, for example, six pituitary adenomas in the tumor group, only one of which failed to be detected. Given the size and diversity of the clinical population, these results

TABLE 5.6
Accuracy of Discriminant Function Computed from AER Features

		AER discriminant score		
Category	N	Abnormal	Normal	%
Tumors	34	28	6	82
Strokes	50	41	9	82
Epilepsy	55	35	20	63
Miscellaneous	38	28	10	74
Totals:	177	132	45	75

establish that computer methods based upon discriminant analysis of AER symmetry features are an extremely accurate diagnostic tool.

2. *Comparison of Effectiveness of AER Symmetry, EEG Symmetry, and Conventional EEG*

In the neurological cases studied in these evaluations of the discriminative utility of neurometric indices extracted from the EEG and the AER, conventional EEG examinations were also performed. All EEGs were studied by three electroencephalographers and classed as *abnormal* if any one of the three judges so concluded. The percentages of detection by conventional methods in these cases were about what might be expected on the basis of existing statistics (Gibbs & Gibbs, 1964).

It is extremely interesting to compare the accuracy of discriminant analysis of EEG symmetry (D_1), AER symmetry (D_2) and conventional EEG examinations, and to estimate the accuracy to be expected were these various techniques to be combined. This comparative analysis was performed by Harmony for the 150 neurological cases in which all three kinds of data were available. Patients for whom any one measure was lacking were excluded from this analysis (Harmony, unpublished observations, 1976).

The results of this comparative analysis are summarized in Table 5.7. Note that there exist slight discrepancies between the detection rates presented in Table 5.7 and those reported elsewhere because data on one or another symmetry measure were lacking from some patients.

Examining Table 5.7, we see that all three methods agree on roughly half the patients in each group. About 15% of the time, both discriminant functions agreed, while a false negative was reported by conventional EEG. Conventional EEG detected abnormality in only 7 cases missed by both discriminant functions, of a total of 150. One or both discriminant functions detected abnormality in 38 cases erroneously classified as negative by conventional EEG criteria, *more than 25% of the sample.* In 16 cases out of the total. AER features detected abnormality that EEG symmetry features failed to identify; in 28 cases, EEG symmetry features detected abnormality that AER features failed to detect. Thus, the two sets of features represented by the two discriminant functions are complementary. A significant increase in accuracy results from using both measures. Abnormality was detected in 79% of the cases by EEG symmetry features alone, but in 90% of the cases when both measures were used. In contrast, the accuracy of conventional EEG, overall accuracy rises to over 95%. The overall relative accuracy of the three methods is presented in Table 5.8, for each disease category.

Particularly in view of the unselected nature of these cases, the overall detection accuracy of the neurometric method must be considered extremely impressive. As a screening method, either AER or EEG symmetry features,

TABLE 5.7

EEG Symmetry Features (D_1)[a], AER Features (D_2), and Conventional EEG Examination

Pattern of results obtained by the three methods			Tumors (n = 31)		Strokes (n = 53)		Epilepsy (n = 40)		Misc. (n = 26)	
EEG symmetry (D_1)	AER symmetry (D_2)	EEG	*N* Positives	%	*N* Positives	%	*N* Positives	%	*N* Positives	%
+	+	+	15	47	26	49	17	43	10	38
0	+	+	1	3	3	5.5	4	10	4	15
+	0	+	4	13	3	5.5	8	20	2	8
0	0	+	2	7	0	0	4	10	1	4
+	+	0	5	17	11	21	2	5	5	19
0	+	0	1	3	2	4	0	0	1	4
+	0	0	2	7	6	11	2	5	1	4
0	0	0	(1)[b]	3	(2)[b]	4	(3)[b]	7	(2)[b]	8
Total positives with the three methods:			30	97	51	96	37	93	24	93
Total positives with symmetry EEG, AER:			28	90	51	96	33	83	23	88

[a]Subjects were considered + if they were assigned to any of the pathological groups.

[b]False negatives.

quantitatively assessed and utilized in a multiple discriminant function, was significantly more accurate than the group judgment of three electroencephalographers working with what was perhaps more than routine meticulousness. Yet such neurometric assessment is more rapid and can be evaluated automatically, without need for a highly trained electroencephalographer. When both neurometric methods were combined, accuracy was far superior to that achieved in routine EEG examinations. Finally, when all available methods were combined, accuracy for the electroencephalographers operating in conjunction with the neurometric analyses was almost perfect.

B. Neurometric Indices Derived by Varimax Factor Analysis of AER Waveshapes

In a collaborative study between our laboratory and that of Harmony, the utility of regression factor analysis for the discrimination between AER waveshapes from normal subjects and from patients with various neurological disorders was investigated. (See Chapter 4 for discussion of these methods.) All normal and pathological data for these factor analysis studies were gathered in the National Scientific Research Center of Cuba (CENIC) and analyzed in the Brain Research Laboratories of the Department of Psychiatry at New York Medical College, to which they were shipped. These data were selected from the much larger sample used by Harmony *et al.* (1975) to include the most difficult diagnostic cases, especially *all* instances of negative findings with the conventional EEG. In each case, conventional EEGs and AERs to flash stimuli were obtained from 10 derivations (C_3/C_4, T_5/T_6, O_1/O_2, C_3O_1/C_4O_2, and T_5O_1/T_6O_2). EEGs were evaluated in the conventional fashion by a team of three electroencephalographers at CENIC, and only the summary reports of findings were sent to New

TABLE 5.8
Relative Overall Accuracy of EEG Symmetry, AER Symmetry, and Conventional EEG for Each Disease Category

	EEG symmetry		AER symmetry		EEG	
Positive cases in:	*n*	%	*n*	%	*n*	%
Tumors	26	84	22	71	22	71
Strokes	46	87	42	79	32	60
Epilepsy	29	73	23	58	33	83
Miscellaneous	18	69	20	77	17	65
Totals:	119	79	107	71	104	69

York. In all cases of neurological disease except epilepsy, independent confirmation by angiography, pneumoencephalography, and in some cases surgery was obtained before data from the patient were included in the sample. In the case of epilepsy, clinical observations were accepted as adequate evidence for the presence of the disease.

1. Determination of the "Normal AER Space"

Using data from 50 normal subjects, three neurometric indices were computed: (1) r – the Pearson product moment correlation coefficient between AERs from bilaterally symmetric derivations; (2) regression (%) – the percentage of each individual AER, accounted for by the set of Varimax factors used for 93% of the energy in the "normal" signal space containing these 500 AERs (10 derivations/subject × 50 normal subjects); (3) the regression symmetry, or *RA*. For each pair of symmetric electrodes, the individual factor loading asymmetry, ΔF, was obtained by calculating the absolute difference in loading coefficients representing the relative contributions of *each* factor to the two AER waveshapes. The total factor loading asymmetry, $\Sigma\ \Delta F$, for the pair was obtained by adding the individual factor loading asymmetries for all the factors required to reconstruct the two waveforms to the criterion accuracy. The regression asymmetry, or $\overset{5}{\Sigma}(\Sigma \Delta F)$, was obtained by summing the total factor loading asymmetries, $\Sigma\ \Delta F$, across the five sets of derivations. $\overset{5}{\Sigma}(\Sigma\ \Delta F)$ constitutes a single measure for the cumulative asymmetry of the entire head of the subject, precisely accounting for the various factors significantly contributing to any AER, and will be referred to as *RA*.

Examining the distribution of the resulting values for each of these neurometric indices, first approximations of the criteria for normal values were established. Any value falling outside these limits, or an abnormal finding in the EEG, was scored as "positive." Any individual whose data included one positive finding was considered "at risk," (?), while any two positive findings in an individual were considered "abnormal," (AB). Individuals with no positive findings were classified as "normal." (–).

By these criteria, three members of the normal group were erroneously categorized as at risk, and one was classified as abnormal (2%).

2. Regression Factor Analysis of AERs from Patients with Tumors, Strokes, or Epilepsy

Next, the data from four test groups were submitted to the same criteria: Group I, 25 normal controls; Group II, 25 patients with brain tumors; Group III, 25 patients with strokes; Group IV, 25 patients with epilepsy. As mentioned before, data from the 75 neurological patients were selected from a much larger population in order to include cases which were particularly difficult to diagnose, so as to devise a particularly stringent challenge for our methods; 25 of these cases were apparently normal by conventional EEG criteria.

For these four test groups, the same neurometric indices were computed, except that the normal Varimax factors *previously* obtained were used in the regression factor analysis, as described in Chapter 3. The data were then classified using the previously defined criteria. The results are presented in Table 5.9.

The overall accuracy of these methods was good. If any single neurometric index was found to be positive (+), the patient was considered *at risk for neuropathology*, or "doubtful," (?). If any two neurometric indices were positive (2+), the patient was classified as "abnormal," (AB). In order to resolve the classification of patients placed in the "doubtful" category on the basis of neurometric assessment alone, the EEG evaluation was taken into consideration. If the EEG evaluation was positive as well, the patient was classified as abnormal. This procedure can be considered as analogous to the use of the regression factor analysis as the first step in a two-stage screening procedure: patients classified as abnormal on the basis of a 2+ neurometric evaluation would be referred directly for treatment. Patients classified as doubtful would be referred for a conventional EEG examination. The results of this two-stage procedure are presented in Table 5.10.

In evaluating Table 5.10, and in comparing it with the results reported using features derived from AER and EEG symmetry presented in the previous sections, the reader should bear in mind that the three patient groups evaluated by this regression factor analysis represent a special sample deliberately selected to include the most difficult diagnostic problems in the available sample. Only 50 of the 75 neurological patients in these groups were considered abnormal on the basis of EEG criteria alone. Of the 75, 60 were classified as abnormal by combining neurometric and EEG evaluation with the requirement that two positive findings be obtained, while 8 more were classified "doubtful" (+). A total of 42 of the 75 patients were classified as abnormal on the basis of neurometric indices alone (2+), while 21 more were classified as doubtful (+).

Among the tumor patients, 20 were classified abnormal by computer indices alone (80%), while combined neurometric and conventional EEG criteria classified 22 as abnormal (88%). EEG abnormalities were only found in 15 of these patients (60%), 2 of whom seemed normal by neurometric criteria. Nine tumor patients considered normal by EEG criteria were classified abnormal by this neurometric method. Twelve of the tumor patients were correctly identified by both conventional and neurometric methods, while one was misclassified by both methods (this patient had a pituitary adenoma).

Among the stroke patients, 15 were classified abnormal by neurometric indices alone (60%), while combined neurometric and EEG criteria classified 20 as abnormal (80%). EEG abnormalities were only found in 15 of these patients

TABLE 5.9

Results of Regression Factor Analysis of Normal Control, Tumor, Stroke and Epileptic Data Sample (% Regression and Regression Asymmetry, or RA) Compared with Correlation Coefficient and Conventional EEG Evaluation

	1	2	3	4	5	6	7	8	9	10	11	12	13	14	15	16	17	18	19	20	21	22	23	24	25	*N* Detected by each index alone (Total: Diagnosis, all measures)
1. Normal control																										
Regression (%)	–[b]	–	+	–	–	–	–	+	–	+	–	–	–	–	–	–	–	–	–	+	–	–	–	–	–	4
RA	–	–	–	–	–	–	–	–	–	–	–	–	–	–	–	–	–	–	–	–	–	–	–	–	–	0
r	–	–	–	–	–	–	–	–	+	–	–	–	–	–	–	–	–	–	–	–	–	–	–	–	–	1
EEG	–	–	–	–	–	–	–	–	–	–	–	–	–	–	–	–	–	–	–	–	–	–	–	–	–	0
Diagnosis	–	–	?	–	–	–	–	?	?	?	–	–	–	–	–	–	–	–	–	?	–	–	–	–	–	0 abnormal 5 doubtful 0 false positives
2. Tumors																										
Regression (%)	–	+	+	+	–	+	–	+	–	+	–	+	–	–	+	+	+	+	–	+	+	+	–	+	+	16
RA	+	+	+	+	–	+	+	+	–	+	+	+	+	–	+	–	+	+	+	+	+	+	+	+	+	21
r	+	+	+	+	–	+	+	–	–	+	+	–	+	–	–	–	+	–	+	+	+	+	–	–	+	15
EEG	–	+	–	–	+	+	+	+	+	+	–	–	+	–	+	+	+	–	+	–	+	–	+	+	–	15
Diagnosis	ab	ab	ab	ab	?	ab	ab	ab	?	ab	ab	ab	ab	–	ab	ab	ab	ab	ab	ab	ab	ab	ab	ab	ab	22 abnormal 2 doubtful 1 false negatives

3. CVA																										
Regression (%)	+	+	+	+	–	+	–	+	+	+	–	–	–	–	–	–	+	+	+	–	–	–	–	+	–	12
RA	+	+	+	+	+	+	+	+	+	+	+	+	–	–	–	+	+	+	+	+	+	+	–	+	+	21
r	+	+	–	–	–	–	+	–	+	–	–	+	–	–	–	–	–	+	–	–	–	–	–	–	+	7
EEG	–	–	+	+	+	+	–	–	–	+	–	–	–	–	–	+	+	+	+	+	+	+	+	+	+	15
Diagnosis	ab	ab	ab	ab	ab	ab	ab	ab	ab	ab	?	ab	–	–	–	ab	ab	ab	ab	ab	ab	ab	?	ab	ab	20 abnormal 2 doubtful 3 false negatives
4. Epilepsy																										
Regression (%)	+	+	–	–	–	–	+	–	–	–	+	+	–	+	–	+	+	–	–	–	–	–	+	–	–	9
RA	–	–	+	+	–	+	–	+	–	+	+	–	+	–	+	–	+	–	–	+	+	+	+	+	–	14
r	–	–	+	+	–	+	–	–	–	–	–	–	–	–	+	–	–	–	–	–	–	–	–	–	–	4
EEG	–	+	+	+	–	+	+	+	+	–	+	+	+	+	+	+	+	–	–	+	+	+	+	+	+	20
Diagnosis	?	ab	ab	ab	–	ab	ab	ab	?	?	ab	ab	ab	ab	ab	ab	ab	–	–	ab	ab	ab	ab	ab	?	18 abnormal 4 doubtful 3 false negatives

[a]Criteria: RA 2.1. Percentage regressed: C<.31; O<.41; T<.25; OT<.20; r_C<.57; r_o <.52; r_T<.26; r_{CO}<.81; r_{OT}<.16.

[b]Key: – Negative; ? doubtful; + positive (if neurometric only one +, conventional EEG necessary). For positive diagnosis, two different plus values needed. CO not used.

TABLE 5.10
Comparative Accuracy of Detection Using (1) Any Single Positive Neurometric Index (+), (2) Any 2 Positive Neurometric Indices (2+), (3) Any Single Positive Neurometric Index (+) *and* Positive EEG Findings, or (4) Positive EEG Findings Alone

	Normal controls (n=25)	Tumor (n=25)	Stroke (n=25)	Epilepsy (n=25)
1. Single neurometric index				
Doubtful (+)	5	22	21	20
Normal (–)	20	3	4	5
2. Three neurometric indices				
Abnormal (2+)	0	20	15	7
Doubtful (+)	5	2	6	13
Normal (–)	20	3	4	5
3. Single neurometric index (doubtful) plus EEG				
Abnormal (2+)	0	2	5	11
Doubtful (+)	5	2	2	4
Normal (–)	20	1	3	3
4. EEG alone				
Abnormal (+)	0	15	15	20
Normal (–)	25	10	10	5
5. Overall neurometric plus EEG				
Abnormal (2+)	0	22	20	18
Normal (–)	25	3	5	7

(60%), 1 of whom seemed normal by neurometric indices. Six stroke patients considered normal by EEG criteria were classified abnormal by neurometric criteria. Only 3 patients were misclassified by both methods.

Among the epileptic patients, 7 were classified abnormal by neurometric indices alone (35%), while combined neurometric and EEG criteria classified 18 as abnormal (72%). EEG abnormalities were found in 20 cases. Although the neurometric method scored relatively poorly with epileptic patients if we look at definitive abnormal classifications, yet in only two cases did the method fail to classify as "at risk" an epileptic patient with abnormal EEG findings. Conversely, the neurometric labeled 2 patients at risk who showed no positive EEG findings.

These results indicate that these neurometric methods could serve a valuable screening function, in addition to supplementing the conventional EEG. Routinely applied to patients with complaints of possible neurological origin, one might expect regression factor analysis of AERs to provide unequivocal identification of abnormality in tumor or stroke victims with accuracy equal to or surpassing the conventional EEG, while labeling few normals erroneously as abnormal. Were all patients classified "at risk" by the neurometric method subsequently referred for an EEG evaluation, the combined neurometric and conventional examination would reject all those who had been thus misclassified, while providing substantially higher detection levels for tumors and strokes than achieved by EEG methods alone. The detection levels for epilepsy would be about the same. Note that even if patients considered normal by the neurometric screening method were not further examined, the incidence of false negative findings would be no higher than if exclusive reliance were placed upon the conventional EEG, and might be substantially lower for certain disorders.

3. Multidimensional Scaling Applied to Discrimination between AERs from Normal Subjects and Patients with Tumors

The method of multidimensional scaling was described in Chapter 3. In a pilot study (Schwartz & John, unpublished observations, 1976), we explored the utility of this neurometric method for separation of AERs obtained from occipitotemporal derivations in 20 normal subjects and 20 patients with brain tumors. The data were a subset of those used in the regression factor analysis just described. In previous unpublished studies, we had found this method remarkably effective for the classification of single evoked potentials elicited by neutral stimuli in differential generalization tests using trained cats (Schwartz, Ramos, & John, 1976). Classifications thus obtained correlated with behavior better than did classification of the same data using sorting methods (see Chapter 10, Volume 1; Chapter 3, Volume 3).

The clustering evident in Fig. 5.1, obtained by multidimensional scaling applied to these human AER data, show impressive separation of the two groups. Since the AERs from the normal subjects were highly symmetric, the

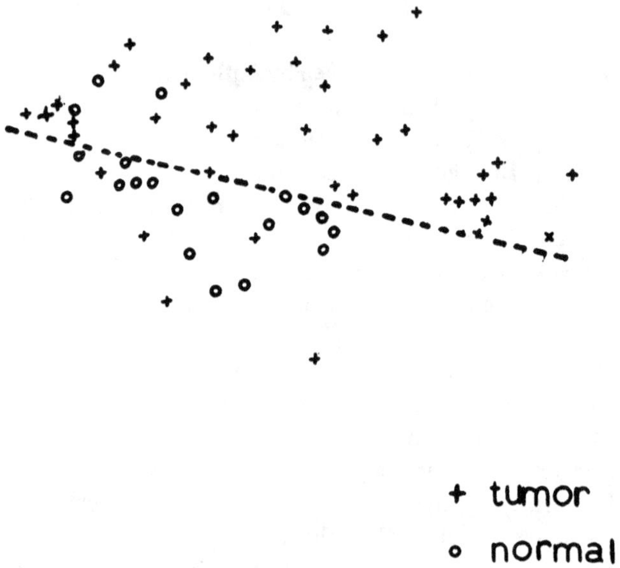

FIG. 5.1 Results of applying multidimensional scaling to classification of AERs from occipitotemporal derivations of 20 normal subjects (open circles) and 20 tumor patients (crosses). Since bilateral AERs from normals were usually almost superimposed, only one point was plotted for those cases.

two points representing the pair of AERs from each normal subject were usually almost superimposed. Accordingly, only one point is plotted in Fig. 5.1 for each normal subject. Since tumor patients displayed asymmetric AERs, both members of the pair from each patient were plotted in the figure.

Examination of the figure shows that AERs from three of the normal subjects lie in the abnormal domain (top portion of figure), while five AERs from tumor patients lie in the normal domain. Actually, these five AERs came from only three tumor patients, two of whom produced pairs of AERs both members of which fell in the normal domain. Thus, this neurometric method classified 15% of the normal subjects as "false positives," while 15% of the tumor patients were classified as "false negatives." The 85% accuracy achieved in this initial trial of the multidimensional scaling method is comparable to the accuracy obtained with the discriminant analyses of features reflecting AER or EEG symmetry, as well as the results of regression factor analysis, and particularly noteworthy since data from only one derivation were used. The results are better than obtained on the same patients using conventional EEG evaluation. These results are quite promising and efforts are in progress to refine and simplify the method to make it a practical tool for routine evaluation of waveshapes.

III. CONCLUSION

A large variety of different neurometric methods for quantitative evaluation of electrophysiological measures obtained from neurological patients has been surveyed in this chapter. These methods can be considered as the first steps toward the application of *numerical taxonomy* (Sokal & Sneath, 1963) for the classification of brain disease. All of these neurometric methods, however disparate their computational details or raw data, share the feature that they develop an objective, quantitative criterion for abnormality and evaluate data from the individual patient relative to this criterion value. Every neurometric method examined equals or outperforms the conventional EEG examination in accuracy. The features of the electrophysiological activity of the brain upon which the indices herein discussed were based included the spectral analysis of the spontaneous EEG, the symmetry of the spontaneous EEG, the symmetry of the AER elicited by blank flashes, and the actual morphology of the AER waveshape. Evidence was presented, and knowledge reviewed in Volume 1 further indicates that some of these neurometric indices reflect relatively independent aspects of brain function. It seems reasonable to expect that once the extent of redundancy of these indices is established by correlative studies, a substantial improvement over the accuracy of any single feature measurement can be attained by combination of these neurometric indices into a single electrophysiological examination carried out and evaluated by computer. Such neurometric evaluations of brain integrity will obviously, on the basis of the mass of evidence presented in this chapter, be more rapid, more accurate, and more economic than evaluations by contemporary methods of visual examination of data by a trained specialist in electroencephalography. Where such specialists are available, neurometric methods should constitute an invaluable screening procedure, enabling paramedical personnel to achieve an enormous increase in the efficiency not only of the electroencephalographer but of the neurologist. The patient who is "caught" by the meshes of such a screening procedure is objectively, replicably abnormal in some aspect of the electrical activity of his brain. Thus, the "sick to healthy" patient ratio in the neurologist's waiting room will be greatly enhanced. Where such specialists are not available, which is in the great majority of communities in the world, a computer installation or a remote terminal connected by telephone or radio to a computer installation, can effectively make the consultation of an exceptionally accurate electroencephalographer available to the pediatrician, psychologist, psychiatrist, or general practitioner.

For the detection of neuropathology, such neurometric evaluations will be enormously useful. Further, it is our hope and belief that the application of similar techniques, but buttressed by the manifold challenges which can be devised and which are encompassed in the quantitative neurometric test battery, NB, defined in Chapter 4, will permit the extension of automatic neurometric assessment to aspects of electrophysiological activity reflecting sensory, perceptual, and cognitive processes. The possibility of such extensions is examined in Chapter 6. Chapter 7 describes current results obtained by using neurometric methods for numerical taxonomic classification of the normal and senile elderly patient, while Chapter 8 describes preliminary results obtained by applying these techniques to children with learning disabilities.

6

Neurometric Assessment of Sensory, Perceptual, and Cognitive Processes

I. INTRODUCTION

In Chapter 5, a body of evidence was presented that established the power and practicability of objective quantitative detection of abnormal electrophysiological activity in patients with a wide variety of neurological diseases, and the feasibility of establishing numerical criteria for abnormality caused by such diseases. We believe that it may well be possible to extend such evaluations into the area of more subtle sensory, perceptual, and cognitive processes, which can be disturbed by brain damage or dysfunction.

In our opinion, it is indisputable that the population of children suffering from learning disorders is not homogeneous (Conners, 1973). Some of these children suffer from the accumulated consequences of inadequate stimulation, some suffer from unfavorable home or school environments, some suffer from psychological disturbances or behavioral disorders, and some have brains which are not working properly because of structural damage or dynamic dysfunction. In order to help these children intelligently and effectively, methods for rapid, economic, and accurate evaluation of electrical activity providing indices of maturation and reflecting brain processes mediating sensory, perceptual, or cognitive functions are sorely needed. Such methods not only facilitate the separation of children with an organic basis for learning disorders from those whose problems have an experiential, environmental, or emotional origin, but provide the basis for differential diagnosis and identification of a wide variety of types of brain dysfunction. Many different kinds of neurophysiological and/or neurochemical dysfunctions may express themselves through the same "final common behavioral path," difficulty in learning. It is a serious error to believe that because all of these children share the same behavioral symptoms, or even the same psychometric profile, they will all benefit from the same treatment.

The problem of learning disabilities, or so-called "minimal brain dysfunction" is discussed in greater detail in Chapter 8.

At the other end of the developmental scale, numerous elderly persons suffer from impairment of cognitive processes. Many of these individuals have undergone personality deterioration due to depression, anxiety, and emotional stresses, while others have discrete or diffuse changes in brain function. The problem of distinguishing between these two classes of etiologies of cognitive impairment is difficult. Effective treatment obviously depends not only upon the accuracy of the differential diagnosis required for this relatively gross distinction, but also upon the evaluation of more subtle differences within the class of elderly patients whose intellectual deterioration has an organic basis, the so-called "organic brain syndrome" or senile dementia. There can be many different organic bases for these behavioral syndromes. The problem of cognitive impairment in the elderly is discussed in greater detail in Chapter 7.

Finally, there exist a large number of mature adults who display cognitive disorders due to neurological disease, psychiatric disease, or brain injury. Diagnosis, selection of treatment, and accurate prognosis of outcome in these individuals can also be facilitated by accurate evaluation of brain functions related to sensory, perceptual, and cognitive processes. Some of these problems are also discussed in Chapter 7.

These three categories, the elderly patient suffering from intellectual deterioration, the victim of head trauma suffering from cognitive impairment, the child with learning disabilities, constitute a far larger population of patients requiring precise and differential evaluation than the sum of the victims suffering from all the various types of neuropathology for which electroencephalographic methods have been found useful or essential adjuncts to accurate diagnosis and effective treatment. (See Table 2.1, Chapter 2) There exists a very substantial literature dealing with the evaluation of sensory, perceptual, and cognitive processes by electrophysiological measurement of brain activity. We attach special importance to the objective assessment of these subtle processes because the capability to do so would offer hope that substantially more people will benefit from neurometrics than only the victims of drastic neurological diseases. For this reason, this literature was deliberately not included in the review provided in Chapter 2, but a general review of these topics was placed separately in this chapter.

After our initial exploration of the effectiveness of neurometric methods in the evaluation of neuropathology, as established in Chapter 5, we turned to the endeavor of extending these methods into the more subtle areas of cognitive function. We chose to deal first with the problem of the organic brain syndrome because the population at risk can be readily identified (the elderly) and the extent of impairment can be relatively well ascertained.

In Chapter 7, we will survey the literature specifically relevant to the electrophysiological assessment of elderly patients suffering cognitive impairment from organic brain syndrome, and present the results of using neurometric methods to compare a group of elderly individuals institutionalized because of cognitive impairment with a carefully matched peer group institutionalized for somatic disorders not involving intellectual deterioration. Results obtained with a group of elderly outpatients showing early and more subtle signs of cognitive impairment will also be presented.

In the final chapter of this book, Chapter 8, we turn to the most difficult and complex of all the problems we have herein identified, that is, the evaluation of children with learning disabilities. In that chapter, we survey the literature specifically relevant to the assessment of organic bases for learning disability, and present some results of applying neurometric methods. The first results obtained with the neurometric test battery (NB) are presented in that chapter. Let us now turn to a general review of the electrophysiological measures reflecting sensory, perceptual, and cognitive processes.

II. AER ASSESSMENT OF SENSORY ACUITY

Adequate sensory input is a prerequisite to the development of adaptive behavior. The most extensive clinical use of AER techniques has been in the evaluation of sensory acuity. AER audiometry and optometry are being used increasingly to obtain objective measures of acuity in children and in patients in whom verbal cooperation is difficult to achieve. Numerous workers have demonstrated that it is possible to estimate auditory thresholds and the focusing of retinal images with adequate precision for diagnostic purposes (Barnet & Goodwin, 1965; Barnet & Lodge, 1966, 1967; Cody & Bickford, 1965; Copenhaver & Perry, 1964; Davis, 1965, 1968; Davis, Hirsch, Shelnutt, & Bowers, 1967; Eason *et al.*, 1970; Engle & Young, 1969; Harter, 1971; Harter & Salmon, 1971; Harter & White, 1968, 1970; Keidel & Spreng, 1965, 1970; McCandless, 1967; Rapin & Graziani, 1967; Rapin, Ruben, & Lyttle, 1970). These methods have been limited by the requirement that the presence of an AER be visually recognized by the examiner, and that particular components be assessed, a process requiring critical judgment by a sophisticated examiner and inevitably generating ambiguous results. Early detection of deafness in children is of particular importance because of the relationship of the defect to difficulty with language acquisition (O'Gorman, 1962) and the consequent need for early treatment (Gellis & Kagan, 1970). The relevant literature on evoked response audiometry has recently been reviewed by Barnet (1972), Graziani and Weitzman (1972), and Regan (1972).

A special-purpose statistical analyzer that indicates the significance of the difference between two AERs by computing the t test at each point in the analysis epoch circumvents such difficulties (Neurodata Statistical Analyzer, Model VT-100). The t test, as may be recalled, is defined as the difference between two mean values divided by the square root of the sum of the variances of the two means, adjusted for sample size.

The statistical analyzer has been used to construct a neurometric screening test for visual acuity. Successive blocks of stimuli are presented to the subject until the significance of the AER difference between two successive blocks fails to exceed the .02 level. Each block consists of 32 presentations of bright flashes at the rate of one every 2 sec, transilluminating a 50% transmission grid with 50 lines/in 1 m from the subject. This grid is perceived as a homogeneous gray field. Failure to achieve a significant difference between two successive AERs indicates that the response has been reasonably well replicated. The number of successive blocks required for stabilization of response depends upon the individual subject, but is usually two or three. Once stability has been achieved, the stimulus transilluminated by the flash is replaced by a 50% transmission grid with 27 lines/in, which will be perceived as a checkerboard by a subject with 20/20 vision. The visual acuity of the subject is judged to be 20/20 if the difference between the AERs to the two different visual gratings is significant at the .01 level or better. A grid with 7 lines/in is used to ascertain whether the subject has at least 20/70 vision, the acuity deficit arbitrarily selected by us as sufficient for concern.

The results in Fig. 6.1 indicate that this technique is sufficiently accurate to serve as a rapid screening procedure. The measurement can usually be completed in about 6 min. With finer gradations of the spatial grid, it is possible to obtain a more precise estimate of the image focusing capability of the subject. This technique is based upon numerous observations that if a blank visual field is replaced by a pattern, a change occurs in the shape of the visual evoked response, the change most commonly consisting of a reversal in polarity of a component between 100 and 150 msec in latency (Harter & White, 1968; Jeffreys, 1969; John, 1974b; MacKay & Jeffreys, 1971; Regan, 1972; Rietveld *et al.*, 1967; Spehlmann, 1965; Van Der Tweel *et al.*, 1970). Preliminary results suggest that the same technique can be applied to evoked response audiometry. These methods can, of course, be implemented in any general-purpose computer.

III. AER ASSESSMENT OF PERCEPTUAL CAPABILITY

AER methods not only provide a way to evaluate the general responsiveness of a sensory system, but permit more specific evaluation of the ability of that system to encode and transmit certain kinds of information. For example, AERs with reproducibly different waveshapes are elicited by two different colors (Burkhardt & Riggs, 1967; Clynes, 1965; Riggs & Sternheim, 1969). Although such assessments would seem to provide information of potential clinical value in the evaluation of color vision in young children, they have not been used for such

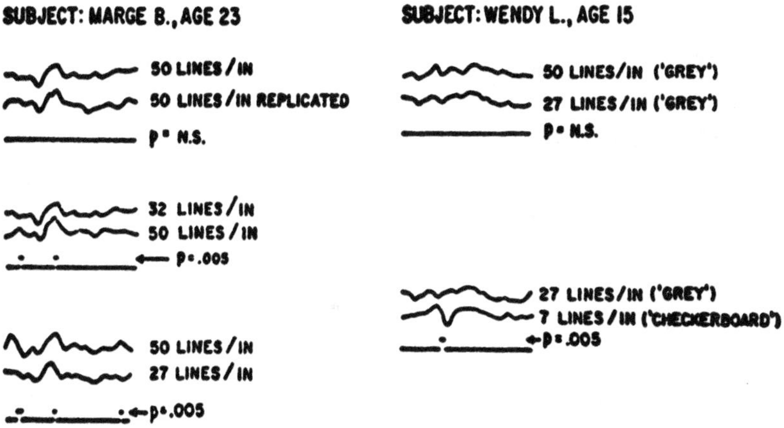

Computer Assessment of Visual Acuity (N=10)

	Different stimuli	Same stimulus
Differences not significant (P ⩾ .02)	2	30
Differences significant (P ⩽ .01)	24	2

The full distribution of these results was as follows:

P level	Different stimuli	Same stimulus
.005	21	0
.01	3	2
.02	2	5
.05	0	9
>.05	0	16

FIG. 6.1 (*Top*) Examples of the assessment of visual acuity by statistical analysis of averaged evoked potentials in two subjects. Both subjects perceived 50 lines/in as a gray field. The subject shown on the left reported that she perceived both 32 lines/in and 27 lines/in stimuli as checkerboards. The subject shown on the right was tested without her glasses and perceived the 27 lines/in stimulus as gray. (From John, 1974.) (*Bottom*) This table summarizes the results obtained by applying the *t*-test method to assess visual acuity in 10 subjects.

purposes. We therefore used the statistical procedure just described to ascertain whether routine neurometric screening tests for perceptual discrimination could be developed. This method provides a sufficiently accurate assessment of color vision to be of practical use (John, 1974).

An example of the use of this method for evaluation of color vision is provided in Fig. 6.2.

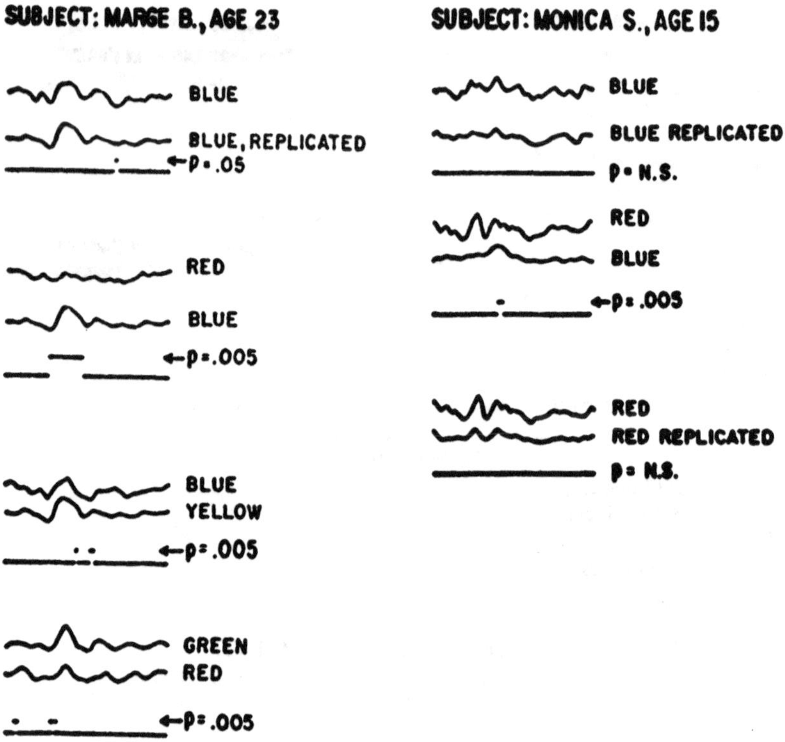

Computer Assessment of Color Vision (N=11)

	Different stimuli	Same stimulus
Differences not significant (P ⩾ .02)	1	13
Differences significant (P ⩽ .01)	14	1

The full distribution of these results was as follows:

P level	Different stimuli	Same stimulus
.005	10	1
.01	4	0
.02	0	0
.05	1	5
>.05	0	8

FIG. 6.2 (*Top*) Examples of the assessment of color vision by evoked potentials in two subjects. (From John, 1974b.) (*Bottom*) This table summarizes the results obtained by applying the *t*-test method to assess color vision in 11 subjects.

IV. AER ASSESSMENT OF SHAPE PERCEPTION

Some years ago, we reported that the waveshape of the VER changed when stimuli of different geometric form but equal area were presented (John, Herrington, & Sutton, 1967). Those findings have been confirmed by Clynes, Kohn, and Gradijan (1967); Pribram, Spinelli, and Kamback (1967); Herrington and Schneidau (1968); and Fields (1969). Of greater interest than the fact that such differences were found, which might have been predicted from our knowledge of brain mechanisms involved in processing information about contours, was our finding that stimuli of the same shape produced similar waveshapes independent of stimulus size. This observation, which is discussed in Volume 1, suggests that the similar evoked potential waveshapes in this instance reflected the activation of the same abstract concept by different sensory stimuli. This observation has been confirmed by Clynes *et al.* (1967) and by Hudspeth (unpublished observations, 1971).

Clearly, to recognize that large and small squares share the same geometric form is an acquired rather than innate ability, corresponding to the well-known psychological concept of shape invariance. This observation, together with many cited in earlier chapters, indicates that evoked potential measures may provide a neurometric index of the development of perceptual capability. Such estimates might be useful in assessing the cognitive maturation of a child. For these reasons, we undertook to develop neurometric methods for rapid assessment of EP correlates of shape perception (John, 1974b).

An example of results obtained by use of the same *t*-test strategy, described above for assessment of visual acuity and color perception, but applied to shape perception, is provided in Fig. 6.3.

V. AER ASSESSMENT OF COGNITIVE PROCESSES

Although neurometric assessment of neuropathology (Chapter 5), sensory integrity, sensory acuity, and perceptual discrimination all have important clinical applications, perhaps neurometric techniques might be most usefully applied to the objective assessment of cognitive processes, particularly in children, since precise measurements of these aspects of brain function are so difficult to achieve.

A. Control of Afferent Input

A critical difficulty of children with learning disabilities or minimal brain damage may consist of difficulty in structuring figure–ground relationships, particularly when cross-sensory interactions are involved (Birch, 1964). A variety

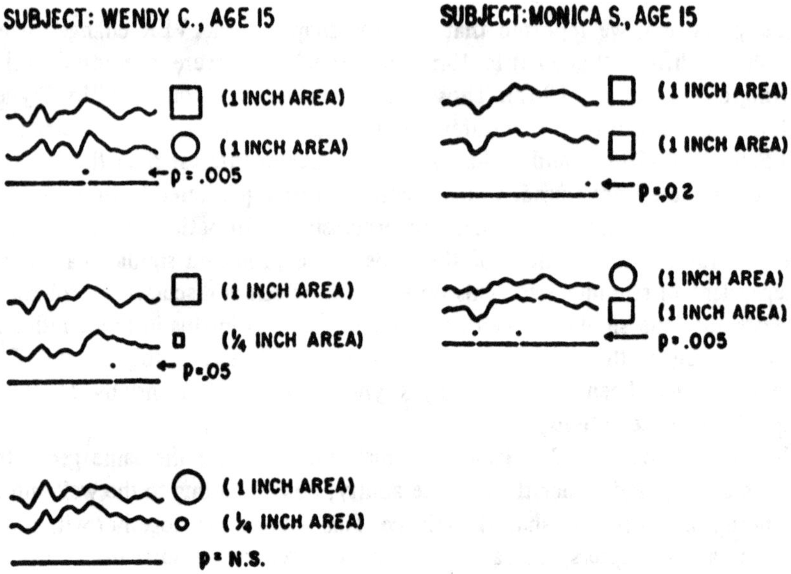

Computer Assessment of Shape Perception (N=6)

	Different stimuli	Same stimulus
Differences not significant (P ⩾ .02)	1	3
Differences significant (P ⩽ .01)	14	0

The full distribution of these results was as follows:

P level	Different stimuli	Same stimulus
.005	12	0
.01	2	0
.02	1	1
.05	0	1
>.05	0	1

FIG. 6.3 (*Top*) Examples of the assessment of shape perception by evoked potentials in two subjects. Note the similarity of response to large and small stimuli with the same geometric form displayed by the upper subject. The similarity of these responses would seem to be an electrophysiological correlate of the phenomenon of size invariance. Some of our subjects produced significantly different responses under these circumstances. (From John, 1974b.) (*Bottom*) This table summarizes the results obtained by applying the *t*-test method to assess shape perception in 6 subjects.

of observations suggest that this capacity might be measured quantitatively. These observations can be classified into two groups: (1) those that relate to the dimunition of the response to repetitious inconsequential events, referred to as habituation, and (2) those related to the selective suppression of irrelevant aspects of the stimulus complex (ground) and the enhancement of relevant aspects (figure).

Habituation is a general property of many different functional systems of the brain. The brain possesses powerful cenrifugal mechanisms capable of excluding afferent input related to informationally trivial events (see Chapters 3 and 4, Volume 1). The anatomy and functional characteristics of this system have been studied extensively in a wide variety of systems (John, 1961), and it is generally accepted that short-term or phasic and long-term or tonic aspects of habituation can be distinguished (Sharpless & Jasper, 1956) and related to different anatomic structures, particularly the cortex (Dunlop *et al.*, 1964; Fernandez-Guardiola *et al.*, 1961; Key, 1965; Marsh & Worden, 1964). Hernández-Peón, a pioneer in the study of habituation, was perhaps the first to argue that this phenomenon represented a rudimentary form of learning, requiring sufficient memory to recognize a present event as one which had been inconsequential in the past (Hernández-Peón *et al.*, 1956b). Broad, general limits have been established for the rate of habituation of evoked responses to sensory stimuli in man (Endroczi *et al.*, 1968; Fruhstorfer *et al.*, 1970). In a particularly interesting series of papers, Barnet and her colleagues have demonstrated that children with certain kinds of brain damage show little or no habituation, and can be reliably differentiated from normal children by the age of one year (Barnet & Lodge, 1967) suggesting that these children lack the ability to inhibit irrelevant afferent input and may possibly have defects in short-term memory.

Data related to the structuring of figure–ground relationships have also been obtained from evoked potential studies. In a well-known classic paper, Hernández-Peón and his co-workers (1956a) showed that the response evoked by a click in the cochlear nucleus of a cat was dramatically inhibited when the cat was shown a mouse in a beaker. This observation clearly illustrated that relevant afferent input in one sensory modality can inhibit irrelevant input in another modality. That relevant input (listening to verbal materials) can suppress irrelevant input (clicks), even within the same sensory modality, was demonstrated by Morrell and Morrell (1965). Such measures should reflect the ability of a child to identify significant sources of information around him.

B. P-300 or the Late Positive Component of the Human AER

Since the initial report by Sutton *et al.* (1965), a voluminous literature (see Price & Smith, 1974) has developed concerning the late positive component (LPC) of the human evoked potential, commonly referred to as the P-300. The notation, P-300, is used to designate the polarity and relative latency (≈ 300 msec) of the late positive EP peak. However, LPCs range in latency from 210 msec (Roth, 1973; Roth & Kopell, 1973; Squires *et al.*, 1975) to as late as 450–550 msec (Ritter & Vaughan, 1969; Thatcher, 1976). Therefore, the general class of P-200, P-300, and P-400 processes will be referred to as LPCs.

A large number of different stimulus conditions result in the production or enhancement of LPCs. For example, late positive peaks can be elicited by variables such as the delivery of information and the resolution of uncertainty (Sutton *et al.*, 1965, 1967), stimulus relevance (Donchin & Cohen, 1967; Hartley, 1970), stimulus novelty (Ritter *et al.*, 1968), cognitive acts of decision (Ritter & Vaughan, 1969; Rohrbaugh *et al.*, 1974), signal detection (Cael *et al.*, 1974; Hillyard *et al.*, 1971), and word and sentence processing (Friedman *et al.*, 1975a, b; Shelburne, 1972, 1973; Thatcher, 1976). LPCs can also be elicited by syncopation of a train of habituated tones (Ritter *et al.*, 1968) or to the absence of an expected stimulus (Sutton *et al.*, 1967; Weinberg *et al.*, 1970).

The diversity of situations in which LPCs can be elicited or enhanced poses serious problems in formulating a simple theoretical interpretation of these phenomena. Many different hypotheses regarding the psychological basis of the LPC have been suggested in the studies cited above. Thus far, only a few authors have attempted to look through all of the various LPC studies in an effort to construct a unified explanation. Donchin *et al.* (1974), Posner (1974), Goff (1969), and Smith *et al.* (1970) have suggested that activation of a conscious attention process is the underlying factor. However, this hypothesis, while being general, does not provide an explanation for the variable latencies of LPCs. In contrast, Hillyard *et al.* (1971) and Squires *et al.* (1973) have proposed a somewhat more testable hypothesis that also assumes a common underlying process. This process is representational matching. Like consciousness, this process is probably operating in all of the LPC experiments. Even LPCs to syncopated (Ritter *et al.*, 1968) or absent stimuli (Weinberger *et al.*, 1970) can be explained as a match or a mismatch process. Also, the variable latencies of LPCs can be explained by assuming that matching between afferent input and a representational 'template' occurs at different levels of serial information processes. Future research is needed in order to accept or reject these hypotheses or to generate better ones.

Recent experimenters have used electrophysiological probes to study the various levels of match–mismatch. For instance, Posner *et al.* (1973), Thatcher (1974a, b) and Thatcher and John (1975), have analyzed electrical responses to letters that match or mismatch with the past. The idea behind the latter studies is to challenge the information processing capacity of the subject while monitoring the electrophysiological processes that correlate with the subject's performance. Such an approach to the electrophysiological analysis of cognitive function requires a carefully controlled stimulus procedure. Such a procedure is illustrated in Fig. 6.4. This is a letter match–mismatch paradigm involving the computer generation of a series of random dot displays (control stimuli),

FIG. 6.4 Illustration of trial sequence of computer generated displays in a letter match–mismatch paradigm. There are a variable number of control and ITI displays before and after the first letter (information). All displays are 20 msec in duration and presented at a repetition frequency of 1/sec. Total luminance and retinal area subtended are the same for all displays. (From Thatcher, 1976).

followed by a letter (A, B, C), followed by a second series of random dot displays (intertest interval stimuli; ITIs) followed by a second letter that matches or mismatches the first letter. The display duration is 20 msec and the presentation rate is 1/sec. Total luminance is equated for all displays and match and mismatch conditions are counterbalanced and equally probable. Since there are a variable number of random dot displays preceding and following the first letter the subjects cannot predict exactly when information will occur and therefore are required to attend to all stimuli. Thus, factors such as attention, information delivery (or uncertainty resolution), and cognitive acts of decision are held constant. The subjects are required to move a small lever left or right in order to indicate whether the test stimulus is a match or a mismatch. Examples of averaged evoked potentials elicited by this procedure are shown in Fig. 6.5. The major finding, which confirms that reported by Posner *et al.* (1973), is that the P-300 process to the mismatch stimuli is consistently attenuated in comparison to the P-300 elicited by physically identical match stimuli.

Recently this procedure was advanced to study semantic match and mismatch using synonyms and antonyms. The results of an EP analysis from the latter experiment are shown in Fig. 6.6.

In this experiment, human subjects view an accessory oscilloscope upon which are presented computer generated displays. The displays are a group of random dots that then become a word (each display is 20 msec in duration presented 1/sec and all displays are equated for luminance) and are followed by another series of random dots which are then followed by a second word that is either an antonym or synonym of the first word. If the second word is a synonym, subjects move a lever to the left (during a 5-sec rest period); if the second word is an antonym, subjects move a lever to the right. There are a variable number of control random dot displays preceding and following the first word so that subjects cannot predict precisely when the words will appear. The experiment is counterbalanced so that for one half of the trials a group of twelve words are synonyms with respect to the first word, and for the other half of the trials the same twelve words are antonyms with respect to the first word. The top row of waves in Fig. 6.6 show averaged evoked potentials elicited by first words. The second row of waves were elicited by the random dot display immediately

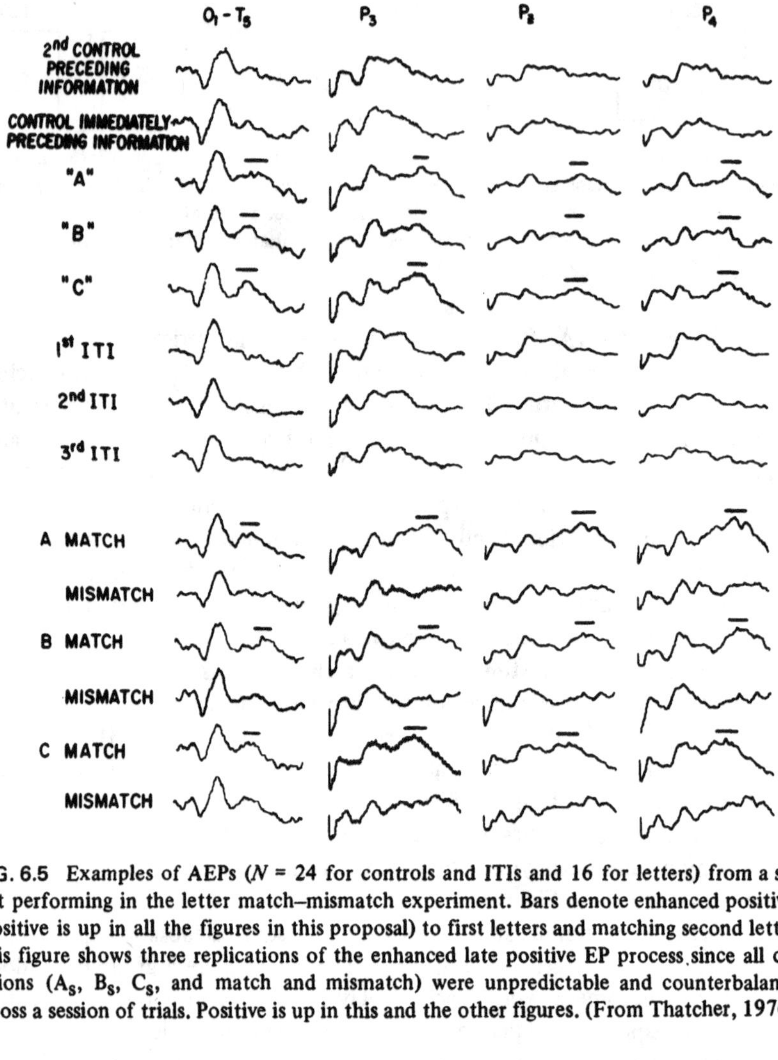

FIG. 6.5 Examples of AEPs (N = 24 for controls and ITIs and 16 for letters) from a subject performing in the letter match–mismatch experiment. Bars denote enhanced positivity (positive is up in all the figures in this proposal) to first letters and matching second letters. This figure shows three replications of the enhanced late positive EP process since all conditions (A_S, B_S, C_S, and match and mismatch) were unpredictable and counterbalanced across a session of trials. Positive is up in this and the other figures. (From Thatcher, 1976.)

preceding the synonym and antonym. The third row were elicited by the random dot display immediately following the synonym or antonym. The fourth row were elicited when the words were synonyms and the fifth row were produced when the words were antonyms. The important comparison is between the top and bottom two rows. Note that the left temporal lobes (T_5 and T_3) exhibit an early enhancement of the evoked potential to the synonym and antonym (but not the right T_6 and T_4) and that a P-300 or LPC (latency approximately 440 msec) is widely distributed but strongest in the posterior regions. These data, in conjunction with the results of match–mismatch experiments involving letters (Posner *et al.*, 1973; Thatcher, 1974a; Thatcher & John, 1975), geometric forms (Thatcher, unpublished), and words, indicate that the P-300 is a general wave process reflecting the completion of a "fit" between external sensory information and internal representational systems (see Thatcher, 1977). The data also emphasize the practicability of developing a cognitive test battery that measures the latency for the appearance of the so-called "match" component. The latency of the match component (see bottom two rows of waves in Fig. 6.6) may reflect the amount of time required to make a "meaning" comparison in order to fit the second word to the synonym or antonym category.

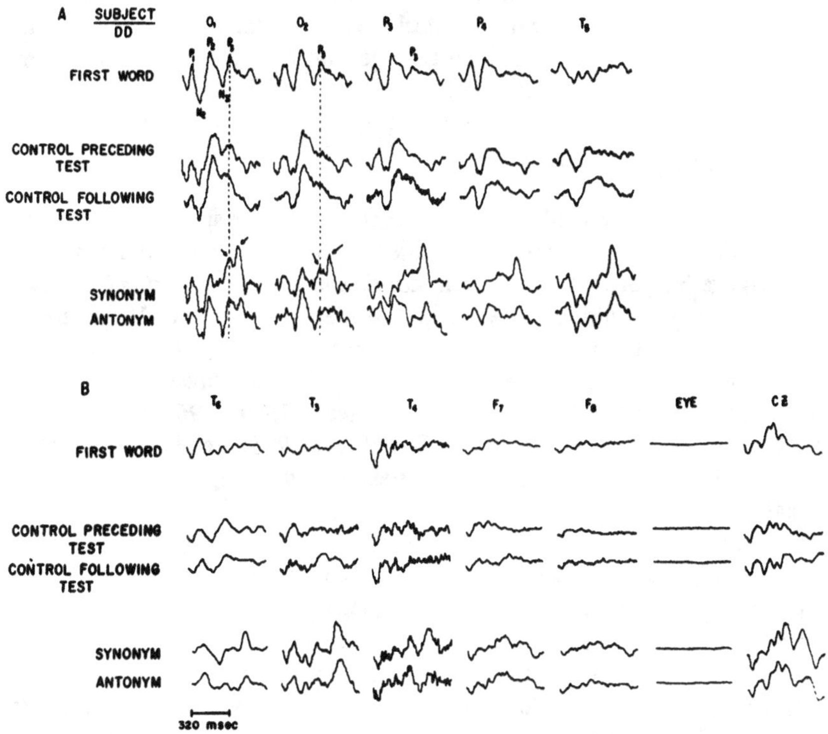

FIG. 6.6 Averaged evoked potentials recorded monopolar (linked ear reference) from (A) O_1, O_2, P_3, P_4; (B) T_5, T_6, T_3, F_7, F_8, Cz, and a bipolar transorbital eye electrode. Top row shows evoked potentials elicited by the first word of a synonym–antonym pair (see text for experimental design). Second row are evoked potentials elicited by a random dot control display which immediately preceded the synonym or antonym. Third row are evoked potentials elicited by random dot control following synonym or antonym. Fourth row are evoked potentials elicited by words which, for one half of the trials, were synonyms to the first word. Synonyms and antonyms were presented in a counterbalanced manner and all displays were equated for luminance. (From Thatcher, 1976.)

In this laboratory, we have demonstrated that the waveshapes elicited in cats by neutral stimuli in a differential generalization situation are not determined by the actual stimulus presented, but correspond to the waveshapes normally elicited by the differential conditioned stimuli appropriate to the behaviors subsequently performed (see Volume 1, Chapters 9 and 10, or John, 1963; John *et al.*, 1969, 1973; Ruchkin & John, 1966). The brain can, therefore, produce a facsimile of the usual response to an absent stimulus. Analogous phenomena occur in human subjects; apparent AERs appear in man at the time of expected but absent events (e.g., Barlow *et al.*, 1967; John, 1967a; Klinke *et al.*, 1968; for other references, see Volume 1, Chapter 10). We have referred to such endogeneously produced potentials as "readout processes," while others have termed them "emitted potentials" (Weinberg *et al.*, 1970). These findings establish that the brain can produce an accurate electrical facsimile of the usual response to an absent stimulus. These facsimiles must be generated by neural readout from memory, representing the absent event.

C. Contingent Negative Variation

When a subject is forewarned of an event that requires a response, then a slow negative shift in the EEG is frequently observed during the prestimulus period. This slow negative potential is called the *contingent negative variation* (or CNV) (see Walter, Cooper, Aldridge, McCallum, & Winter, 1964). The CNV appears to be correlated with both sensory and motor processes and develops primarily when a subject is expecting a salient stimulus within a defined time interval. Changes in the amplitude (McCallum, 1969; Tecce & Scheff, 1969) and distribution (Donchin, Tueting, Ritter, Kutas, & Heffley, 1975; Weinberg & Papakostopoulos, 1975) of the CNV have been correlated with changes in attention and information processing. In addition, the CNV has been correlated with stimulus readiness (Karlin, 1970), motivation (Irwin, Knott, McAdam, & Rebert, 1966), cognition (Low, Borda, Frost, & Kellaway, 1966), arousal (Tecce, 1971), and memory retrieval (Roth, Kopell, Tinklenberg, Darley, Sikora, & Vesecky, 1975).

Several authors have suggested that the appearance of the P-300 or LPC is functionally linked or dependent on the occurrence of the CNV (Donchin & Smith, 1970; Karlin, 1970; Näätänen, 1967, 1970; Wilkinson & Lee, 1972; Wilkinson & Spence, 1973). However, many studies show that the CNV and P-300 can be dissociated. For example, P-300 amplitude has been demonstrated to change while CNV amplitude is invariant (Donald & Goff, 1971; Lombroso, 1969; Tueting & Sutton, 1973). Also, LPCs can be elicited in paradigms which preclude subject anticipatory activity (Donchin & Cohen, 1967; Eason *et al.*, 1969; Harter & Salmon, 1972; Hillyard *et al.*, 1974; Thatcher, 1977). Recently, the CNV and P-300 have been shown to have different anatomic distributions (Donchin *et al.*, 1975). Although data indicate that the CNV and LPC reflect independent processes, these processes can interact (see Tueting & Sutton,

1973). Currently, however, the precise nature of the neuroelectrical generators giving rise to these processes and the parameters controlling their interaction is poorly understood.

D. Differential Anatomic Distribution of Exogenous and Endogenous Processes in Man

In closely related studies (Grinberg & John, unpublished, 1974), we have found that the waveshape of the AER elicited by a visual stimulus from certain head regions depends primarily upon the physical characteristics of the stimulus (exogenous), while in other head regions the AER waveshape depends primarily upon how the stimulus is perceived (endogenous). These studies gathered data from scalp electrodes at various 10/20 positions from subjects engaged in two tasks. In the first task, subjects watched a block of tachistiscopic presentations of the numbers "1" and "2," followed by a block of presentations of the letters "I" and "J." The same physical stimulus was used to represent either the number 1 or the letter I. The subject was instructed to "look at the numbers" before the first block and "look at the letters" before the second block.

Figure 6.7 shows typical results obtained from parietal (P_4) and occipital (O_1) derivations in two of the subjects in this experiment. For each derivation, the upper waveshape shows the AER computed in response to 100 presentations of the number "1," while the second waveshape shows the AER elicited in response to 100 presentations of the letter "I." The third row of data shows the difference obtained by subtracting the second AER from the first, and the fourth row shows the *t* test for the significance of the difference between the two AERs at each point along the analysis epoch. All subjects in the experiment, like the two illustrated, showed a statistically significant difference in the AERs recorded from the parietal but not the occipital cortex under the two conditions.

In the second task, the subject viewed four stimuli: a big "E," a little "e," a big "A," and a little "a." When AERs elicited by presentation of big and little versions of the same letter were compared, significant differences were found in occipital but not parietal cortex, as seen in the upper half of Fig. 6.8. When AERs elicited by presentation of two different letters of the same size were compared, significant differences were found over parietal, occipital, and temporal cortex, as seen in the lower half of Fig. 6.8.

Thus, these data suggest that the parietal reflected changes related to the different *meaning* assigned to the symbol "1" under the two viewing conditions, while the occipital cortex showed no difference in response because the *physical* stimulus was the same. When two versions of the same letter with different sizes were viewed, the parietal cortex showed no differences because the same meaning was attributed to both stimuli, while the occipital cortex reflected the physical differences between the two stimuli. Finally, when two different letters

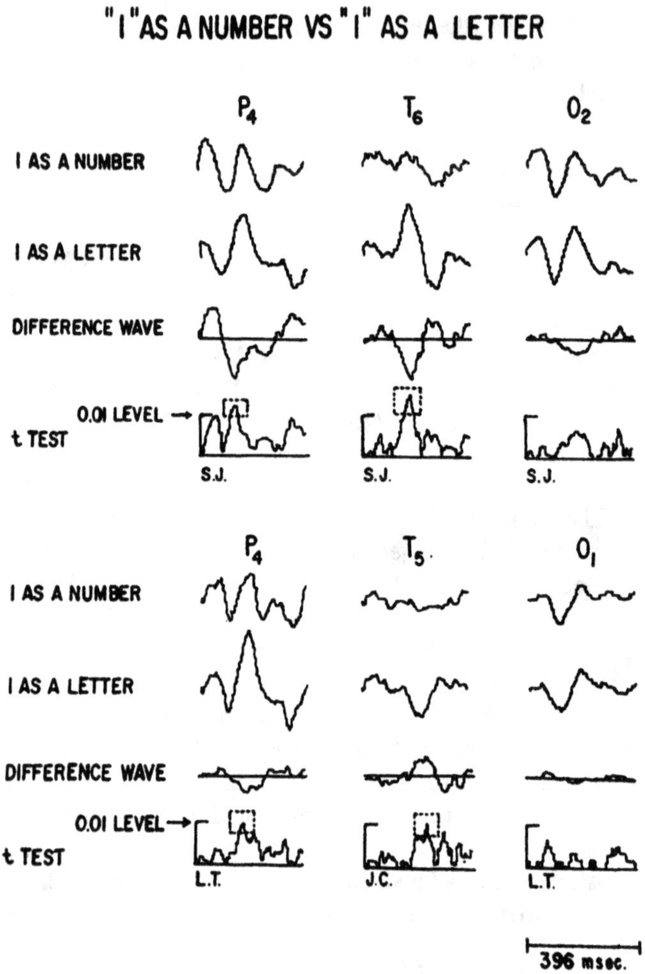

FIG. 6.7 Results obtained in two subjects (L.T. and S.J.), when "I" as a number and "I" as a letter were presented. The first and second row shows 100 samples of averaged evoked potentials recorded from parietal and occipital derivations. The third row depicts the difference wave; and the last row, the *t* test results. Notice that only the parietal leads reach significant differences at approximately 160 msec of latency.

equated for size were compared, differing in physical features as well as meaning, significant differences were obtained from both parietal and occipital, as well as temporal, cortex.

These results are closely related to findings that differences in the amplitudes of late components of the AER can be demonstrated between good and poor readers. These differences are particularly manifested in posterior parietal and temporal regions.

These results suggest that electrodes located over occipital regions detect processes primarily reflecting exogenous influences, while electrodes located over parietal regions reflect endogenous processes related to the interpretation of the sensory input, or perception. These interpretative processes clearly play a major role in defining the significance of the afferent input for the person, and must involve reference to memory about previous experience for such evaluation. Presumably, those memories in man are mediated by an anatomically distributed system of the sort discussed in Chapter 10 of Volume 1. The parietal electrodes simply reflect activity in a cortical region participating in that representational system. Findings such as these suggest that valuable insights into the development of reading skills and the mechanisms involved in certain reading disorders might be forthcoming from analysis of the emergence of characteristic topographic distributions of certain AER features and changes in AER features with maturation, as a function of systematic increase in the structure of visual form from featureless flashes to structured geometric pattern to patterns with verbal connotations.

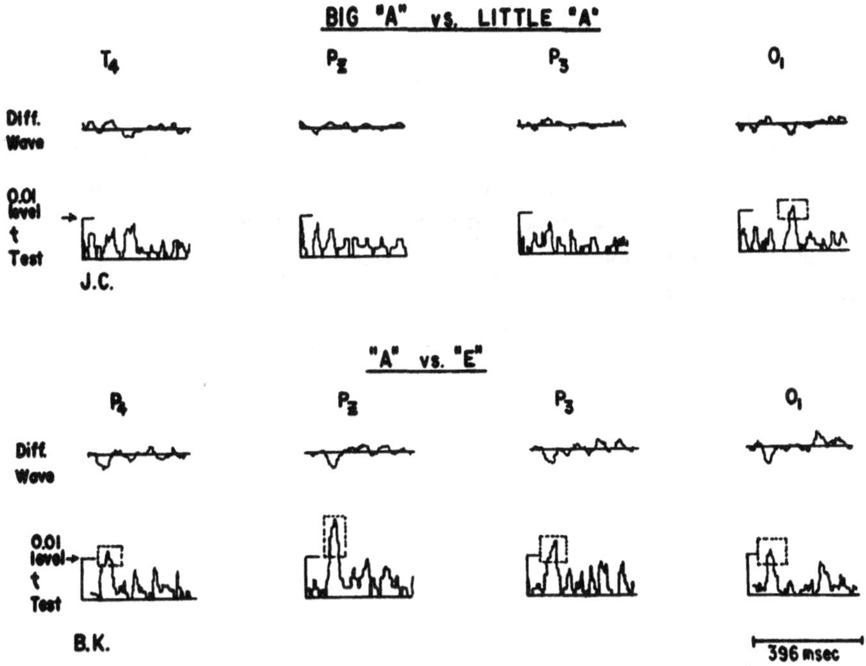

FIG. 6.8 The upper part of the figure depicts the difference wave and the *t*-test results between 100 samples of averaged potentials evoked by a big "A" versus a little "A." Notice that only the occipital derivation shows significant differences. The lower part of the figure depicts the same as the upper, but between 100 samples of averaged potentials evoked by an "A" letter versus an "E" letter. Notice that all the derivations show significant differences.

These various findings indicate that the shape of the AER is not solely determined by the actual physical stimulus, but also reflects processes related to short-term and long-term memory, and the influence of recent experience, expectations, and set. Standardization and quantification of these various aspects of the AER might be expected to provide indices of brain function that not only reflect the gross structural integrity of the brain but provide insight into more subtle functions related to sensory, perceptual, and cognitive processes. We have presented the details of a methodology for objective quantification of these features and have described a set of functional challenges intended to provide quantitative measures of such functions. In this chapter, we have reviewed much of the evidence that indicates electrophysiological measures provide meaningful estimates of sensory acuity, color vision, shape perception, and processes intimately related to cognitive functions such as control of afferent input by habituation, differential suppression of irrelevant input, and changes in AER waveshape as a function of experience, of possible utility in evaluation of brain function in the learning disabled child or the adult with impairment of cognitive functions.

These findings provide a substantial basis for the hope that careful numerical evaluation of electrophysiological features in individuals with learning disabilities or cognitive impairment might show significant deviations from normal patterns of brain electrical activity. Before we dared set foot onto what was obviously extremely uncertain ground, we undertook to test our measuring instruments on the relatively simple activity reflecting gross neuropathology. Encouraged by the effectiveness of neurometric evaluation of electrophysiological abnormalities in patients suffering from confirmed neurological disorders, reported in Chapter 5, we began to examine the utility of such methods for the assessment of children with learning disorders, elderly individuals, and mature adults with impairment of intellectual function. The literature most specifically relevant to these problems is reviewed and our experimental results thus far are presented in the next two chapters.

7
Assessment of Brain Dysfunction in Elderly Patients with Cognitive Impairment

I. COGNITIVE DETERIORATION IN THE ELDERLY: THE ORGANIC BRAIN SYNDROME

As our ability to manage medical crises with life support systems and drugs has improved, the average life expectancy in many countries has increased slowly but steadily. Although we have as yet not learned how to achieve a significant shift in the upper limit of the human life span, more and more people are surviving until old age. More than 30 million people, or about 12% of the United States population, are over 60 years old (United States Bureau of the Census, 1971). It seems reasonable, as medical techniques advance and become more generally available, to expect that in the relatively near future 20 to 25% of the people in many societies will belong to this age group and will be in relatively good physical condition. A substantial proportion of these individuals and their families will derive little enrichment from this extension of their lives because of cognitive impairment or senile deterioration, often referred to as organic brain syndrome (OBS). At least one-tenth of the elderly, or about 3 million people at present, suffer from significant cognitive impairment, ranging from severe memory loss to confusion and disorientation (Kay, 1972; Redick, Kramer, & Taube, 1973; Slater & Roth, 1969), and more than half the aged in nursing homes are classified as senile (United State Department of Health, Education, and Welfare, 1969). In addition, a much larger number will experience sufficient cognitive decline to impede their activities and compound the normal problems of aging. The cost to society, in terms of skilled labor, nursing and medical care, institutional requirements, and so forth, is difficult to estimate. The cost to individuals and their families is incalculable and often tragic.

The physiological basis for such cognitive impairment is poorly understood. In regard to etiology, two possibilities, diffuse cell death and cerebrovascular insufficiency, underlie the two traditional classifications, senile brain disease (senile dementia) and cerebral arteriosclerosis (arteriosclerotic dementia) (Slater & Roth, 1969). Although many senile patients show severe cerebral arteriosclerosis, this condition is commonly found post-mortem in elderly persons who displayed no cognitive impairment (Obrist, 1951; Sheridan, Yeager, Oliver, & Simon, 1955; Sokoloff, 1976; Terry & Wisniewski, 1976). While patients with severe cerebral atrophy are usually seriously impaired, such impairment does not invariably occur. Further, comparable intellectual dysfunction often occurs in the absence of cerebral atrophy (Deisenhammer, Hofer, & Jellinger, 1968). Thus, no morphological changes have been found that constitute a necessary and sufficient condition for cognitive impairment. Other possible causes of age-related cognitive decline may include electrolyte imbalance, decreased availability of certain neurotransmitters, and changes in RNA and protein synthesis, since these processes all relate to brain mechanisms of information processing, storage, and retrieval (John, 1967a). These five potential etiologies are not the only possibilities, nor are they necessarily independent or mutually exclusive.

Thus, it seems highly unlikely that there is a single common factor responsible for all of the clinical symptoms of intellectual deterioration in the elderly. The population of aged individuals with cognitive impairments must be heterogeneous. The full set of causes has not yet been identified, nor have existing diagnostic techniques been capable of differentiating specific subgroups or clusters within this population.

Typically, efforts at evaluation and differential diagnosis have relied upon neurological examinations, clinical assessments of mental status and behavior, and standardized psychological tests. The major utility of the neurological examination has been to rule out specific neurological diseases as the basis for deterioration. Clinical evaluations have relied upon interviews, including standardized questions such as the Mental Status Questionnaire (Kahn, Goldfarb, Pollack, & Peck, 1960) or rating scales (Salzman, Kochansky, & Shader, 1972). The psychological tests used most often are standard tests of organicity such as the Bender–Gestalt (Bender, 1938) and tests of memory span (Wechsler, 1945). These clinical and psychological evaluations provide objective methods to estimate the general extent of impairment, but do not yield much insight into the relative impairment of different specific aspects of cognitive function. Perhaps more important is that these assessment methods are purely behavioral. A particular behavioral dysfunction, especially in the complex domain of cognitive processes, may be the result of a wide variety of underlying organic causes. It is difficult, if not indeed impossible, to draw reliable and accurate inferences about the actual cause for the cognitive impairment of a particular individual from behavioral observations alone.

The limitations of present assessment techniques, together with the present lack of understanding of the underlying etiologies, have imposed severe constraints upon the search for effective treatments of cognitive deterioration in the elderly. Although there have been a wide variety of such endeavors, ranging from attempts to increase cerebral blood flow or oxygen supply through trials of compounds specifically suggested to improve cognitive functions to the treatment of particular behavioral symptoms with tranquilizers, antidepressants, and stimulants, the present consensus is that cognitive deterioration in the elderly is currently untreatable (Goldfarb, Hochstadt, & Jacobson, 1972; Jarvik & Milne, 1976; Jennings, 1972; Rao & Norris, 1972; Sathananthan & Gershon, 1976; Stotsky, 1976; Thompson, 1975).

Without direct measurement of different aspects of brain functions related to sensory, perceptual, and cognitive processes, it seemed unlikely that it would be possible to identify different subgroups within the heterogeneous population of aged persons with cognitive impairment. Since such differential diagnosis seemed to be an indispensable prerequisite for the development of effective individualized therapeutic interventions, we undertook to explore the effectiveness of neurometric techniques for the identification of specific patterns of brain dysfunctions in cognitively impaired elderly individuals. The results of these endeavors are presented later in this chapter. First, however, it is desirable to review previous electrophysiological studies of this problem.

II. ELECTROPHYSIOLOGICAL STUDIES OF CHANGES WITH AGING, WITH SPECIAL RELEVANCE TO ORGANIC BRAIN SYNDROME

The EEG and the evoked potential provide unique insights into the structural and functional integrity of the brain, into the dynamic interactions between different brain regions, and into the physiological processes involved in the encoding, transmission, and analysis of information by the brain. A steadily growing body of literature reviewed in earlier chapters documents the sensitivity and utility of these electrophysiological measures for the diagnosis of brain damage or dysfunction. A portion of this accumulated knowledge describes the changes that have been found in these measures with aging. Naturally, as it was realized that the EEG and the evoked potential changed with maturation and aging, it became of interest to search for features of brain electrical activity that might distinguish between normal old persons and patients with senile dementia or chronic brain syndrome.

A. EEG Studies

1. General Changes with Aging

Many investigators have studied the changes in the EEG that occur during aging. Some discrepancies appear between the extent and frequency of incidence of such changes as described by different workers. These discrepancies are probably to be attributed to whether the upper limit of the theta band is defined as 7 Hz, 8 Hz, or even 8.5 Hz, and to the difficulty of attempts to quantify the frequency composition of the EEG by visual inspection and manual measurements.

Nonetheless, numerous reports support the general conclusion that aged subjects tend to show slowing of the alpha rhythm and a diffuse increase of activity in the delta and theta bands in various head regions (Berger, 1929; Muller & Grad, 1970; Muller, Grad, & Kral, 1971; Obrist, 1951, 1954; Obrist & Busse, 1960; Surwillo, 1961, 1963, 1964).

Dominant low-amplitude theta–delta activity with occasional spikes in the left temporal region was reported by Silverman, Busse, and Barnes (1955) in 30% of a group of normal elderly volunteers. These findings were confirmed by Obrist and Busse (1960) who reported 29% incidence of such temporal theta–delta in a group of elderly normals as against 3% in a group of young controls. In an attempt to correlate changes during aging with alteration of cerebral metabolism, Obrist (1954), Obrist and Busse (1960) reported 19% abnormal EEGs among a group of healthy males over 65 years of age, with a mean age of 72.3 years. Focal slowing outside the temporal areas in normal aging occurs in less than 4% of community-living elderly normals and in 10% of normal institutionalized elderly persons.

Diffuse nonfocal abnormal slowing, predominantly bifrontal, has been reported in 15–20% of aged normal controls by Gibbs and Gibbs (1950), and Maggs and Turton (1956). A lower estimate of the incidence of diffuse slowing was provided by Mundy-Castle, Hurst, and Beerstrecher (1954), who found diffuse excessive theta activity in only 8% of normal elderly volunteer subjects. Obrist and Busse (1960) observed only 2% of the normal elderly who displayed diffuse delta activity, an incidence comparable to that found in normal younger control subjects.

Several authors have described an increase in fast beta activity during aging in normal subjects (Gibbs & Gibbs, 1950; Mengoli, 1952; Mundy-Castle, 1951). The proportion of such activity increases with aging, reaching 25% in normal controls between 60 and 70 years, becoming variable between 70–80 years, and generally decreasing after age 80. The latter observation has been confirmed by Silverman, Busse, and Barnes (1955).

Simultaneous aging processes are taking place in many other systems of the body, which undoubtedly contribute to some of the observed changes in EEG activity. Anoxia and hypoglycemia have been correlated with slowing of the spontaneous EEG activity (Brazier, 1948; Davis, Davis, & Thompson, 1938). Circulatory changes also produce slowing in the EEG, probably on the basis of a hypoxic and glucose "fuel" deficiency, interfering with the citric acid or Krebs cycles. Diseases of the cardiovascular system, such as congestive heart failure, have been reported to produce diffuse theta and delta (Ewalt & Ruskin, 1944; Stuhle, Cloche, & Kartun, 1952), while focal slowing is an obvious consequence of local vessel thrombosis (see Chapter 2).

Sheridan and his co-workers (1955) have presented evidence showing that slowing of the EEG in the aged is usually correlated with arteriosclerotic changes. Deisenhammer *et al.* (1968) found that many aged patients with EEG slowing displayed cerebral atrophy. In general, persistence of fast activity in the EEG was correlated with good performance in the more difficult psychological and cognitive tasks.

2. Studies of Patients with Organic Brain Syndrome

A relatively small subset of the EEG literature on changes accompanying aging has been concerned with features specifically related to cognitive impairment and the senile psychoses. Mundy-Castle and his colleagues (1954) reported a diffuse increase in delta activity in the senile psychoses,but failed to demonstrate any distinctive difference between this increase in slow waves and the increase usually found to accompany aging. Psychiatric patients diagnosed as moderate to severe senile deterioration have shown a greater incidence of abnormal EEGs (Barnes, Busse, & Friedman, 1956). Some authors believe that the extent of EEG abnormality bears some relationship to the severity of symptoms in orientation, memory, and intellectual function (McAdams & Robinson, 1956). Organicity of mental symptoms is reportedly suggested by dominant 6–7-Hz theta activity and an overall decrease in alpha abundance (Luce & Rothchild, 1953; Mundy-Castle *et al.*, 1954).

Some correlations have been reported between slowing of the EEG in elderly patients with cognitive deterioration and post-mortem evidence of arterio-sclerotic brain disease (Sheridan *et al.*, 1955) but comparable severity of cerebral arteriosclerosis was found in post-mortem cerebral examinations of normal elderly patients whose deaths were from other causes, such as accidents. This study revealed that most of the patients with cerebral arteriosclerosis had diffuse slowing as well as focal abnormalities. It is important to note that in a clinical study, a significant shift from alpha to the theta and delta range was found in patients with cardiovascular disease even when acute circulatory disturbance was not demonstrable. Further, overtly abnormal EKG and/or X-ray evidence of cardiomegaly was frequently accompanied by abnormal changes in the EEG (Obrist & Bissell, 1955).

Referring to blood pressure and the abnormally slow EEG, Obrist, Busse, and Henry (1961) reported an inverse relationship in elderly psychiatric patients. Once again, this finding did not discriminate between the abnormal and the normal aged with chronic hyper- or hypotension. Records of six elderly senile patients with dementia and reduced cerebrovascular oxygen consumption revealed abnormal EEGs in all, five of which had slow wave activity (Lassen, Munch, & Totty, 1957). This was observed only in the resting wakeful record but did not persist during sleep and/or mental activity.

Drachman and Hughes (1968) reported correlations between cognitive impairment in the aged and asymmetries in the EEG recorded bilaterally from the temporal lobes. These workers asserted that aged patients with such temporal lobe abnormalities were more likely to show cognitive defects than aged subjects with EEG slowing in other regions or with normal EEGs. The findings of Obrist

(1951, 1954) tend to contradict this assertion. In healthy old subjects, Obrist found little correlation between EEG variables and psychological or psychiatric assessments. However, subjects with diseases of the cardiovascular system consistently showed such relationships, displaying alpha slowing and diffuse slow activity in the delta band. He reported that the severity of intellectual deterioration was related to the abundance of this diffuse slow activity, with the highest correlations found in patients with organic brain syndrome. In contrast with the findings of Drachman and Hughes, Obrist found that sharply localized EEG abnormalities were rarely associated with senescent intellectual deterioration. Slow waves restricted to anterior temporal regions, although prevalent among patients with cerebrovascular disease, were not per se indicative of psychological or neurological impairment in the elderly.

This review of the literature on EEG changes in the elderly reveals a consensus that aging is accompanied by a shift toward slower frequencies in the composition of the spontaneous activity. These changes are probably associated with cerebral arteriosclerosis and/or cardiovascular diseases. The slowing is diffuse, but is often most prominent in the temporal regions, where asymmetries appear. However, these EEG changes are commonly found in the elderly, whether they display marked cognitive impairment or retain normal intellectual function. No evidence emerges from the literature to support the suggestion that EEG slowing, whether diffuse or focal, is uniquely found in elderly patients with organic brain syndrome. At the same time, there is rather general agreement that aged individuals with cognitive impairment are extremely likely to display diffuse slowing of the EEG. Thus, it seems reasonable to conclude that a diminution of alpha activity and the emergence of increased theta and delta activity in widespread regions of the brain are necessary but not sufficient conditions for cognitive deterioration with aging. It may be worthwhile to point out that these studies have largely been confined to comparisons between the extremes of normal and markedly abnormal patients, and reveal little about the vast number of early symptomatic and minimally to mildly symptomatic patients.

B. AER Studies

1. General Changes with Aging

When special-purpose average response computers capable of providing a picture of the typical or average features of a set of evoked potentials became readily available little more than a decade ago, a series of studies was carried out exploring changes in the averaged evoked potential during aging, by workers such as Cohn (1964), Kooi and Bagchi (1964), Dustman and Beck (1969), Tamura, Lüders, and Kuroiwa (1972), and Straumanis, Shagass, and Schwartz (1965). These studies primarily were concerned with changes in the morphology of the averaged evoked response, or AER, in response to visual or to somatosensory stimuli with no functional significance, delivered to the passively receptive subject. The changes in morphology of the visual evoked responses found with aging were not dramatic. In general, slight increases in amplitude of one or two components were reported, with some increases in latency with age, especially in the later components usually attributed to the influence of nonsensory-specific pathways. The typical, long latency rhythmic afterdischarges usually seen in response to bright visual stimuli diminished or disappeared with age.

The description of the visual evoked response waveshape that has been employed most often was provided by Ciganek (1961). The response consists of a series of waves. Ciganek designated their peaks by successive Roman numerals. Latency of onset of the response averaged 28 msec. The initial peak was negative, with mean latency of about 40 msec. Mean latencies of waves II–VII were 53, 73, 94, 114, 135, and 190 msec, respectively.

As must be apparent from the discussions in previous chapters and especially the factor analytic and clustering results reported in Chapter 5, there is great interindividual variability in the morphology of the visual evoked response. The "prototypic" waveshape and peak latencies described by the nomenclature proposed by Ciganek represent an oversimplification or idealization of the response. Marked deviations from this average picture can be encountered in perfectly normal individuals. Many investigators, however, prefer to use this nomenclature for convenience in discussing changes in waveshape.

Following the initial complex of the seven major peaks in Ciganek's scheme, there is a rhythmic afterdischarge. This afterdischarge has been termed "ringing" by Walter (1962). Kooi and Bagchi (1964) have also presented comparable normative data for the visual response.

The resemblance between the rhythmic afteractivity and the "spontaneous" alpha rhythm in form and frequency has led some investigators to believe that these two kinds of activity may be identical and to attempt to verify this view. Ciganek considered the afterrhythm frequency to be equal to that of the alpha rhythm, but Barlow (1960), using more accurate instruments, found that although the frequencies were close they were not identical. Cohn (1964) noted that the afterrhythm was dependent on the presence of spontaneous rhythmic EEG activity around the alpha frequency band. In view of the apparent close relation between alpha rhythm and the visual afterrhythm, Cohn proposed an explanation for the latter based on the effects of the light flash on the ongoing alpha activity. He suggested that the incoming flash first terminates the alpha activity and that then, following the flash, there is an interval during which the alpha frequency waves are reinstated. The light would thus set the phase of the alpha frequency, in a sense triggering new alpha activity after a delay. The alpha rhythm thus triggered would be roughly time-locked to the flash and be seen as the afterrhythm. The issue of the identity between the rhythmic afteractivity and EEG alpha activity does not yet appear to be definitely settled, although there is agreement that they are probably intimately related.

Viewed in this context, it seems reasonable to suggest that the diminution or disappearance of the rhythmic afteractivity from the visual evoked response generally reported to occur in old age simply reflects the diminution of alpha activity in the spontaneous EEG. These observations, while they corroborate the conclusions drawn from studies of EEG changes with aging, provide little additional insight.

Tamura, Lüders, and Kuroiwa (1972) found no significant alterations in the somatosensory evoked response during the aging process, failing to confirm slight latency increases previously reported by Lüders (1970).

2. *Studies of Patients with Organic Brain Syndrome*

Remarkably little research has been published directly comparing the morphology of the AER in healthy old subjects with the responses of aged patients showing severe cognitive impairments due to such deterioration as cerebral arteriosclerosis. One of the few evoked potential studies making this kind of comparison was published by Straumanis, Shagass, and Schwartz (1965). In this study, 20 elderly patients with severe cognitive impairment due to cerebral arteriosclerosis, 18 healthy "old" controls, and 18 healthy "young" controls were compared with respect to the characteristics of their visal evoked responses. Healthy young and old controls differed primarily with respect to the earlier portions of the AER, with larger amplitudes and longer latencies in the older subjects. The elderly patients with organic brain syndrome differed from healthy old controls mainly with respect to AER components longer than 100 msec in latency. Latencies of these components were prolonged and the rhythmic afterdischarge was reduced. The amplitude differences were significant at the .05 level and the late component latency differences were significant at the .01 level. The strongest statistical effects were with respect to the diminished afterdischarge, significant at the .001 level.

These findings suggest that the reactivity of the nonsensory-specific system may be lower in the patients with cognitive impairment, as reflected in the longer latency of the late components. The significantly greater diminution of the rhythmic afterdischarge seen in these patients may reflect greater resistance of the alpha generating system to "triggering" of the sort speculated about earlier.

III. NEUROMETRIC FEATURES THAT DISCRIMINATE BETWEEN NORMAL ELDERLY SUBJECTS AND PATIENTS WITH COGNITIVE IMPAIRMENT (OBS)

As mentioned at the end of Chapter 6, our results in using neurometric assessment of EEG and AER features in discriminating between normal subjects and patients with various forms of neuropathology encouraged us to attempt to extend those methods into the more difficult domain of impaired cognitive functions. The remainder of this chapter will describe the successful results achieved by application of neurometric methods to the problem of discriminating between normal and cognitively impaired elderly subjects.[1]

The research reported here had several goals: (1) to apply digital computer techniques to the neurometric evaluation of a wide variety of spontaneous EEG and AER parameters in a population of aged subjects; (2) to obtain a precise description of changes in these neurometric indices as the subjects were challenged with situations requiring processing or attending to complex visual and auditory information; (3) to compare the values of these neurometric measures in two populations of elderly patients with and without cognitive impairment, but carefully equated for age, socioeconomic status, amount of education, and length of hospitalization. All patients with overt neurological abnormalities were excluded from the study; and (4) to correlate the amount of cognitive impairment as assessed by a battery of psychometric tests yielding a derived composite psychometric score with specific neurometric features.

A. Patient Selection[2]

1. Normal Controls

Normal controls were selected from healthy community volunteers. We used a number of criteria to exclude subjects from this control group, as specified in Footnote 2. There were 56 persons selected above age 60, all of whom were partially or completely gainfully employed as one evidence of their cerebral competence. Other evidences of normality were: (a) a score of "not present" in

[1] The original research to be described was carried out in collaboration with Dr. Irvin M. Gerson, Frank Bartlett, and Veronica Koenig and was partially presented at the annual meeting of the American Medical EEG Society in April of 1975. We acknowledge the assistance of Daniel Brown with the cluster analysis. This research was partially supported by funds provided by Sandoz, Inc.

[2] Criteria for exclusion of subjects from the normal control group were:
(a) Patients having a current diagnosis of psychosis, presenile psychosis, depressions, paranoid features, posttraumatic brain damage, postinfective brain disease, cerebral neoplasm, marked mental deterioration, mental retardation, alcoholic brain syndrome or tardive dyskinesia; (b) patients having a past history of schizophrenia, manic–depressive psychosis, or electroconvulsive therapy; (c) patients having a recent history (i.e., during the three months prior to study) of long-term (i.e., six months or longer) phenothiazine therapy, (d) patients requiring continuous administration of psychotropic drugs, vasodilators, daytime sedative, central nervous system stimulants, or steroids; (e) patients having sufficient symptomatology to justify a symptom cluster of cognitive, emotional, and physical deterioration associated with senile degeneration as specified under "Patient Selection"; (f) patients having conditions such as blindness, deafness, language difficulties, or any other disability which would prevent the patient from participating in all test requirements; and (g) patients having pathological conditions which could be expected to progress, recur, or change to such an extent during the study period that they might bias assessments of clinical and mental status to a significant degree, for example, deteriorating cardiovascular disease, renal insufficiency, and so on.

all seven parameters of impairment in an "Assessment of Clinical Status," conducted by psychiatric interview. These parameters included confusion, impaired mental alertness, impaired recent memory, disorientation, mood depression, emotional lability, and inadequate capacity for self-care, as defined in Table 7.1. Psychometric tests were administered in each case to standardize the scoring of each parameter in the "Assessment of Clinical Status." These tests are described in detail in Gerson, John, Koenig, and Bartlett (1976); (b) none of the prodromal, transitional, or manifest symptoms described in Table 7.2 evaluated during the interviews; and (c) negative routine medical and neurological exam with reference to this age group, including laboratory tests.

The normal control group was matched by sex, age, race, level of education, and ethnic background with the group of cognitively impaired patients below.

2. Cognitively Impaired Group (OBS)

A selection of 62 patients aged over 60 years was made, comprised of geriatric patients suffering from cognitive, emotional, and physical symptoms associated with senile mental deterioration, or OBS. Each patient exhibited at least a mild to moderate degree of severity (a 3, 4, or 5 numerical rating on the "Assessment of Clinical Status" on at least two key symptoms) and had a minimum score of 11. Approximately half of the OBS group had prodromal and transitional symptoms while the other half had symptoms that were manifest (Table 7.2), as assessed in a brief psychiatric interview.

These patients were recruited from a private chronic care center serving an upper middle-class socioeconomic group and a community geriatric center serving a middle-class and lower middle-class socioeconomic group. This provided a fairly equal mix of professionals, laborers, and housewives, with respect to previous occupations.

Laboratory tests evaluated any chemical or metabolic imbalance which could influence the EEG. All patients had currently normal SMA-12 values.

B. Psychometric Evaluation[3]

As indicated above, the psychometric tests provided estimates of parameters such as: disorientation, confusion, mental alertness, memory, depression, emotional lability, self-care, and socialization. An additional parameter, schizoid tendencies, was added to eliminate any elderly schizophrenics who might or might not be diagnosed as seniles. This last score (ScZ) was not used in calculating the final "Assessment of Clinical Status." A psychologist evaluated the scores on each subtest and assigned a contribution to the overall assessment, using a rating of 1 = not present, 2 = very mild, 3 = mild, 4 = mild to moderate,

[3] Dr. Gerson was responsible for patient selection, psychiatric, psychological, and medical evaluations and electrophysiological data acquisition, assisted by Ms. Koenig. Data analyses were carried out by Dr. John, Mr. Bartlett and Mr. Brown.

TABLE 7.1
Assessment of Clinical Status[a]

Behavior	Description
1. Confusion	Lack of proper association for surroundings, persons, and time—"not with it." Slowing of thought processes and impaired comprehension, recognition, and performance; disorganization. Rate on patient response and behavior of interview and on reported episodes since last interview.
2. Mental alertness	Reduction of attentiveness, concentration, responsiveness, alacrity, and clarity of thought, impairment of judgment and ability to make decisions. Rate on structured questions and response at interview.
3. Impairment of recent memory	Reduction in ability to recall recent events and actions of importance to the patient, for example, visits by members of family, content of meals, notable environmental changes, personal activities. Rate on structured pertinent questions and not on reported performance.
4. Disorientation	Reduced awareness of place and time, identification of persons including self. Rate on response to questions at interview only.
5. Mood depression	Dejected, despondent, helpless, hopeless, preoccupation with defeat or neglect by family or friends, hypochondriacal concern, functional somatic complaints, early waking. Rate on patient's statements, attitude, and behavior.
6. Emotional lability	Instability and inappropriateness of emotional response, for example, laughing or crying or other undue positive or negative response to nonprovoking situations as the interviewer sees them.
7. Self-care	Impairment of ability to attend to personal hygiene, dressing, grooming, eating, and getting about. Rate on observation of patient at and outside interview situation and not on statements of patient.

[a]Rating key: 1 = Not present; 2 = Very mild; 3 = Mild; 4 = Mild to moderate; 5 = Moderate; 6 = Moderately severe; 7 = Severe.

TABLE 7.2
Clinical Impressions from Psychiatric Interview

A. Prodromal
1. Unexplained reactive depression.
2. Poor emotional recovery after physical illness is improved.
3. Limitation of ideas and impairment of capacity for abstract thinking.

B. Transitional symptoms
1. Dislike of environmental change.
2. Reduction of ambition.
3. Reduction of activity.
4. Constriction of interests in the direction of self-centering.
5. Increased difficulty in comprehension.
6. Increase in effort and time consumed in performance of familiar duties.
7. Difficulty in adaptation to new circumstances.
8. Lessened sympathy for new ideas and views.
9. Reminiscence and repetition.

C. Manifest symptoms
1. Loss of recent event memory.
2. Loss of social courtesies and personal hygiene habits.
3. Hostile and anxious dependence.
4. Natural affections become blunted and may turn to hatred.
5. Self-isolation.
6. Defensive reaction to impaired memory.
7. Sexual activities become uncontrolled or at least a mental preoccupation.
8. Hoarding.
9. Delusions of poverty and not being wanted.
10. The behavior patterns are caricatures of earlier personality.
11. Amnesia, confusion, disorientation, nocturnal delirium.

5 = moderate, 6 = moderately severe, and 7 = severe. A patient achieving a total score of 11 or more was considered as abnormal.

C. Neurometric Evaluation

1. Recording Procedures

Each patient was recorded resting in a dimly lit room, using an eight-channel EEG machine with nine standard montages employing the standard 10/20 international placements. Four standard bipolar and five standard monopolar runs were recorded for computer analysis. Data were recorded on FM tape and on paper.

2. Test Conditions

(a) Two minutes of spontaneous EEG were recorded from each derivation (monopolar and bipolar). Frequency analyses of these data provided a spectral profile and measures of the symmetry of the resting activity.

(b) Evoked responses were obtained first for monopolar derivations under the following ten stimulus conditions[4]:

(*i*) Five minutes of click at 1/sec: auditory evoked responses (AER) were computed at the beginning (condition 1) and at the end (condition 2) of this interval ($N = 60$) and were compared to obtain a measure of auditory habituation. AER symmetry and morphology were also evaluated by computation of Pearson correlation coefficients between homologous derivations.

(*ii*) Five minutes of flash at 1/sec: visual evoked responses (VER) were computed at the beginning (Condition 3) and at the end (Condition 4) of this interval ($N = 60$) to obtain a comparable measure of visual habituation. VER symmetry and morphology were evaluated as above.

(*iii*) Five minutes of paired click plus flash at 1/sec were then presented in order to establish sensory–sensory conditioning. No recordings were made during this period.

(*iv*) One minute of flash at 1/sec: change in VER waveshape (Condition 5) was measured as a result of the previous conditioning procedure ($N = 60$).

(*v*) One minute of click at 1/sec: change in AER (Condition 6) was measured as a result of the previous conditioning procedure (N = 60).

Next, figure–ground relationships were examined using a method based on cross-modal inhibition:

(*vi*) A 5-minute color film was presented simultaneously with a 1/sec click. The click sequence was continued for 2 minutes after the film ended. AERs

[4] All evoked response measures were obtained first using monopolar and then bipolar derivations. A Grass Photic Stimulator (Model PS-2) was used to produce the flash as well as the click. These stimuli could be delivered separately or simultaneously. The duration of the stimulus was 100 μsec and was at intensity 4, low scale. The strobe light was 5 ft from the patient's face. The click frequency was approximately 2,000 cycles. The source was 8 ft away from the patient at an intensity of 60 dB. An audiometric test was performed prior to the test to insure adequate hearing. The patient's range was tested between 256 Hz and 4,096 Hz. Patients with more than 10% loss in the 1,000- to 3,000-Hz range were eliminated from the study.

At the beginning of each session, the EEG machine was calibrated in three ways: (1) a square wave calibration to insure proper voltage and alignment of pens; (2) a balance run to insure synchronous recording of the EEG on all tape recorder channels; and (3) a "biocal" montage to insure that each channel receiving data from the patient processed this information through the system in exactly the same way.

The data flow was from patient through the EEG amplifiers to a Bell and Howell Model CPR 4010 Magnetic Tape Recorder/Reproducer using 1-inch high-fidelity, low-noise magnetic tape, (these tapes were previous degaussed) and in parallel to an RCA oscilloscope which monitored the output of the recording head. There was a switch so that any channel could be monitored to ensure proper recording.

Periodically, a centering meter was used to check the alignment of the 8-channel tape heads, since these had to be exactly synchronous for playback to the PDP 11/45 computer used for data analysis.

Since the dynamic range of the tape recorder was 1–4 V, it was necessary to use maximum EEG output, or approximately 2 μV/mm during the tape recording sessions.

during (Condition 7) and after (Condition 8) the film were compared to see whether suppression of auditory response occurred during attention to meaningful visual information (N = 60).

(*vii*) A 5-minute audio recording of a short story was presented simultaneously with a flash at 1/sec (eyes closed). The flash sequence was continued for 2 minutes after the story ended. VERs during (Condition 9) and after (Condition 10) the story were compared to see whether suppression of visul response occurred during attention to auditory information (N = 60).

These ten conditions were then repeated in the same order during bipolar recording and will be referred to as Conditions 11–20.

3. *Data Analysis*

The tape recorded data were shipped from Philadelphia, where they had been gathered, to the Brain Research Laboratories of the Department of Psychiatry at New York Medical College, where they were subjected to computer analysis. All analyses were carried out on the PDP 11/45 computer of the DEDAAS system. In essence, the prerecorded data were played back into the slightly modified DEDAAS system so that the system treated the playback as if it were data coming on-line from a patient. The playback electronics were carefully matched with those in the tape recorder used initially.

a. Data conversion and editing. All analog tape recordings received at New York Medical College were first played back through the analog-to-digital converters of the PDP 11/45 in the Brain Research Laboratories and stored on digital magnetic tape. The segments of data to be subjected to digital analysis were then displayed in analog form and subjected to visual editing to exclude all data contaminated by visible artifacts due to movement. After data analysis, which is described below, additional data were rejected because the quantitative analysis revealed unacceptably high variance in the sample collected from some individuals. Thus, not all of the data which had been recorded in retrievable form were included in the final analyses. The rationale for excluding data with high variance (almost solely from the EP conditions) was that a patient who was restless or irritable might be expected to move about and thereby generate average evoked responses (AERs) with high variance. Senile patients might be expected to display such uncooperative behavior more often than normal subjects. Failure to exclude data reflecting such behavioral differences might result in the detection of electrophysiological differences between the two groups arising from these spurious sources and not truly reflecting differences in brain function. This editing was one of the most time-consuming aspects of the data analysis, but was considered indispensable if conservative and valid data were to be subjected to the quantitative analyses.

The results of editing were to exclude the data from certain subjects under particular conditions, yielding a sample size which varied from condition to condition. The final edited sample size for each condition is presented in Table 7.3.

Exmination of the table shows that the normal patients provided data of approximately equal quality under almost all of the test conditions, showing a slight tendency toward deterioration after the monotonous flash-click pairing condition. The senile patients showed a greater variability, with almost half yielding unusable data during conditions 5 and 6, while over 80% provided usable data during the more interesting conditions 7 and 9, which involved viewing a film or listening to a record. Note that even under the worst conditions (5 and 6), the sample size was quite adequate. It should be pointed out that subjects fluctuated in their cooperativeness throughout the session, so that the overall pattern of usable data varied from subject to subject. In other words, the pattern of missing "test items" was variable.

b. Neurometric indices. EEG. Fast Fourier transform was used to compute the wide-band spectral analysis of the 8 monopolar (T_5, T_6, T_3, T_4, F_7, F_8, O_1, O_2) and 8 bipolar (C_3O_1, C_4O_2, F_3P_3, F_4P_4, $F_{P1}C_3$, $F_{P2}C_4$, C_3T_3, C_4T_4) derivations selected for analysis (the symbols refer to International 10/20 electrode placements). The following measures were computed for every derivation:

Absolute power (microvolts squared): Channel total, $delta_1$ (.5–1.5 Hz), $delta_2$ (1.5–3.5 Hz), theta (3.5–7 Hz), alpha (7–13 Hz), $beta_1$ (13–19 Hz), $beta_2$ (19–25 Hz), and gamma (25–40 Hz).

Relative power (%): percentages of $delta_1$, $delta_2$, theta, alpha, $beta_1$, $beta_2$, gamma.

Power ratio: the ratio of absolute power between symmetrical derivations was computed separately for the $delta_1$, $delta_2$, theta, alpha, $beta_1$, $beta_2$, gamma frequency bands and for the total EEG signal.

TABLE 7.3

	Monopolar			Bipolar	
Condition	Normal	Senile	Condition	Normal	Senile
Spont. EEG	93	51	Spont. EEG	93	51
1	62	38	11	62	38
2	64	42	12	66	37
3	65	45	13	65	38
4	63	39	14	66	37
5	59	31	15	59	26
6	58	26	16	55	28
7	67	43	17	64	42
8	62	38	18	54	29
9	62	44	19	66	42
10	65	35	20	65	37

Cross correlation: the cross correlation coefficient was computed between homologous derivations, in phase and 90° out of phase, separately for the $delta_1$, $delta_2$, theta, alpha, $beta_1$, $beta_2$, and gamma frequency bands and for the total EEG signal.

EP. Average evoked responses (AER) and variance were computed using a 1000-msec epoch, for each of the 16 derivations under each of the 10 conditions. For each of these 160 AERs, the following measures were computed across the whole 1000-msec epoch, and separately for the intervals 50–99, 100–199, 200–500, and 500–1000 msec:

Signal energy: defined as the integral of the voltage squared.

Noise: defined as the average of the variance.

Signal to noise ratio: defined as the signal energy divided by the noise.

Signal energy asymmetry: defined as the average difference between the signal energy of homologous derivations.

Signal waveshape asymmetry: defined as the Pearson cross-correlation coefficient between the two AERs obtained from homologous derivations.

For the spontaneous EEG, these analyses yielded a total of 31 measures × 16 derivations, or 496 measures per patient. For each EP condition, these analyses yielded a total of 320 measures per condition per patient. Since the full battery of electrophysiological tests contained 10 EP conditions plus a frequency analysis of the spontaneous EEG, a grand total of 3696 numerical measures was available from each patient for whom a full sample of data had been collected.

For each of these measures, means and standard deviations were computed separately for the normal and senile population, as well as for the sample population as a whole. Since it was desirable to be able to combine scores from various selected measures, all measures were subjected to the appropriate z transformation; that is, the difference between the mean value and the individual value on each measure was divided by the standard deviation of the whole population for that measure.

Analysis of variance between symmetrically placed leads showed that the senile patients displayed significantly greater homogeneity of variance in frontal pole leads. This was interpreted to reflect a higher incidence of artifact due to eye blink in the senile group. Consequently, data involving frontal pole electrode were excluded from most of the subsequent analyses comparing the two groups.

D. Neurometric Findings

1. *Frequency Analysis of Resting EEG*

Figure 7.1 shows the percentage of energy in four frequency bands for three pairs of bilaterally symmetric electrode derivations. The values for each subject were obtained by taking the average value of fast Fourier analyses on 18 consecutive EEG segments each of 5.12 sec duration, sampled by the computer

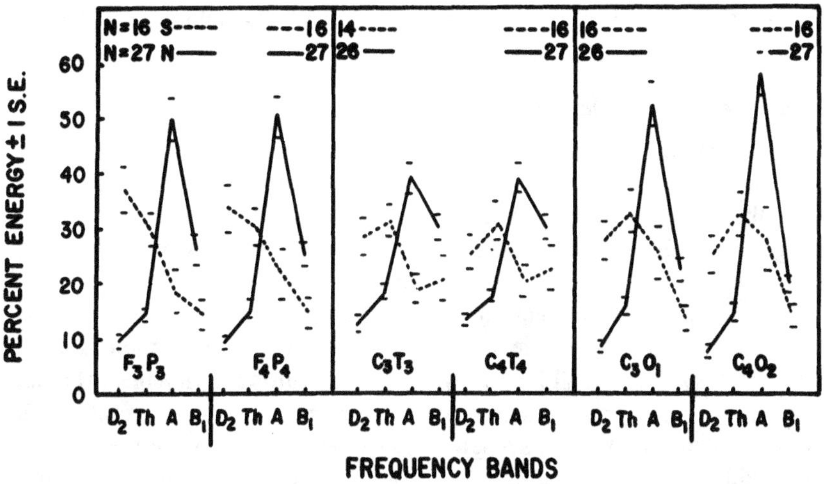

FIG. 7.1 Percentage of total power of EEG in four frequency bands for three pairs of bilaterally symmetric electrode derivations. The values are the average or mean value for each group plus or minus one standard error of the mean. The number of subjects in each group is shown at the top of each panel. The values for the senile group are shown as a dotted line, the normal group as a solid line. Although this figure, and Fig. 7.2, shows data only for about the initial third of our sample, similar findings were obtained from the total group.

at 100 points/sec. The resting EEG was classified into only a $delta_2$ band (1.5–3.5 Hz), a theta band (3.5–7 Hz), an alpha band (7.0–13 Hz), and a $beta_1$ band (13–19 Hz). The percentage of the energy in each of these bands was computed, relative to the total energy in these four bands.

The values shown in Fig. 7.1 are the average or mean value for each group plus or minus one standard error of the mean. The number of subjects in each group is shown at the top of each panel. The values for the senile group are shown as a dotted line, the normal group as a solid line. The left panel compares the frequency distribution of the EEG energy for these two groups in the left (F_3P_3) and right (F_4P_4) fronto-parietal derivations; the middle panel for the left (C_3T_3) and right (C_4T_4) centro-temporal derivations, and the right panel for the left (C_3O_1) and right (C_4O_2) centro-occipital derivations.

In all derivations, the senile group showed significantly more low-frequency activity (delta and theta bands) than the normal group. Despite the marked differences between the groups, there was striking bilateral symmetry of the energy distribution in the four frequency bands within each group. Examination of the general profile of these frequency distributions suggests relatively greater differentiation in the frequency composition of the EEG in the anterior–posterior plane for the senile group than for the normal group. Specifically, the normal group shows an alpha peak in every derivation while the senile group shows a delta predominance in the fronto-parietal, a theta predominance with more beta than alpha in the centro-temporal, and a theta dominance with more alpha than beta in the centro-occipital derivations.

2. *Bilateral EEG Synchrony*

Figure 7.2 shows the bilateral covariance in six frequency bands for the same three pairs of bilaterally symmetric derivations. This measure quantifies the synchronization of the activity in each of these frequency bands for corresponding regions of the hemispheres. Values for each subject were obtained by squaring the correlation coefficient between the EEG activity in symmetric leads, limited within each frequency band using the mean values for 18 consecutive 5.12-sec samples of the resting EEG, sampled at 100 points/second. For this analysis, we used all frequency bands. The values shown represent group means plus or minus one standard error of the mean. The number of subjects in each group is shown in the lower left corner of each panel. As before, the data for the senile group are represented by the dotted line, while the normal group is shown by the solid line. The left panel presents the data for the squared cross correlation, or *covariance*, between bilateral fronto-parietal derivations, the center panel for bilateral centro-temporal derivations, and the right panel for bilateral centro-occipital derivations.

Note that in the fronto-parietal and centro-occipital derivations, the senile group displays higher bilateral coherence or covariance in the slow delta range and lower bilateral synchrony in the alpha range than the normals. For the

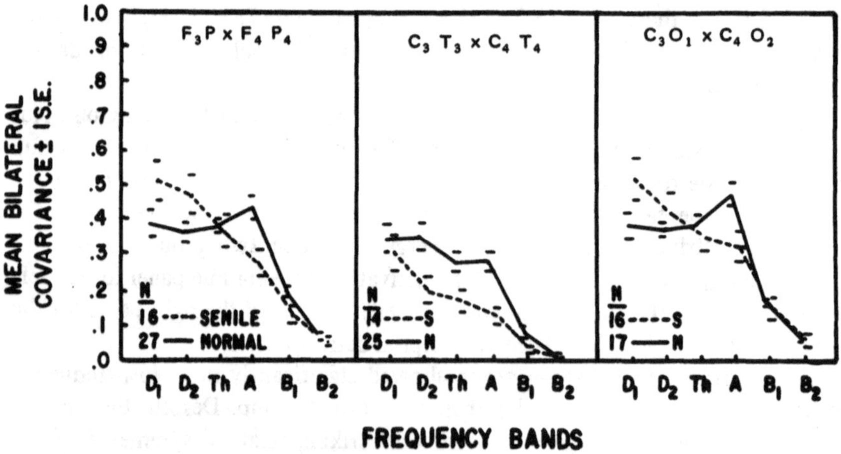

FIG. 7.2 Bilateral covariance of the EEG in six frequency bands for the same three pairs of bilaterally symmetric derivations as shown in Fig. 7.1. The values shown represent group means plus or minus one standard error of the mean. The number of subjects in each group is shown in the lower left corner of each panel. Data for the senile are represented by the dotted line, while the normal group is represented by the solid line. (See text for further details.)

centro-temporal derivations, the senile group has lower bilateral synchrony in all frequency bands. Thus, it appears that the alpha activity of the two cerebral hemispheres is consistently less coherent or synchronized in the senile group than in the normals. Lack of coherence is most marked in the centro-temporal regions, where it occurs for all bands. Note that the higher delta coherence of the senile group in FP and CO derivations is with respect to the slow activity usually considered pathognomonic.

3. Comparison of Overall Interhemispheric Covariance of Resting EEG

Each subject was assigned an overall interhemispheric covariance score as follows: First, the mean correlation (r) was computed, (using 18 consecutive 5.12-sec samples of resting EEG sampled at 100 points/second) between each of four symmetrically placed bipolar derivations (frontopolar-central, fronto-parietal, centrotemporal, and centrooccipital). Second, each of the resulting mean values of r was squared to yield the covariance between each pair of symmetric derivations. Third, the four resulting values of r^2 were averaged to obtain an estimate of the mean covariance of the resting EEG between the two hemispheres. *No* significant differences between the senile and normal groups were found on this measure. The results are shown in Fig. 7.3.

4. Symmetry of AERs from Bilaterally Symmetric Derivations

A comparable analysis of AER symmetry and covariance was carried out, under a variety of stimulus conditions. The results are shown in Fig. 7.4. For each stimulus condition, the average cross correlation (r) between the AER computed from the four pairs of bilaterally symmetric derivations was determined. There were two basic groups of data: AERs to auditory stimuli and AERs to visual stimuli. For each modality of sensory stimulus, there were five different conditions: (1) the response to the sensory stimulus while it was novel (start habituation); (2) the response to the stimulus after a period of familiarization consisting of regular monotonous repetition of the stimulus (end habituation); (3) the response to the stimulus after a period of systematic temporal pairing with a stimulus of the other sensory modality (post sensory-sensory conditioning); (4) the response to the meaningless stimulus during a period of competition with meaningful stimulus patterns in another sensory modality (response to sound during a color cartoon film and response to light during a comic recording); and (5) the response to the stimulus following such cross-modal meaningful stimulation.

The data in Fig. 7.4 show that consistent differences were obtained between the two groups with respect to the symmetry of the AER between corresponding regions of the two hemispheres as assessed by the mean cross correlation. The absolute magnitude of these differences remained essentially constant for the auditory stimuli no matter what stimulus condition was used. However, the

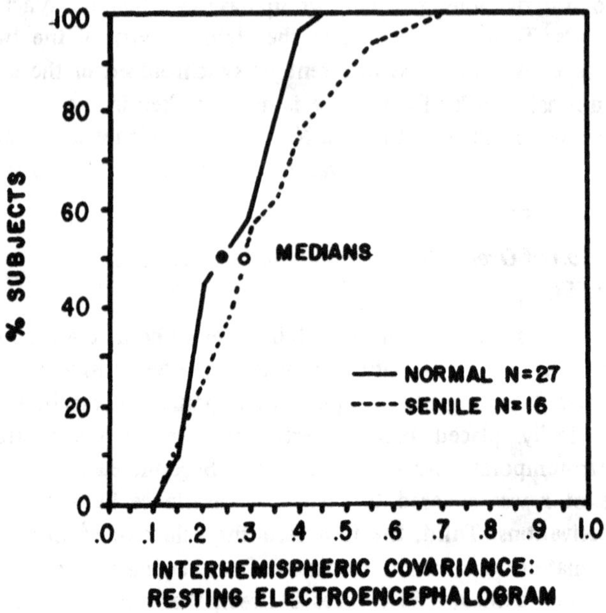

FIG. 7.3 Overall interhemispheric covariance of the EEG from normal (solid line) and senile (dotted line) groups. Data were computed by averaging interhemispheric covariance values (r^2) for four symmetrically placed bipolar derivations (frontopolar-central, fronto-parietal, centro-temporal, and centro-occipital). This figure shows the percentage of subjects that exhibited a particular interhemispheric covariance value. Apparently, the higher alpha covariance of normals obscures the higher delta covariance of seniles in this *overall* measure.

difference between the two groups was not only consistently greater when a visual stimulus was employed, but tended to increase after habituation (regarded as a rudimentary learning experience) or during competition with cross-modal information. Thus, imposition of an information-processing activity might cause an *increase* in AER symmetry in the normals and a decrease in the seniles. Presumably, this would suggest that the states of the two cerebral hemispheres might become more similar in normal aged subjects as an information-processing load is imposed, while the senile patients display decreased similarities of state of the hemispheres under the same conditions.

Analysis of variance of these data revealed a highly significant difference between the two groups on these repeated measures ($p < .001$) and a marked difference in the effect on various derivations ($p < .001$).

5. Comparison of Overall Interhemispheric Covariance of AERs

The conditions which achieved greatest separation between the normal and senile groups were then selected to provide the basis for computation of an

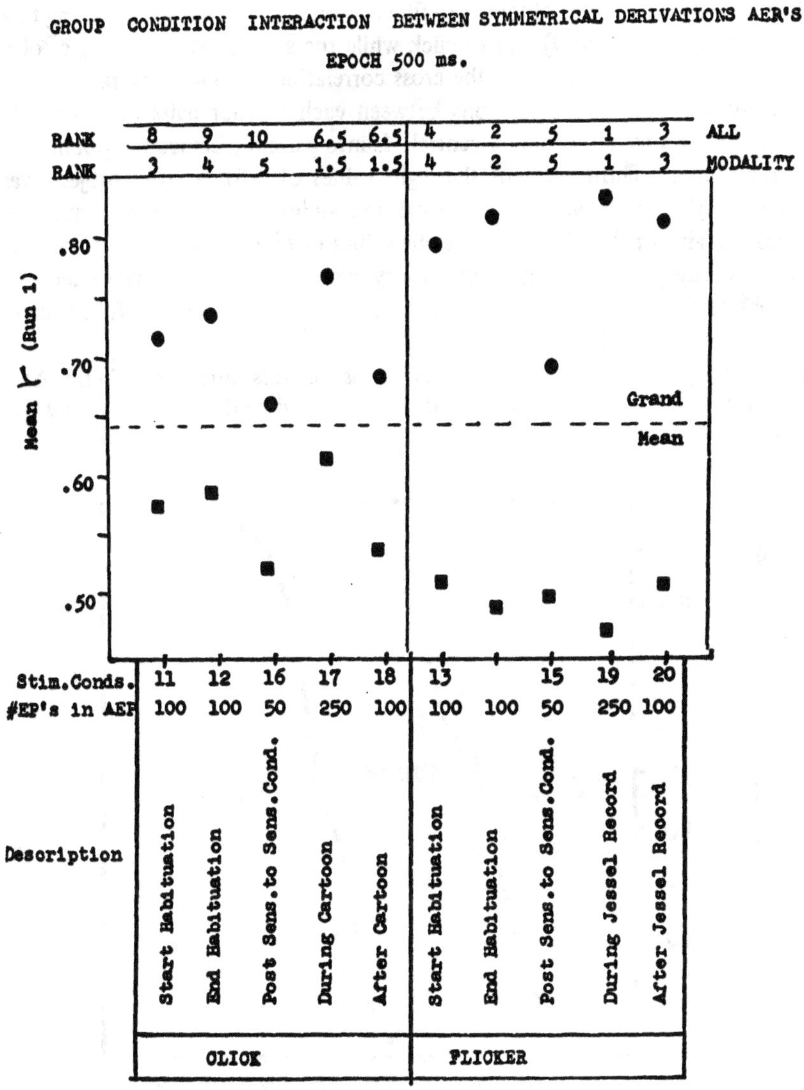

FIG. 7.4 AER correlation during different stimulus conditions. For each stimulus condition the averaged cross correlation (*r*) between AERs recorded from four homologous derivations (frontopolar–central, fronto-parietal, centro-temporal, centro-occipital) was derived. Mean *r* is shown for both click and flicker conditions. Circles represent normals, squares represent seniles.

overall interhemispheric covariance score for AERs analogous to that devised for the EEG. Each subject was assigned an interhemispheric AER covariance score as follows: First, AERs were obtained to flicker while the subject was listening to a comic record (Condition 9) and to click while the subject was watching a color cartoon (Condition 7). Second, the cross correlations r_1 and r_2 were obtained separately for these two conditions between each of four pairs of symmetric bipolar derivations (frontopolar-central, fronto-parietal, centro-temporal, and centro-occipital). Third, each of the eight values of r for a given subject was squared to give the covariance of visual and auditory AERs from symmetric derivations, and finally, fourth, the eight values of r^2 were averaged to yield an estimate of the mean covariance of sensory evoked responses between the two hemispheres under conditions when the subject attended to meaningful events in another sensory modality.

The mean values and standard deviations of this interhemispheric AER covariance were then computed for the normal and senile groups. Figure 7.5

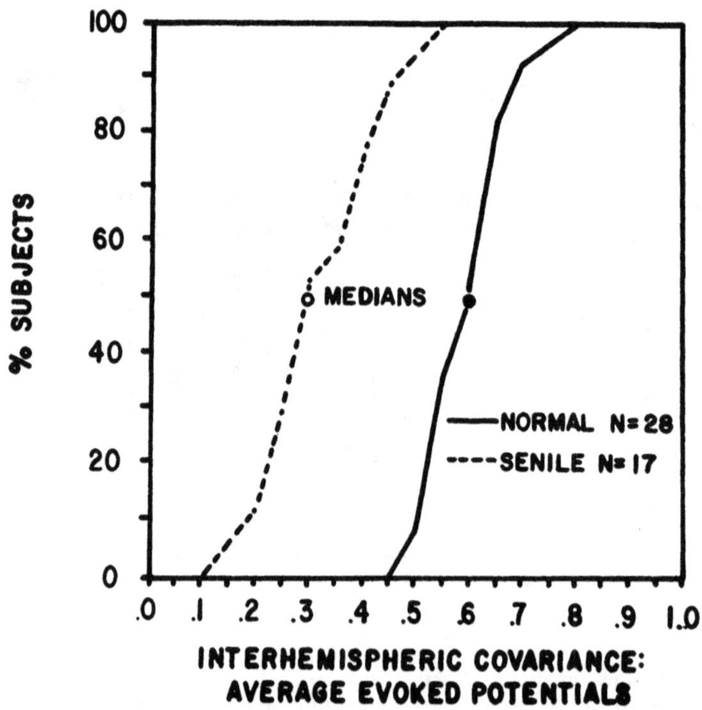

FIG. 7.5 Overall interhemispheric covariance of AEPs for normal (solid line) and senile (dotted line) subjects. Data were computed by averaging interhemispheric covariance (r^2) values for the four homologous derivations described in Fig. 7.3 and 7.4. The cumulative histograms of the two groups are shown, that is, the value of the interhemispheric covariance versus the cumulative percentage of subjects with that value.

shows the cumulative histograms of the two groups, that is, the value of the interhemispheric covariance versus the cumulative percentage of the groups with that covariance value or less. The difference between the two groups is highly significant ($p \ll .001$). It can be seen that this measure almost completely separated the two groups. Only one case of overlap was found between the two subsets of subjects, with one senile subject scoring higher than the two lowest normals.

The data shown in Fig. 7.5 represent values obtained from an unselected subgroup, that portion of our sample for which these data first became available. Subsequent computation of this average interhemispheric correlation for the full population in our sample revealed a separation accuracy of 89.3%. Only minor fluctuations in this accuracy were observed as the stimulus conditions were varied. Under any and all stimulus conditions, the senile population displayed markedly lower interhemispheric correlations between AER waveshapes than the normal subjects. These findings are summarized in Table 7.4.

6. *Discriminant Functions on Individual Conditions*

Using the value obtained from each derivation, it was possible to construct a discriminant function for every test condition in the neurometric battery. These computations were carried out, using the CDC 6600 computer at the Courant Institute, New York University. The results of these computations are presented in Table 7.5 for a selected subset of the measures.

These results show that the normal and senile subgroups displayed pervasive differences which were highly significant in every EP condition as well as in the resting EEG. Under all of these EP conditions, as well as at rest, the senile individual was characterized by a larger amount of electrical energy in both the spontaneous EEG and the late components of the average evoked response from at least some region of the scalp (200–500-msec latency).

Furthermore, the senile individuals had a greater percentage of energy in the 1.5–3.5-Hz ($delta_2$) frequency band in at least some head region(s), and a smaller percentage of energy in the alpha band (7–13 Hz).

Examination of these data shows that many senile patients display a hypersynchronous slow wave in the delta band, in some head region, together with a

TABLE 7.4

		Psychometric classification: Normal	Psychometric classification: Senile
Interhemispheric correlation classification	Normal	66	8
	Senile	4	34

TABLE 7.5

Measure	N (normal, senile)	Discriminant accuracy (%)
S, Condition		
1[a]	62, 38	79
2	64, 42	85
3	65, 45	80
4	63, 39	74
5	59, 31	64
6	58, 26	73
7	67, 43	78
8	62, 38	80
9	62, 44	84
10	65, 35	79
11	62, 38	74
12	66, 37	80
13	65, 38	80
14	66, 37	79
15	59, 26	81
16	55, 28	77
17	64, 42	82
18	64, 29	79
19	66, 42	83
20	65, 37	86
Absolute EEG power, total band, mono	79, 51	80
Relative $delta_2$ power	53, 51	82
Relative alpha power	53, 51	68

[a]S denotes signal energy in the 200–500-msec latency interval. Condition numbers refer to the EP conditions defined earlier.

deficiency of alpha activity. The overall higher amount of absolute power in the EEG is primarily due to this slow activity. The high energy in the late component of EPs under almost every condition in the senile group probably reflects the greater ease of achieving a highly synchronized evoked response when the neural population is already in a hypersynchronous mode. These features of many members of the senile group might be due to underactivation of the midline reticular activating system.

7. Multivariate Analysis

Our primary goal in this study was to learn whether neurometric indices could reliably discriminate between normal and senile subgroups in the geriatric population. Many of the individual measures for which results have been presented here obviously accomplish that separation with quite high sensitivity. Thus, the primary goal was successfully achieved.

However, combinations of measures might provide even more sensitive indices of the cognitive deterioration of the senile patient. Different patterns of abnormal values on these multivariate combinations might characterize clusters of senile patients who share profiles of abnormality of brain function representing different etiologies of the same behavioral syndrome. Cognitive deficits resulting from these different etiologies might respond differentially to particular pharmacotherapy or other treatments.

a. Multiple discriminant function. As an initial step in multivariate analysis, we computed the multiple discriminant function between the normal and senile groups using the signal energy (S) from Conditions 1–6. The result is shown in Table 7.6. Out of 110 subjects 94 were "correctly" classified, that is, the EP and psychometric classifications agreed, for an overall accuracy of 85.5%.

The computation was repeated, leaving out those subjects where the discrimination was classified as "uncertain" (ambiguous F ratios). The results are shown in Table 7.7. Out of 88 subjects 80 were correctly classified, for an overall accuracy of 90.9%.

The computation was repeated, leaving out those subjects whose performance score contradicted their psychiatric classification. The results are shown in Table 7.8. Out of 82 subjects 77 were correctly classified, for an overall accuracy of 96.3%.

b. "Leave-one-out" replication of discriminant function. Since overly redundant data can actually detract from the accuracy of a discriminant function by adding unnecessary variance or "noise" to the computation, we computed a discriminant function using only the signal energy from Conditions 2 and 9. These measures were selected because they yielded the highest accuracy for discriminations based upon single measures in monopolar derivations (see Table 7.5). The initial multiple discriminant function had an overall accuracy of 92%.

A so-called "leave-one-out" replication of this discriminant was then computed. In this ingenious new technique, *replication of a discriminant function is*

TABLE 7.6
Classification of Normal and Senile Subjects by Multiple Discriminant Function

		Psychometric classification	
		Normal	Senile
EP classification	Normal	57	9
	Senile	7	37

TABLE 7.7
Accuracy of Discriminant Function When Cases Classified "Uncertain" Were Omitted

		Psychometric classification	
		Normal	Senile
EP classification	Normal	50	5
	Senile	3	30

accomplished by iterative computations on the initial data sample itself. The principle is simple: On each computation, a different member of the sample is left out while the discriminant function is computed. The resulting discriminant function is then used to classify the omitted subject, and the process is iterated for the next subject.

While computation-costly, the leave-one-out replication is experimentally economical. The discriminant function computed on each iteration is statistically independent of the subject who was "left out." Thus, *leave-one-out replication provides a wholly independent replication of a study without the need to gather any additional data.* Further, it offers the unique advantage that the replicate population is an identical duplicate of the original sample, a condition impossible to achieve by conventional means.

The results of leave-one-out replication of the discriminant whose initial accuracy was 92% is shown in Table 7.9. Out of 81 subjects 68 were correctly classified, for a replicated overall accuracy of 84%. The chi-squared value for this replication was 28.7, with $p < .0001$.

Thus, multivariate analyses, utilizing multiple discriminant functions, yield even more powerful criteria for the separation of normal and senile subgroups.

TABLE 7.8
Accuracy of Discriminant Function When Cases with Ambiguous Behavioral Evaluations Were Omitted

		Psychometric classification	
		Normal	Senile
EP classification	Normal	48	4
	Senile	1	29

TABLE 7.9
'Leave-One-Out' Replication of Discriminant

		Psychometric classification	
		Normal	Senile
EP classification	Normal	54	10
	Senile	3	14

8. *Cluster Analysis*

Examination of the univariate analyses presented thus far made it clear that each of the tests in the neurometric battery which had been administered to these elderly subjects provided the basis for quite powerful separation of the senile and normal subgroups in our sample. Even though the discriminative power of the individual measures was uniformly rather high, they were, nonetheless, substantially nonredundant. This became apparent when multivariate analyses of the various types reported were performed, since appreciable improvements in accuracy were achieved thereby. Thus, it appeared that although the normal and senile subgroups were significantly different with respect to most, if not all, of the neurometric indices extracted from the test battery, *different* senile individuals displayed abnormal values on *different* indices. Shared profiles of abnormal values on selected subsets of neurometric indices would characterize individuals who shared a common pattern of brain dysfunction.

a. Method. In order to search for individuals who shared these hypothetical common profiles, we utilized a method of cluster analysis devised by Daniel Brown and implemented on our PDP 11/45. In brief, this method begins by representing each individual as a point in a neurometric space, in which each neurometric index corresponds to a different dimension. The distance matrix between each individual and every other individual in the neurometric space is next computed. Using prelabelling or other criteria, tentative initial clusters are defined. As will be seen from what follows, these initial clusters could, in fact, be selected arbitrarily, since the same final outcome would ensue but after additional computation. Using the distance matrix, the average distance between the individuals comprising *each* such initial cluster is computed. Next, an average distance is computed between *every* individual and all of the members of *each* initial cluster. An F ratio can now be constructed, relating the average distance between members of the same cluster to the average distance between any individual and all of the members of that cluster. The initial clusters are now "purified" in view of the results of the F ratio computation: that is, individuals who fail to meet a minimum F ratio criterion are removed from the corresponding cluster. Further, individuals who lie so nearly equidistant from two cluster centers that the corresponding F ratios are closer together than some "guard band" defined by ΔF (difference between the F ratios) are categorized as "unclassified". This procedure is iterated until the initial cluster membership stabilizes at some residual set of members. At this point, that unclassified subject who has the largest number of unclassified nearest neighbors is selected as the center around which a new cluster is to be grown. This individual and those nearest neighbors are defined as a new cluster. The purification procedure is then iterated on this new constellation of clusters. The process continues until no new clusters can be grown which satisfy the criteria selected (Brown, unpublished method, 1976).

Neurometric data from 60 elderly subjects considered to be normal on the basis of the clinical assessment, 52 subjects considered to be senile by that criterion, and 8 subjects who were not definitely preclassified because full clinical assessment data could not be obtained (and by that fact alone most probably senile) were subjected to cluster analysis using the method just described. Since so many of the neurometric indices had displayed strong univariate discriminative power, and since excessive redundancy simply introduces noise into the distance computation, a set of neurometric indices was selected which seemed to us to provide a good sample of different types of indices extracted from basically different conditions. Using such subjective criteria, we selected signal energy (S), signal-to-noise ratio (S/N), signal energy asymmetry (PEASY), and signal waveshape asymmetry (PSASY) from the monopolar and bipolar AER to click during the color film (Conditions 7 and 17) and to flicker during the comic record (Conditions 9 and 19) and the absolute power in the delta and beta bands of the spontaneous EEG, recorded monopolar and bipolar.

9. Numerical Taxonomy of the Elderly

Cluster analysis of this sample of subjects was carried out in the neurometric space defined by these indices. The results are shown in Table 7.10 and Fig. 7.6. The 120 elderly subjects classified by this analysis fell into 7 clear clusters that contained 101 subjects while 19 remained unclassified. The normal elderly group was found to comprise three clusters (N_1, N_2, N_3). One of these clusters (N_1) was characterized by low values of delta and beta power in the resting EEG, AER power slightly below and S/N slightly above the population mean on all EP

TABLE 7.10
Groups of Normal (1–3) and Senile (4–7) Found by Cluster Analysis

	N_1	N_2	N_3	S_4	S_5	S_6	S_7	Did not cluster
Normals	30	9	6	2				13
Seniles	1	1		15	13	10	7	5
No global rating (?)				3	2	1	1	1

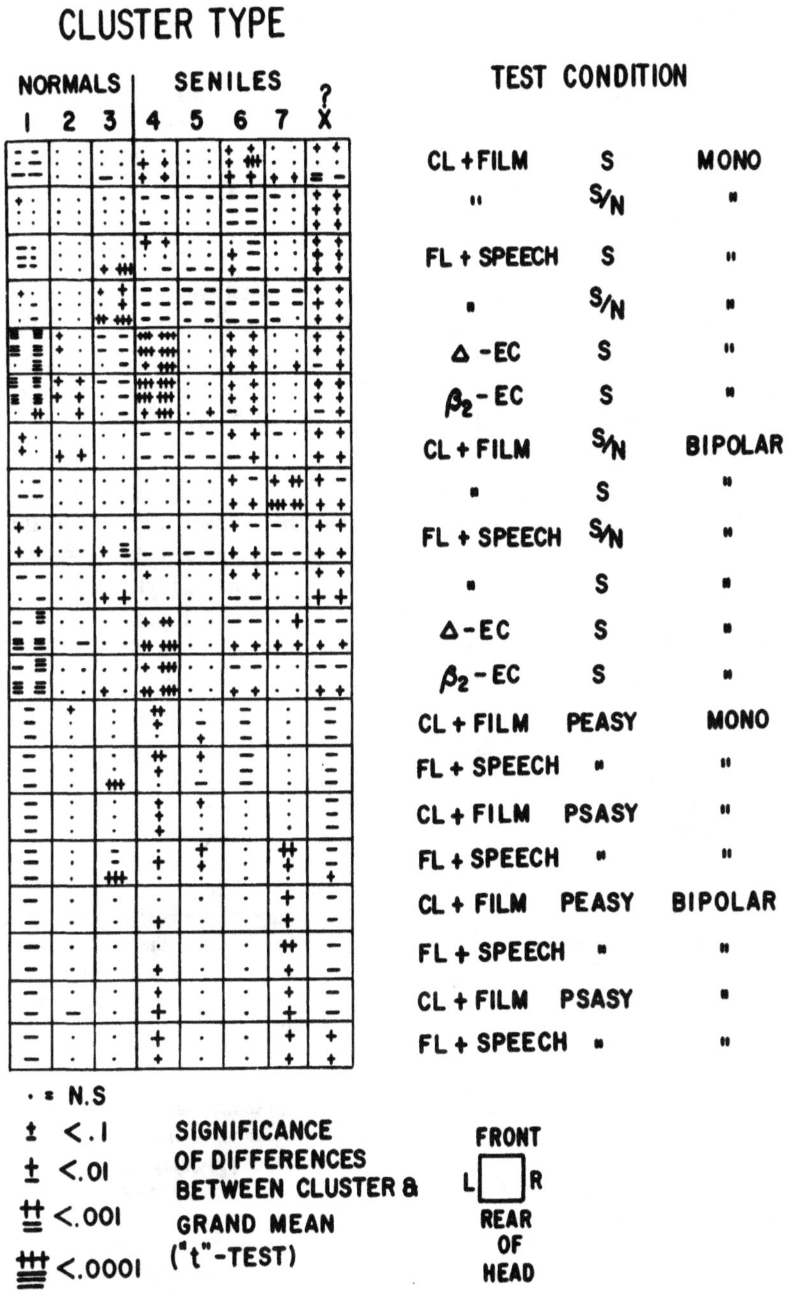

FIG. 7.6 Neurometric features distinguishing seven different clusters found within a sample of 120 elderly patients. Each *column* of rectangles shows the features displayed by members of the cluster indicated at the top of the column. Each *row* of rectangles indicates the values of a particular neurometric index extracted from a specific NB test condition, indicated at the right side of the row.

Each rectangle represents a head viewed from the top, facing upward, as illustrated at the bottom of the figure. Within each rectangle, the location of every entry corresponds to the

conditions, and good bilateral AER symmetry. A second normal group (N_2) differed from the first in having an extremely high amount of power in the beta band and a high amount of power in the delta band of the resting EEG. The third normal group (N_3) displayed the same general EEG profile as group 1, but showed very high S/N to the visual stimuli.

The senile elderly group was found to consist of four clusters (S_4–S_7). The largest cluster (S_4, 33.3%) was characterized by extremely high amounts of delta and beta activity in the resting EEG, recorded either monopolar or bipolar, a poor S/N to visual stimuli while listening to speech, and by a consistent amplitude and shape asymmetry in the AER during sensory-sensory interactions. The second-largest cluster (S_5, 25%) showed a low S/N to visual stimuli while listening to speech, and moderate shape asymmetry in the AER during sensory–sensory interactions but did not display severely extreme values on any index. The third group (S_6, 18%) displayed very high signal power to auditory stimuli while watching a film, poor monopolar but good bipolar S/N during sensory–sensory interactions, a moderate excess of delta and beta activity especially in the monopolar resting EEG and in posterior regions in the bipolar resting EEG. The fourth group (S_7, 13%) was characterized by very high signal power to auditory stimuli while watching a film, an excess of delta but not beta activity in the EEG, and marked asymmetry of AERs especially in bipolar records obtained during sensory–sensory interactions.

FIG. 7.6 (*contd.*)
anatomical location of the electrode or set of electrodes from which the measure was derived. Indices of symmetry are presented along the midline at a point corresponding to the anterior–posterior position of the regions which were compared.

Every entry represents the values obtained by computing the z-transformation of each individual value, relative to the grand mean and standard deviation of the total sample, and then averaging across all the members of that cluster. The resulting average z-transform is presented as a symbol whose sign shows the direction of deviation from the grand mean and whose size is proportional to the statistical significance of the deviation, encoded as illustrated at the bottom of the figure.

On the basis of psychometric and psychiatric assessment of clinical status, 60 of these patients had previously been classified as "normal," 52 as "senile," and 8 as "unclassified" because adequate data for clinical evaluations could not be obtained from the patient. (The mere fact that these "unclassified" patients were incapable of providing any answers to the questions utilized for clinical evaluation suggests that their proper classification would be "senile.") Three of the clusters defined by the application of numerical taxonomic methods to the neurometric data obtained from this total sample contained subjects almost all of whom were normal: Cluster 1 (30 normal, 1 senile), Cluster 2 (9 normal, 1 senile), Cluster 3 (6 normal, 0 senile). Clusters 1, 2 and 3 thus contained a total of 45 normal but only 2 senile subjects. Most of the subjects in four other clusters were senile: Cluster 4 (15 senile, 2 normal, 3 unclassified), Cluster 5 (13 senile, 0 normal, 2 unclassified). Cluster 6 (10 senile, 0 normal, 1 unclassified) and Cluster 7 (7 senile, 0 normal, 1 unclassified). Clusters 4, 5, 6 and 7 thus contained a total of 45 senile but only 2 normal subjects, plus 7 subjects who had been "unclassified." Nineteen Ss (13 normal, 5 senile and 1 "unclassified") could not be located within any of these clusters by our methods, and are denoted as group ?/X.

The overall accuracy of diagnosis achieved by the cluster analytic method must be calculated in at least two ways; (1) consider only those subjects classified definitively by clinical methods and (2) consider the trichotomy which includes the classification "unclassified" (?). By the first criterion, the accuracy of the cluster analytic classification was 95.5%. By the second criterion, the accuracy was 75%. A third criterion, which we consider to be the most correct, classifies the "unclassified" as senile simply because these subjects were incapable of providing data good enough to permit confident classification. Using this criterion, the accuracy was 85% (see Table 7.11). Finally, it is noteworthy that no significant differences were found between the four senile clusters with respect to the Assessment of Clinical Status. These different clusters did not reflect different severities of impairment.

IV. CONCLUSIONS

A number of clear conclusions emerge from this study:

First, with respect to previous studies of the electrophysiological characteristics of the aged in general and the senile in particular, our results indicate that excessive slow waves are neither necessary nor sufficient for clinical manifestations of senility. Although three of the senile clusters showed excessive slow waves, members of one of the normal clusters also did so.

Second, two of the senile clusters showed extremely low interhemispheric symmetry for AERs. Although these senile individuals often displayed poor signal to noise ratios in the AER, the low interhemispheric coherence is not merely a reflection of noise in the system. Amplitude and waveshape asymmetry of the AER were found to be more powerful discriminators between normal and senile clusters than the signal to noise ratio, if one evaluates the relevant univariate F ratios. Further, it was observed that S/N was high in some leads but low in others, and that these variations in S/N did not coincide with values for amplitude and waveshape symmetry. Thus, we conclude that the low interhemispheric correlation between AER waveshapes is not reasonably to be attributed to high variance of the AER in senile individuals, but more probably

TABLE 7.11
Classification of Normal, Senile, and "Questionable" Subjects by Cluster Analysis

Group cluster classification	N	S	?
N	45	2	0
S	2	45	7
?	13	5	1

reflects a real discrepancy in the response of homotopic regions of the two hemispheres to afferent input. It was found that this discrepancy is most marked with respect to late, or endogenous components, rather than short latency or exogenous components. This suggests that the processes related to interpretation of the meaning of afferent input are not well coordinated in the two hemispheres. Since callosal fibers interconnect corresponding regions of the two hemispheres, this finding suggests that in the senile individual each hemisphere sends an interpretation of the significance of incoming information to the other hemisphere that is likely to contradict the interpretation transmitted by that hemisphere. The resulting informational inconsistencies might be expected to have major functional implications. The data suggest that in most senile individuals, cognitive impairment is as appropriately to be attributed to this specific interference with information processing as to the general consequences of cerebral arteriosclerosis as reflected in slowing of the dominant EEG rhythms.

Finally, and most important, our results establish unequivocally that the population of senile individuals thus far considered to be homogeneous is, in fact, heterogeneous. Subsequent studies with larger sample sizes and new neurometric indices may confirm the existence of only four clusters, as tentatively identified in the relatively small sample examined in this study, or may reveal the existence of additional groupings. In either case, our results demonstrate that the method of numerical taxonomy, applied to adequate neurometric data from elderly patients, not only permits the accurate separation of normal from senile individuals but clearly reveals the existence of homogeneous subgroups within the heterogeneous population hitherto subsumed under the category of "senile." Once these homogeneous subgroups are recognized and their differential profiles specified, the search for individualized prescriptive therapies becomes possible. It now would seem only a matter of time and effort until we can resolve the questions of whether or not treatment is possible for the members of any cluster, and what the optimal treatment might be.

8

Neurometric Assessment of Brain Dysfunction in Children with Learning Disabilities

I. MINIMAL BRAIN DAMAGE

Examining the literature on minimal brain dysfunction, or MBD, one cannot help but be reminded of the novel *Catch-22:* minimal brain damage might whimsically be defined as brain damage so slight that it cannot be detected. Once definitely detected, it is no longer minimal. The concept of MBD is vague. It stands for a complex of disturbances of learning and behavior that can sometimes be correlated with demonstrable functional disorders of the brain. Yet, children without any abnormal neurological signs often are considered to suffer from MBD on the basis of behavioral observation alone. MBD is considered by some to be the single most common disorder seen by child psychiatrists, yet its existence is often unrecognized (de la Cruz *et al.,* 1973; Wender, 1971).

The impoverishment of life experiences by the victims of this disorder, the unhappiness and tension caused in their families, the burden they place on an already overtaxed school system, the welfare costs resulting from the difficulties these individuals encounter in supporting themselves and from the antisocial behavior to which failure, frustration, and alienation must lead many of them, amount to an incalculable but undoubtedly staggering price in both human and economic terms. This price becomes doubly tragic when one realizes that present knowledge permits the early identification and successful treatment of many children with MBD, and that more diversified and effective therapies must inevitably ensue if precise evaluation of the functional consequences of this disorder in a given individual was readily available. Conversely, failure to treat MBD results in increased risk for psychopathology later in life, ranging from poor social adjustment, underachievement and scholastic failure, to juvenile delinquency and even psychosis (Menkes *et al.*, 1967; Weiss *et al.*, 1971).

A. The MBD Syndrome

The syndrome of MBD includes some or all of the following symptoms (after Wender, 1971): (1) a high activity level (*hyperkinesis*) and impaired coordination (*dyspraxia*); yet, some of these children are *hypoactive* and listless; (2) short attention span; (3) poor ability to concentrate or pay attention, often accompanied by "forced responsiveness," by which is meant undue attention to minute details. Perseverative stereotyped performance of trivial activities is often displayed; (4) difficulty in school, often found in children of normal intelligence, "underachievers." The most common manifestation is difficulty in learning to read; (5) poor "impulse control," by which is meant inability to inhibit behaviors known to result in undesirable outcomes, as well as low frustration tolerance, physical recklessness, and defective sphincter control; (6) difficulty in interpersonal relations, characterized by resistance to social demands, and excessive dependence or bravado suggestive of reaction formation; (7) inappropriate emotional behavior, evidenced as increased lability, increased aggressiveness, altered reactivity, diminished or exaggerated pain responses, or overexcitement; (8) *dysphoria,* by which is meant reduced ability to experience pleasure, depression, low self-esteem, and anxiety. Surveying the foregoing list of symptoms, the reader must begin to understand how easy it would be to conclude that a neurotic or psychologically disturbed child suffered from minimal organic brain damage, or vice versa.

1. "Learning Disability" and "Learning Disorder"

There is considerable inconsistency in the literature concerned with the evaluation of children who display learning problems. As will become clearer during the subsequent discussions in this chapter, much of the apparent inconsistency and contradiction between findings can probably be attributed to failure to recognize that the population of children who share these behavioral signs contains many different subgroups. Children who display some or all of the behavioral syndrome described above usually are identified as suffering from minimal brain dysfunction. This term is misleading and inadequate. Hitherto, it has been used because unequivocal signs of brain dysfunction have been very difficult or impossible to detect in most of these children. However, if we succeed in our endeavor to develop methods whereby clear signs of brain dysfunction can be demonstrated in a substantial proportion of such children by neurometric assessment, brain dysfunction will no longer be "minimal." Therefore, we consider it useful to make a distinction between "learning disability" and "learning disorder." By learning disabled, or LD, we refer to those children who have some detectable abnormalities in brain function that are correlated with certain impediments in learning. By learning disordered, we refer to those children whose brain function shows no such abnormality and who nonetheless also display learning problems, perhaps because of experiential, emotional, or other difficulties. At present, and perhaps inherently, it may well be impossible to distinguish learning disability from learning disorder on the basis of behavioral observations alone. Although the ensuing discussion of the various bodies of relevant literature uses the terminology as it appears in the original articles, it is our ultimate purpose to learn how to distinguish between the disabled and the disordered subgroups within the population, and further to separate those clusters of children within the learning disabled subgroup who suffer from distinctly different types of brain dysfunction.

2. *Heterogeneity of Etiology*

It is intuitively clear, and most students of learning agree, that learning and performance depend upon a wide variety of factors. Among the more salient of these are the existence of an adequate state of arousal, the ability to focus and maintain attention, the presence of appropriate motivation, the sensory acuity required to resolve relevant cues, the perceptual capacity to distinguish relevant signal or "figure" from background events, the capability to perform required discriminative behaviors, the functioning of the various neurochemical systems involved in registration of sensory input, the maintenance of some representation of the neural consequences of that sensory input sufficiently long for consolidation of memory to occur, the ability to establish and sustain the macromolecular syntheses probably required for long-term storage of the experience, and effective access to stored information when related events subsequently take place. Our present understanding of the many complex functions enumerated above is far from complete, but it establishes unequivocally that many different functional systems and neurochemical reactions in the brain participate in an important way in the processes of learning and memory. In view of such knowledge on the level of basic neuroscience, as discussed in Volume 1 of this work, it seems obvious that a particular brain dysfunction may manifest itself in a variety of different aberrations of learning and behavior, that a particular behavioral peculiarity may be attributed to diverse causes, and that a wide variety of different factors may cause one or another type of "learning disability."

For example, Conners (1973) used a population of 267 children between 6 and 12 years of age referred to the Child Development Laboratory of the Massachusetts General Hospital for evaluation and treatment of behavioral and/or learning disorders and obtained a number of measures, including the Wechsler Intelligence Scale for Children (WISC), the Bender Visual Motor Gestalt Test, the Porteus Mazes, the Wide Range Achievement Test, the Frostig Test of Visual Perception, The Harris–Goodenough Draw-A-Man Test, a continuous performance test, and a visual paired-associates learning test.

The scores on these tests were then factor analyzed, using a Varimax rotation. Five major factors were obtained, which were arbitrarily labeled: (1) general IQ, (2) achievement, (3) rote learning, (4) attentiveness, and (5) impulse control. The profiles of each subject on these factor scores were then constructed, and the set of profiles subjected to cluster analysis in order to identify separate patterns or "cluster types" of test profiles. (See Chapter 3 for a discussion of cluster analysis.) Six groups with different characteristic profiles emerged from this analysis: Group 1 was characterized by good impulse control, low IQ, and learning difficulty; Group 2 was inattentive, underachieved, and learned poorly; Group 3 showed poor impulse control but was attentive, although they learned poorly; Group 4 had good impulse control, high IQ, and learned quickly; Group 5 was attentive and learned well; and Group 6 had poor impulse control and low IQ, but learned well.

The functional validity of the clustering procedure was then assessed in several ways. The six groups differed significantly from each other on measures of motor development (Lincoln–Oseretsky test). These groups also showed significantly different responses to dextroamphetamine, methylphenidate, magnesium pemoline, and placebo, as assessed by changes in response on the Porteus mazes.

Finally, evoked responses to visual and auditory stimuli were recorded separately and while the children were attending separately to each modality. The groups differed significantly from one another with respect to various parameters. Group 1 displayed significantly higher amplitudes of late components of the occipital VER on the left side relative to the right, while Group 2 had significantly longer latencies for some of the VER components. Hemispheric asymmetry of the averaged evoked response was a parameter that discriminated between groups on the basis of their psychological test profile.

These findings provide strong support not only for the contention that the population of children with learning disabilities displays marked heterogeneity in terms of behavioral capability, but that this heterogeneity is reflected in differential electrophysiological features and differential responsiveness to various drugs. Other evidence, to be presented later, confirms this conclusion.

Thus, one is compelled to conclude that children who display difficulties in learning must do so for many different reasons. Learning disability is not a homogeneous diagnostic entity with a single etiology, but a term subsuming a heterogeneous collection of different problems. Some of these problems have primary organic bases (learning disabilities) reflecting intrinsic inadequacies of brain structure or function, while the others (learning disorders) have their primary origin in features of the environment or previous experience, even though such features may themselves produce a secondary organic basis. The attempt to achieve optimal remediation of the individual child depends upon the ability to distinguish between learning disabilities and learning disorders, and to identify the specific differential cause of the learning disability or disorder as precisely as possible.

Unfortunately, the diagnostic tools available to help in this endeavor have had severe limitations. Classifications are often based upon vague, intuitive observations of behavior of ambiguous origin, or upon psychometric measures of product that fail to make process explicit. There are all too many ways to produce an inappropriate behavioral test response. The techniques of classic neurological examination serve well for the identification of certain defects in sensory acuity, integration, or motor coordination, but are not optimal for the analysis of subtle dysfunctions in information processing, storage, and retrieval. The conventional electroencephalogram, or EEG, evaluated by traditional methods of visual inspection, has yielded suggestive but inconsistent findings of abnormal features in children displaying difficulties in learning. More recently, studies of the sensory averaged evoked response (AER) have suggested that such measures of brain function may reflect aspects of information processing or cognitive functions. However, most AER studies of learning disabled children have focused upon gross sensory reactivity rather than indices of information processing and have further suffered from the limitation of qualitative visual evaluation.

Our development of the automatic digital electrophysiological data acquisition and analysis system (DEDAAS) and the neurometric test battery (NB) was intended to overcome some of these shortcomings and to enable fuller advantage to be taken of the potential utility of neurometric assessments for the differential diagnosis of learning disabilities in children. After further review of the relevant literature and discussion of some of the conceptual and methodological issues that must be recognized, results thus far obtained by applying these new methods to the evaluation of children with learning disabilities will be presented later in this chapter.

3. *Estimates of Prevalence*

Using the term "learning disability" in a much broader sense ("disordered" as well as "disabled"), many studies have attempted to assess the incidence of MBD within this general category. Prevalence estimates vary from less than 1% to well over 15% of the school-age population (ages 5–19), an indication of either the looseness of the criteria and/or the heterogeneity of the category. Estimating its prevalence is made even more difficult by variations in epidemiological factors, such as urbanicity (Eisenberg, 1966), socioeconomic status, and geography, in criteria (from reading problems to generalized underachievement, to psychological, neurological, and pediatric criteria), in measurement instruments (reading tests, achievement tests, teacher and administrator records or estimates, and extrapolations by large agencies), and in variations in conceptual frameworks among the various disciplines. (These problems are summarized by Minskoff, 1973.) The highest estimates (7% and up) probably represent very relaxed criteria and may include many "nuisance problems" with the hard-core learning disabled. The lowest estimates, below 3%, are likely to include only the very specific learning disabilities (Silverman & Metz, 1973). The HEW Report (1969) included estimates from 7.5% to 14.7% of learning disability sufficient to require remediation. More conservative conclusions were drawn by Myklebust and Boshes (1969) who found that only half of the underachievers defined in their study were learning disabled, representing a 7.5% figure. Thus, 3.0–7.5% might represent a reasonable range of estimates for the present.

At least 2 million children in the United States under the age of 15, and perhaps as many as 10 million, are therefore suffering from some form of brain dysfunction. Such findings recently led a past president of the American Neurological Association to state that MBD constitutes the most urgent problem currently faced by neurologists (Carter, 1971).

The high level of variance observed in estimation attempts suggests both a complexity and heterogeneity of the behavior. This would underscore the danger of overgeneralization and miscategorization of individuals. Of equal seriousness is the failure to detect a dysfunction where the subsequent tension produced within the family over poor school performance could be devastating to the child and likely to lead to serious personality disorders, even to the point of psychosis (Kirk, 1972). It is also possible that our current lack of accuracy in detection and classification permits evaluations to be heavily influenced by socioeconomic or other factors, which are not directly related to the real nature of the pathology. Most workers in the field confess great difficulty in making sense of the variety of learning problems (Clark, 1968; Gallagher, 1973; Keogh, 1971). It has become clear that a more precise methodology for measuring and a better diagnostic taxonomy are needed.

B. Relevant Electrophysiological Measures

Electrophysiological assessment can be used in a variety of ways to assess the neonatal status, maturational development, and functional integrity of a child. Some of the major dimensions of such an assessment are indicated below.

1. Neonatal Status

a. EEG measures. A good review of EEG and AER anomalies in children with hereditary, congenital, and perinatal diseases can be found in Dreyfus-Brisac and Ellingson (1972). Monod and Dreyfus-Brisac (1972) have evaluated a body of evidence showing that the neonatal EEG is useful as an estimate of the severity of perinatal distress and provides information relating to the prognosis for full-term babies, while Monod *et al.* (1972) has presented statistical studies and estimates of the prognostic value of the EEG in premature as well as full-term babies. They classify abnormal infant EEGs into the following major classes: (A) *Permanent diffuse abnormalities.* These abnormalities are further subdivided into five types:

1. Inactive EEGs, defined as records in which the voltage is continuously less than 5 μV.

2. Paroxysmal EEGs, defined as interruption of an "inactive" background by irregular, asynchronous, and asymmetric bursts of spikes or slow waves.
3. Slow EEGs, defined as continuous, diffuse low voltage delta waves, present in wakefulness as well as sleep. Theta activity is absent, and these records reflect no reactivity or spatial organization.
4. Low voltage EEGs, defined as records displaying voltages between 5 and 15 μV, with some lability and reactivity.
5. Discontinuous EEG in quiet sleep, which differs from the paroxysmal record in that discontinuity only appears in quiet sleep.

These workers have further identified four other major classes of abnormalities:

(B) *Localized abnormalities*, consisting of electrical seizures, spikes, or rhythmic activity in different frequency bands.
(C) *Diffuse slow waves*.
(D) *Abnormal spatial or temporal organization*, especially asymmetry, asynchrony, low voltage occipital activity, and absence of lability.
(E) Abnormalities of sleep cycle organization.

Dreyfus-Brisac and Monod (1972) reach the following general conclusions about the prognostic value of the neonatal EEG:

1. A normal neonatal EEG has strong prognostic value. No cases of severe neurological sequelae have been observed in surviving babies who showed normal neonatal records, and a very small percentage of infants with normal neonatal EEGs die.
2. With respect to the prognostic value of EEG abnormalities, transitory abnormalities are generally less alarming than persistent ones.
3. A paroxysmal EEG in a comatose baby, and sometimes even in a conscious one, is consistently of very grave prognostic significance, as is an inactive EEG after the first day of life.
4. If the EEG recorded in a comatose infant does not display one of the classes of permanent diffuse abnormality listed above, or diffuse slow waves or gross focalized abnormalities, there is a good chance of normal development despite a severe neurological state.
5. Rapid normalization generally implies a good prognosis, while persistence or augmentation of abnormalities implies a poor prognosis.
6. Multiple abnormalities are more serious than only one.

These conclusions indicate that the neonatal EEG has considerable prognostic value and that it may be useful as an adjunct to evaluate the influence of treatment upon the status of the high-risk newborn infant.

b. AER measures. By conventional clinical methods, it is extremely difficult to evaluate neonatal cerebral function, brain maturation, and sensory perception. Most neonatal neurologic reflexes depend mainly upon functions in structures other than the cerebrum and are difficult to evaluate reliably, while scalp-recorded AERs are related to cerebral events, including activity at the cortical level. Pointing this out, Graziani and Weitzman (1972) have recently reviewed the maturational features of neonatal sensory evoked responses and discussed their clinical utility in early infancy. Barnet (1972) has reviewed the evidence that visual and auditory defects are extremely important sequelae of prenatal infections, and stressed the diagnostic importance of AER measures in high-risk infants. Barnet discussed specific infections and their effects on the CNS, relating these diseases to EEG and AER measures, and focused particularly upon the utility of evoked response audiometry, while Graziani and Weitzman discussed the characteristics of auditory, visual, and somatosensory AERs in the neonatal period and their various clinical applications.

The waveshapes of AERs to various modalities of sensory stimuli have been extensively studied in full-term and premature infants. Clear changes take place in various AER features with age, which might serve as indices of maturation and postconceptual age (Akiyama, Schulte, Schultz, & Parmelee, 1969; Barnet & Goodwin, 1965; Creutzfeldt & Kuhnt, 1967; Desmedt & Manil, 1970; Dustman & Beck, 1969; Ellingson, 1964, 1967a,b; Engel & Benson, 1968; Engel & Fay, 1971; Ferris, Davis, Dorsen, & Hackett, 1967; Fogarty & Reuben, 1969; Groth, Weled, & Batkin, 1970; Hrbek, Hrbkova, & Lenard, 1968, 1969; Hrbek & Mares, 1964; Lenard, von Bernuth, & Hutt, 1969; Lodge, Armington, Barnet, Shanks, & Newcomb, 1969; Monod & Garma, 1971; Umezaki & Morrell, 1970; Vanzulli, Wilson, & Garcia-Austt, 1964; Weitzman, Fishbein, & Graziani, 1965; Weitzman & Graziani, 1968). These studies support the suggestion that analysis of sensory AERs may provide an estimate of neonatal condition as well as postconceptual age.

Harter and Suitt (1970), Karmel, Hoffmann, and Fegy (1974), and Karmel and Maisel (1975) have shown that visual patterns elicit reliable developmental changes in component amplitudes of infant AERs. Further, Karmel and Maisel show how these changes positively correlate to development of visual attention over the first five months of life. The data might directly link pattern AER development with potential cognitive impairment, since Sigman and Parmelee (1974) have shown that high-risk infants differ from normal neonates in visual attention, while Miranda and Fantz (1974) have shown that Down's syndrome infants developmentally lag normal infants in their visual attention.

In general, a decrease in the latency of AER components, together with changes in waveshape and altered topography occur with increasing age. In premature infants, AERs are not only of longer latency, but tend to show smaller amplitude and greater variability. Similar changes with maturation have been described for both visual and auditory AERs. These findings support the suggestion that neurometric analysis of AER waveshapes in the newborn, using methods of pattern recognition discussed in Chapter 3 and illustrated in Chapters 5, 6, and 7, might permit identification of a variety of types of newborn

AER waveshapes and the establishment of reliable correlations between particular parameters or features of the AER and neonatal status and postconceptual age. Routine assessment of neonatal status using quantitative EEG and AER features might eventually be of substantial utility as a prognostic index for certain aspects of subsequent development.

The little evidence presently available on the relationship between neonatal AERs and subsequent "longitudinal" development is particularly interesting. Butler and Engel (1969) demonstrated a high correlation between gestational age, birth weight, neonatal AER latency, and mental test scores at age 8 months. A clear inverse relation was found between visual AER latency and test scores, independent of gestational age or birth weight. Engel and Fay (1971) showed that certain aspects of language skill were better in 3-year-olds who had relatively short neonatal AER latencies. Jensen and Engel (1971) showed that there was a correlation between short neonatal AER latencies and the ability of an infant to walk unaided at 1 year of age. In a particularly important study in this series, Engel (1971) reported no relationship between neonatal AER latencies and IQ test scores at 4 and at 7 years of age. These results suggest that neonatal AER latency by itself is a measure with substantial predictive value for certain aspects of subsequent development, but that such predictive value diminishes with increasing age.

These fascinating studies suggest a number of conclusions:

1. Neonatal AER latency has some utility for short-term but less utility for long-term prediction of language and motor development.
2. Long-term development is influenced by a multitude of variables, of which brain state at birth is merely one.
3. Brain state at birth may not be adequately characterized by the single measure of AER latency. Multivariate description of neonatal brain state and longitudinal studies of substantial numbers of children over a longer period might reveal that certain patterns of neonatal electrophysiological activity were meaningfully related to significant aspects of brain function later in life.
4. Certain aspects of brain dysfunction or damage related to hereditary, congenital, or perinatal diseases may not be immediately reflected in features of the postnatal EEG or AER, but may exist as a latent defect requiring the passage of time before it becomes manifest.

Longitudinal prospective and retrospective studies of large populations, as yet unavailable, will be necessary before some of these questions can be properly answered.

2. *Indices of Maturation*

a. EEG measures. Normative baselines for the frequency composition of the EEG from birth to maturity have recently been published (Hagne, Persson, Magnusson, & Petersén, 1973; Matoušek & Petersén, 1973b). These norms reveal a systematic evolution of the EEG with maturation, and therefore permit objective determination of whether or not a particular child shows a maturational lag in the spectral composition of his EEG. Characteristic profiles of EEG spectra from different head regions in children from 1 to 21 years of age are illustrated in Fig. 8.1. These normative data have been adapted by us from those published by Matoušek and Petersén (1973b). Their original data were presented as the absolute amount of energy in each frequency band, while those values have here been transformed into *relative* values by calculating the mean percentage and standard deviation of energy in each frequency band. In our opinion, both types of information are important, but the representation of data in a normalized form is more intuitively comprehensible and perhaps more meaningful for comparisons across laboratories and samples of children. Because of their great utility as a baseline for the changes in EEG spectrum in the normal individual with age, the original data of Matoušek and Petersén (1973b) are reproduced (with their permission) in Appendix 8.1, Table A1; the transforms into percentages corresponding to Fig. 8.1 are to be found in Appendix 8.1, Table A2; while Table A3 of the Appendix shows the percentages of energy in each band representing deviations from normal spectral composition significant at the .05 and .01 levels.

It seems clear that the EEG can provide useful insights into perceptual and cognitive processes in children. An extensive literature, reviewed earlier in this volume as well as by John (1967a), Regan (1972), and Leiman and Christian (1973), has established unequivocally that EEG and AER measures provide sensitive and powerful measures of sensory information processing and show clear changes during learning. Hughes (1968) has reviewed the evidence showing relationships between EEG measures and the results of a variety of psychological tests. Many studies have revealed a high incidence of abnormal EEG activity in children who were underachievers, or who had speech problems, dyslexia, or other learning or behavioral disorders. The neurological examination of such children is usually within normal limits and they are often of average intelligence (Benton & Bird, 1963; Capute, Niedermeyer, & Richardson, 1968; Cohn & Nardini, 1958; Gergen *et al*., 1965; Hughes, 1968; Hughes & Myklebust, unpublished results, 1966; Hughes & Park, 1966; Klinkerfuss, Lange, Weinberg, & O'Leary, 1965; Klove, 1959; Muehl *et al*., 1965; Pavy & Metcalfe, 1965; Ritvo, Ornitz, Walter, & Haley, 1970; Ryers, 1967; Sorel *et al*., 1965; Stevens, Sachdev, & Milstein, 1968; Torres & Ayers, 1968; Wikler, Dixon, & Parker, 1970).

Posterior slow waves: Excessive slow wave activity is seen with high incidence (14%) of unselected referrals, and has been reported to be commonly associated with a pattern of hyperactivity and behavioral learning difficulties (Capute *et al*., 1968; Cohn & Nardini, 1958; Sorel *et al*., 1965; Wikler *et al*., 1970). Pavy and Metcalfe (1965) found bilateral occipital slowing in many children with speech and language impairment. Hughes and Park (1966) found abnormal posterior slow waves to be the *sole* abnormality in 19% of a group of dyslexic children. Muehl *et al*. (1965) found a high incidence of EEG asymmetry, as well as slow

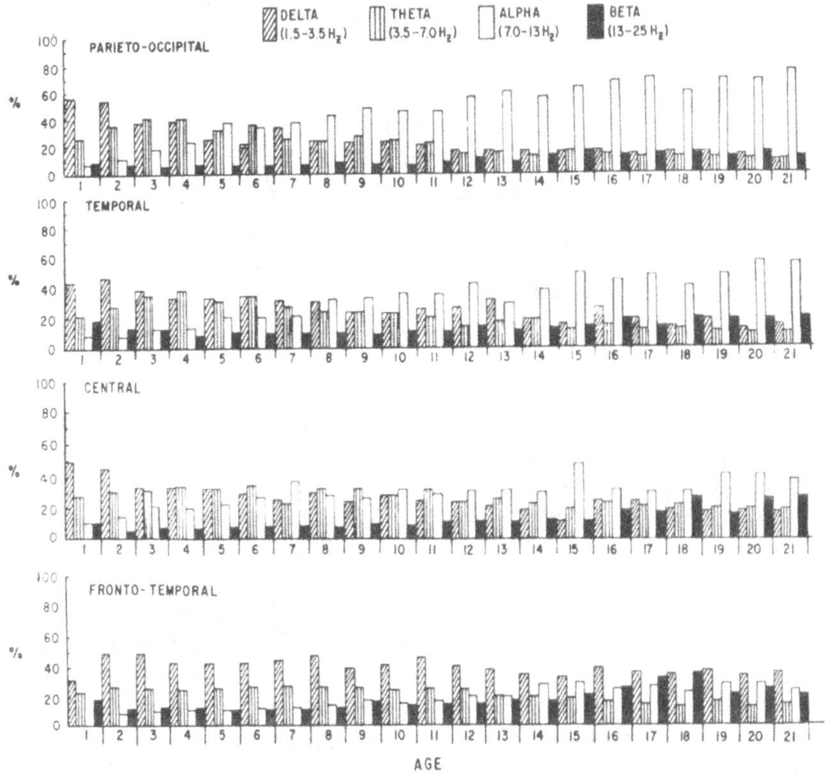

FIG. 8.1 Distribution of spontaneous EEG energy into different frequency bands plotted for different head regions as a function of age. Different shading of the vertical bars refers to delta (1.5–3.5 Hz), theta (3.5–7.0 Hz), alpha (7–13 Hz), and beta (13–25 Hz) bands. See key at top of figure. Data for each age group based on samples from 20 to 50 children. (Adapted from Matoušek and Petersén, 1973b.)

waves, in dyslexic children. Gergen *et al.* (1965) reported a high incidence of asymmetry in the resting EEG, and with photic driving, in underachievers. Hughes and Myklebust (unpublished results, 1966) found a high incidence of posterior slowing in children with learning disabilities. Hughes (1968) concluded that excessive posterior waves can often be found in learning disabled children and stressed the need to determine the amount of slow wave found in asymptomatic children of various ages in a quantitative way. That need, well expressed by Hughes, has now been met by the normative data laboriously compiled by Matoušek and Petersén (1973b) and presented by us in normalized form in Fig. 8.1 and in the appendices. Surely all workers in this field owe Matoušek and Petersén a vote of thanks.

b. Age-dependent EEG quotient and maturational lag. These findings show widespread agreement that there is an unusually high incidence of excessive slow wave activity, as well as spikes and EEG asymmetries, in children with learning or behavioral disorders who seem neurologically normal. In view of this consensus, it is particularly interesting that Christian and April (1974) have shown that obstetrical methods which increase the probability of head trauma are correlated with a higher incidence of excess slow EEG activity. Considering these findings in the context of Fig. 8.1, one might conclude that the excessive slow activity represents a maturational lag in the development of the brain, reflected in an electrophysiological parameter. This suggests that the Age-Dependent Quotient of Matoušek and Petersén (1973a), which was discussed in Chapter 5 from the viewpoint of neuropathology, might reasonably be expected to reveal abnormalities in children with learning disabilities. Data bearing on this question are presented in Sections II and III of this chapter.

c. AER measures. The changes in AER features which take place with age, discussed above as indices of neonatal status, also might serve as indices of maturation and of post-conceptual age (Dustman & Beck, 1969; Ellingson, 1967; Rose & Ellingson, 1970b). Several of the changes that occur during maturation seem particularly relevant for the assessment of cognitive function:

1. The waveshapes obtained from severely retarded children differ in amplitude, complexity, and variability from those observed in normal children (Barnet & Lodge, 1967; Bigum, Dustman, & Beck, 1970; Lelord, Laffont, Jusseaume, & Stephant, 1973). Along similar lines, AERs that were poorly formed (few components) and/or showed long latency or asymmetry have been found in children with minimal brain dysfunction or learning disabilities (Buchsbaum & Wender, 1973; Conners, 1971; Shields, 1973). Further, reading disabled children have significantly smaller component amplitudes of P-300 than either same-aged normals matched on IQ or younger children matched on the basis of reading level when words are flashed for viewing (Preston, Guthrie and Childs, 1974). Satterfield (1973) found the amplitude of P-200 to be significantly smaller in hyperkinetic children compared to normal children. Prichep, Sutton, and Hakerem (1976) found the amplitude of P-200 to be significantly smaller and of N-250 to be significantly larger in hyperkinetic children than in normal children.

2. Studies show that evoked as well as spontaneous activity in anterior temporal derivations (corresponding to the location of cerebral areas involved in language functions) becomes asymmetrical with maturation, apparently reflecting the lateralization of neural processes relating to language acquisition (Giannitrapani, 1969; Rhodes, Dustman, & Beck, 1969). A relationship between hemispheric lateralization and AER asymmetry is further supported by several studies which report that AER amplitudes of subjects of normal or above normal intelligence are larger over the right hemisphere than over the left hemisphere (Perry & Childers, 1969; Rhodes, Dustman, & Beck, 1969; Richlin, Weisinger, Weinstein, Giannini and Morgenstern, 1971; Schenkenberg & Dustman, 1970; Vella, Butler, & Glass, 1972). Appropriate hemispheric differentiation may be

reflected in EP characteristics since retardates are known to possess different hemispheric patterns of EP responses than normals (Bigum *et al.*, 1970; Rhodes *et al.*, 1969). Buchsbaum and Wender (1973) found that the changing pattern of hemispheric differentiation in response to sine wave stimulation associated with increasing age in normal children was different from that found in children with MBD.

Doyle, Ornstein, and Galin (1974) have measured hemispheric EEG asymmetries in normal subjects performing various cognitive tasks. Spectral analysis of the EEG during task performance indicated that verbal and arithmetic tasks produce significantly larger right/left power ratios from homologous temporal (T_4/T_3) and parietal (P_4/P_3) regions than do spatial tasks. Power differences were shown primarily in the alpha band, less so in the beta and theta bands, but not in the delta band. Evoked responses to signals representing primarily verbal or spatial concepts may also become differentiated (Begleiter & Platz, 1969; Buchsbaum & Fedio, 1969, 1970; Shelburne, 1972). Such findings suggest that the lateralization of verbal and spatial functions observed in the "split-brain" human patient after commissurectomy may well be experientially acquired rather than innate, perhaps serving to minimize confusion and cross-talk between neural regions engaged in qualitatively different modes of abstraction.

3. We and others have reported that the waveshape elicited by a visual form of specified shape contains features that are independent of size (Clynes *et al.*, 1967; John *et al.*, 1967), as discussed in Chapter 6. These stable aspects of disparate stimuli reflect a physiological concomitant of the psychological phenomenon of size constancy; a square is a square regardless of its size. It seems reasonable to believe that certain forms of perceptual constancy are not innate, but are acquired by maturation and experience, and reflect the evolution of an individual's ability to construct abstract generalizations from the idiosyncratic experiences that impinge upon him.

C. Electrophysiological Findings in Pharmacotherapy of MBD

Satterfield (1972, 1973) confirmed the observations of others that the most common type of clinical EEG abnormality found in hyperactive children is excessive slow waves. He found that hyperactive children responded better to treatment with a stimulant (methylphenidate) when the EEG showed excessive slow waves than when it was normal ($p = .001$).

Satterfield, Cantwell, Lesser, and Podosin (1972) compared small groups of hyperactive children who showed good responses and poor responses to methylphenidate treatment. It was found that good responders, as compared with poor responders, had low levels of CNS arousal as demonstrated by a low resting skin conductance level, a high amount of power in low-frequency bands of the EEG, and a high amplitude of auditory AERs. Following methylphenidate treatment, the good responders showed a greater increase in these electrophysiological measures of arousal level than poor responders.

On the other hand, Burks (1968) found that medication was more effective on MBD children with normal EEGs. Similarly, Gross and Wilson (1974) have shown that children showing excessive slow waves respond poorest to drug treatment. Thus, at present, no clear relationship between EEG characteristics and subsequent drug response has been demonstrated.

Several investigators have studied the AER in children with learning difficulties in endeavors to find correlates of differential drug responses (Buchsbaum & Wender, 1973; Halliday, Rosenthal, Naylor, & Callaway, 1974; Satterfield *et al.*, 1972). These studies have proved more fruitful than the attempts to predict drug response from the EEG, as described above. Some of the findings suggest that characteristics of the AER before treatment may be correlated with therapeutic outcome and, furthermore, that the differential clinical response to particular drugs seen in responders and nonresponders may be associated with different changes in the AER.

One example of such findings is a recent paper by Conners (1973). Conners assumed that some children with MBD might suffer from defects in arousal mechanisms, while others have more specific defects in the ability to attend selectively. On the basis of other studies, he hypothesized that magnesium pemoline and dextroamphetamine might have different effects on arousal and selective attention. Accordingly, 70 children with MBD were randomly assigned to three treatment groups: magnesium pemoline, dextroamphetamine, or placebo. Visual and auditory evoked responses were obtained from monopolar derivations O_1, O_2, P_3, and P_4 before and after drug treatment, under two conditions of attention: (1) an "auditory relevant" condition in which behavioral responses were required to certain auditory stimuli while bright irrelevant flashes were presented, and (2) a "visual relevant" condition in which behavioral responses were required to certain visual stimuli, while interspersed clicks were irrelevant. Evoked responses were averaged separately for auditory relevant, auditory irrelevant, visual relevant, and visual irrelevant stimuli. Average responses were compared before and after treatment, which was over an 8-week period.

For auditory stimuli, the attending condition produced AER effects which were larger on the left side and more pronounced on the parietal than the occipital region. For visual stimuli, the largest effects were on the right side in the parietal region, but consisted of decreases in amplitude under the attending or relevant condition.

Magnesium pemoline decreased latencies whereas dextroamphetamine increased latencies of AERs on the left side to auditory stimuli. Magnesium pemoline increased the amplitude of the second negative peak on the right side. Magnesium pemoline increased the amplitude of the AER to visual stimuli under attending conditions, as did dextroamphetamine, but reduced the amplitude under nonattending conditions. This differential effect was not found for dextroamphetamine. Conners concluded that magnesium pemoline seemed to produce a facilitation of selective attention, if one assumes that enhanced attention is reflected by higher amplitudes of the averaged evoked response.

Magnesium pemoline seemed to agument visual information processing when a behavioral task required visual discrimination, while reducing attention to irrelevant visual stimuli. However, attention per se reduced the amplitude of AERs for visual stimuli on the right parietal area.

The interpretation of these results is confounded by habituation effects due to inadequate balancing of relevant and irrelevant visual and auditory stimuli in the experimental design. However, the results might indicate that magnesium pemoline may have more effects on selective attention than dextroamphetamine, whereas both drugs act to increase the general level of arousal. Unfortunately, these differential drug effects were not related either to the extent of behavioral improvement or to the features of the pretreatment AER in the individual subjects. Thus, although this study provided suggestive evidence that two drugs which are each effective in the treatment of children with MBD act in a differential manner, no data were provided which might assist in a priori identification of those children who would benefit more from one of these drugs than from the other.

Among the papers suggesting more definitive correlations between the characteristics of the pretreatment AER and the therapeutic outcome of treatment with a particular drug is a recent work by Saletu, Saletu, Simeon, Viamontes, and Itil (1975). These workers randomly assigned 62 hyperkinetic children to 8 weeks of treatment with either placebo, thioridazine, or dextroamphetamine. Behavioral evaluations of the children were derived from ratings on the Global Clinical Impression Rating Scale, the Parents' Questionnaire (Conners, 1970), and the Teachers' Questionnaire (Conners, 1969). Average evoked responses to blank flashes were obtained from the left and right occipital regions, recorded monopolar against ipsilateral earlobe references.

The latencies of each peak in the AER and the peak-to-peak amplitudes were measured and subjected to statistical analysis. This quantitative evaluation showed that there was a significant increase in latencies and decrease in amplitudes of many components of the AER after thioridazine treatment, while treatment with dextroamphetamine tended to increase amplitudes as well as latencies of a somewhat different set of AER components. Thioridazine caused a significant decrease in "inattentive–passive behavior," while dextroamphetamine decreased hyperactivity. Using regression and correlation analysis between the pretreatment AER component latencies and peak-to-peak amplitudes and changes in symtomatology as a function of the 8 weeks of a particular treatment condition, it was found that short latencies and small amplitudes of the AER before treatment [especially with response to peaks 7–8, according to the nomenclature of Ciganek (1961), found between about 250 and 300 msec] were predictors of good therapeutic outcome with subsequent thioridazine treatment. Short latencies and high amplitudes of the pretreatment AER, especially for peaks 3–4 at about 80–100 msec, were predictors of good response to dextroamphetamine. Finally, these authors reported that the greater the drug-induced increase of component latencies, the greater the clinical improvement.

Certain aspects of these findings are similar to results reported previously by Buchsbaum and Wender (1973) and Satterfield *et al.* (1972). These results provide some support for the contention that hyperactive children can be divided into subgroups with different clusters of behavioral symptomatology, that certain electrophysiological features may be differentially characteristic of membership in these clusters, and that particular features of the pretreatment AER may serve as predictors of differential responsiveness to different medications.

Two reservations can be voiced about the findings of Saletu and his colleagues. First, the symptom of hyperkinesis is but one of the eight possible components of the MBD syndrome as identified by Wender (1971). Second, paradoxically and ironically, treatment with placebo produced *a greater* improvement in seven of the eight symptom clusters of the hyperkinetic children than did thioridazine. Similarly, placebo produced essentially as great improvement as dextroamphetamine in six of the eight symptom clusters, the exceptions being "perfectionism" and "muscular tension." However, placebo produced no consistent, statistically significant changes in the AER. Thus, although Saletu and his co-workers went on to demonstrate that it was possible to discriminate between placebo, thioridazine and dextroamphetamine using stepwise-discriminant analyses of AER features, the behavioral findings reported in this work raise grave doubts as to whether either of the "active" compounds studied is preferable to placebo for the treatment of hyperkinetic children. Since placebo produced no significant changes in the AER, and achieved essentially as good or better behavioral improvement as thioridazine or dextroamphetamine, the relationships reported in this work between pretreatment AER features and subsequent differential responsiveness to the two drugs seem to have no functional significance at all.

One should not be discouraged by this example of significantly different findings between the electrophysiological effects of behaviorally ineffective or marginally effective compounds. While we confess that the example was selected somewhat with tongue in cheek, nonetheless the evidence as a whole, even including that reported in the example, supports the contention that the population of children with learning problems is heterogeneous and that accurate identification of the features of the various subgroups or clusters within this population will be a prerequisite to optimum treatment and evaluation of treatment efficacy.

This review of the present status of the field of assessment and remediation of learning disabilities in children led us to draw several conclusions:

1. The EEG and the AER, assessed under various conditions of information processing, potentially provided a fairly direct insight into brain functions related to sensory, perceptual, and cognitive processes likely to be relevant to learning and performance.

2. In order to take optimal advantage of these indices, it was necessary to devise a neurometric technology that would construct precise quantitative measures of the salient features of the EEG and the AER under a set of conditions instituting "challenges" of a wide variety of brain functions devised to reflect such informational processes.

3. These electrophysiological measures of brain function should be obtained in conjunction with a wide spectrum of behavioral and functional measures, as assessed by psychometric tests and evaluation of academic performance.

4. The measures of brain function and behavior should be reduced to numerical representations permitting the use of mathematical and statistical methods for analysis of the data and classification by numerical taxonomic methods.

With these goals in mind, we undertook an extensive research program. This program, now in its third year, consisted of five stages:

1. Development of a new technology (DEDAAS) permitting rapid automatic acquisition of a standardized set of EEG and AER measures of brain function reflecting sensory, perceptual, and cognitive processes (NB). These measures were devised to reflect dynamic processes, or functional challenges, as well as static measure, or conditions.

2. Precise quantification of the essential features of these electrophysiological measures and their reduction to numerical representation (neurometric indices).

3. Construction of a psychometric test battery, including a wide variety of standardized instruments generally considered to reflect perceptual or cognitive disorders of organic origin.

4. Gathering these neurometric and psychometric data from a large group of children consisting of two subgroups, one (normal controls or N) considered behaviorally normal and the other (learning disabled or LD) considered at risk for minimal brain damage because of the nature and amount of difficulty displayed with learning tasks.

5. Analysis of these data by various statistical techniques to identify significantly different subgroups within the sample.

The results thus far obtained using DEDAAS and the full NB to compare samples of normal and learning disabled children will be presented in Section III of this chapter. As a pilot study, we examined the utility of a small subset of NB measures for the assessment of children with learning disorders who were referred to the Neurophysiology Clinic of the Department of Psychiatry at New York Medical College. This subset of NB measures was selected because substantial amounts of normative data on those measures had already been provided by other laboratories. Therefore, we could begin to explore their utility while we proceeded with the task of gathering norms on the full set of measures contained in the NB.

II. NEUROMETRIC ASSESSMENT OF 50 CONSECUTIVE CHILDREN WITH LEARNING DIFFICULTIES REFERRED TO THE NEUROPHYSIOLOGY CLINIC

This section presents preliminary findings obtained from a group of 50 children, between the ages of 6 and 14 years, who were examined consecutively in our clinic. This pilot study suffered from severe shortcomings:

1. No objective assessment of learning disability was obtained. Children were considered as learning disabled if they were referred to the clinic by a qualified psychologist, pediatrician, neurologist, or psychiatrist to establish whether an organic basis for an apparent learning disability could be found. The conclusion that such learning disability did in fact exist rested upon the opinion of the referral source, with no independent confirmation.
2. Electrophysiological measurements were restricted to spectral analysis and symmetry analysis of the spontaneous EEG and to determinations of the symmetry of the AER from homologous derivations in response to blank flashes and patterned visual stimuli.
3. Although normative data for spectral analysis across the ages represented in the group of 50 children studied were available from Matoušek and Petersén (1973a), we *assumed* that those norms, constructed from a Swedish population with possible peculiarities due to genetic characteristics, nutrition, and factors reflecting child-rearing practices of that society, were legitimately applicable to a heterogeneous population of children from New York City and its environs. In the more comprehensive study reported in Section III, we found that those norms deviated slightly from the values we obtained, at least for 9-year-olds.
4. We assumed that normative values of EEG and AER symmetry, available for adults in the literature, were applicable to young children. Our later findings indicate that some of those values will require correction for age, and cannot be considered quite precise until such age-dependent norms have been established.

Nonetheless, we consider our findings sufficiently interesting to warrant presentation here. The basic features of our findings are in good agreement with the literature surveyed earlier in this chapter, and are easily reconciled with one's expectations based upon that literature. A substantial portion of our EEG data is essentially congruent with Matoušek and Petersén's (1973a) normative spectral analyses and Harmony *et al.*'s (1973a) norms for EEG symmetry, and much of our AER data fits well with the norms of AER symmetry in adults published by Harmony *et al.* (1973b). Thus, the data to be discussed show features basically corresponding to theoretical predictions based upon the literature about both normal and abnormal electrophysiological activity. The data display a high incidence of deviations from the expected normal patterns, as will be presented in what follows.

A. Deviation from EEG and AER Norms

The normative data for the frequency composition of the EEG, illustrated in Fig. 8.1, reveal a systematic evolution of the EEG with maturation and provide the basis for objective determination of whether an individual child shows a maturational lag manifested electrophysiologically. Many authors have provided qualitative evidence that children with MBD often display an apparent excess of slow waves in the EEG, as reviewed above. For these reasons, we carried out spectral analyses on the EEG from bilateral fronto-temporal (F_7T_3/F_8T_4), temporal (T_3T_5/T_4T_6), and parieto-occipital (P_3O_1/P_4O_2) derivations of 50 children between 7 and 14 years of age, most of whom had been referred to our clinic on the basis of psychological test results suggesting an organic basis for their learning difficulties. In the great majority of cases for which EEG evaluations had been performed, the findings had been negative, or "borderline."

Since we felt that an objective criterion for abnormality was highly desirable, indeed essential for our purposes, we decided to utilize conventional statistical guidelines. For spectral analyses of spontaneous EEG, the normative data provided by Matoušek and Petersén (1973b), normalized to percentage energy, were used to define means and standard deviations for each frequency band as a function of age and derivation; for PCC and SR analyses of the symmetry of the spontaneous EEG, means and standard deviation were obtained from Harmony, Otero, *et al.* (1973a); for AER symmetry, means and standard deviations were based on the norms provided by Harmony, Ricardo, *et al.* (1973b). All of these normative data presented shortcomings, about which we had reservations; the spectral norms had been gathered on a Swedish population that could conceivably differ from a population of children in the United States because of genetic or nutritional factors; the PCC, SR, and AER symmetry norms had been gathered for a group of adults and might well differ from normal values for children. In view of these reservations, we tried to construct a composite "abnormality index" that did not rely on any single feature. Significance of deviations from norms were classified as + (.05 level), ++ (.01 level), and +++ (.001 level) for each feature, and an abnormality index was defined as the total number of pluses (positive findings) across the set of measures for each child.

Table 8.1 presents the total body of data obtained from 50 children who were examined consecutively in the Neurophysiology Clinic in 1974, all referred because they were considered at risk for minimal brain dysfunction underlying their learning disorders. Of these 50 children, 49 showed one or more abnormal EEG features, defined as a statistically significant deviation from normal percentage of delta, normal percentage of theta, PCC, or SR values of the sponta-

TABLE 8.1
Values of % Delta, % Theta, PCC, SR, and AER Symmetry for Parieto-occipital, Temporal, and Fronto-temporal Derivations in 50 Children with Learning Disorders[a,b]

		Parieto-occipital							Temporal							Fronto-temporal						
		Delta (%)		Theta (%)					Delta (%)		Theta (%)					Delta (%)		Theta (%)				
Age	Patient	L	R	L	R	PCC	SR	AER symmetry	L	R	L	R	PCC	SR	AER symmetry	L	R	L	R	PCC	SR	AER symmetry
7	1	39	25	35	25	.3	1.0	+	35	39	31	25	.3	1.2	+	4.4	59^{++}	16	12	.4	1.2	+
7	2	39	39	11	25	.3	1.4	ND	44^{+}	31	21	25	.5	1.0	ND	69^{+++}	59^{++}	15	14	ND	1.2	ND
8	3	16	31^{+}	16	16	.4	1.2	+	25	51^{+}	16	13	.2^{+}	1.2	+	69^{+++}	69^{+++}	13	11	.3	1.0	–
8	4	34^{++}	39^{+++}	35	39^{+}	.5	1.1	.72	39	39	28	35	.3	1.1	.82	44	44	35	31	.1^{++}	1.1	.74
8	5	39^{+++}	69^{+++}	23	15	ND	1.4	–	51^{+}	51^{+}	25	23	ND	1.2	–	ND	ND	ND	ND	ND	ND	–
8	6	25	35^{++}	25	21	.2^{+}	1.3	+	31	44	25	25	.2^{+}	1.3	+	44	39	28	21	.3	1.4	–
8	7	28	25	11	12	.34	1.2	–	39	51^{+}	17	15	.2^{+}	1.2	+	51^{+}	44	11	15	.3	1.0	+
8	8	44^{+++}	51^{+++}	11	11	.5	1.7^{+}	+	51^{+}	51^{+}	11	11	.3	1.0	+	59^{+++}	59^{+++}	11	13	.4	1.0	+
8	9	31^{+}	25	16	13	.4	1.5^{+}	.86	59^{+++}	51^{+}	14	13	.3	1.5	.86	69^{+++}	69^{+++}	11	10	.3	1.3	.71
8	10	25	31^{+}	16	21	.4	1.3	+	35	44	16	15	.2^{+}	1.0	ND	ND	ND	ND	ND	ND	ND	ND
8	11	25	25	15	21	.36	2.0^{++}	.21^{+++}	44	39	11	11	.2^{+}	1.6^{+}	.38^{+++}	83^{+++}	69^{+++}	6	8	.2^{+}	1.0	.63
8	12	53^{+++}	56^{+++}	28	25	.5	1.0	.95	46^{+}	45	36	34	.4	1.1	.66^{++}	ND	ND	ND	ND	ND	ND	ND
9	13	31	59^{+++}	19	10	.1^{+}	1.2	+	39^{+}	44^{+}	14	15	.1^{++}	1.4	–	59^{+++}	59^{+++}	8	8	.2^{+}	1.2	ND
9	14	39^{++}	44^{+++}	35	39	.1^{+}	1.4	+	39^{+}	35	21	31	.3	2.0^{++}	+	59^{+++}	34	9	28	.2^{+}	2.0^{++}	+
9	15	31	35^{+}	8	25	.5	1.7^{+}	+	44^{++}	39^{+}	21	21	.2^{+}	1.2	+	44	59^{+++}	11	13	13	1.0	ND
9	16	25	39^{++}	31	25	.4	1.5^{+}	+	35	51^{+++}	25	25	.3	1.7^{+}	+	51^{++}	51^{++}	25	21	.4	1.0	+
9	17	37^{++}	33^{++}	32	36	.1^{+}	ND	.53^{++}	34	44^{++}	32	29	.0^{++}	ND	.87	46^{+}	48^{+}	34	35	.3	ND	.87
9	18	20	14	23	19	.6	ND	.88	27	21	22	21	.6	ND	.80	47^{+}	40	30	28	.2^{+}	ND	.87
9	19	24	30	22	20	.4	ND	.42^{+++}	25	29	18	20	.4	ND	.65^{++}	42	38	24	26	.1^{++}	ND	.95
9	20	33	11	24	17	.5	ND	.36	39^{+}	29	28	25	.6	ND	.65^{++}	60^{+++}	61^{+++}	20	23	.4	ND	.83
9	21	26	39^{++}	27	25	.0^{++}	ND	.93	39^{+}	39^{+}	34	32	.1^{++}	ND	.86	52^{++}	39	35	39	.1^{++}	ND	.49
9	22	34^{+}	25	24	20	.6	ND	.95	24	20	32	19	.4	ND	.75	35	61^{+++}	16	21	.2^{+}	ND	.92
9	23	18	23	15	20	.3	ND	.29^{+++}	57^{+++}	41^{+}	21	22	.3	ND	–.16^{+++}	66^{+++}	59^{+++}	23	27	.2^{+}	ND	.43

9	24	30	ART	22	ART	ND	ND	.93	24	ART	21	ART	ART	ND	$.54^{+++}$	36	37	27	25	.6	1.0	$.05^{+++}$
9	25	40^{++}	ART	36	ART	.6	ND	.93	36	ART	36	ART	$.1^{++}$	ND	$.16^{+++}$	57^{+++}	56^{+++}	26	24	$.1^{++}$	ND	.41
9	26	42^{+++}	44^{+++}	26	25	.6	ND	.92	28	26	27	35	.9	ND	$.40^{+++}$	57^{+++}	38	27	33	.3	ND	.57
9	27	40^{++}	32	28	29	.4	ND	.93	36	27	33	31	.7	ND	.87	57^{+++}	61^{+++}	28	34	$.1^{++}$	ND	.71
9	28	31	38^{++}	17	27	$.1^{+}$	ND	$.29^{+++}$	38^{+}	33	14	19	.9	ND	$.43^{+++}$	34	37	23	26	$.1^{++}$	ND	.92
9	29	21	34^{+}	20	24	$.1^{+}$	ND	ND	27	31	16	37	$.2^{+}$	ND	ND	43	51^{++}	41	37	$.1^{++}$	ND	ND
9	30	21	25	13	13	.5	ND	.75	33	33	18	23	.8	ND	$.30^{+++}$	61^{+++}	53^{++}	21	22	$.1^{++}$	ND	.60
9	31	27	32	42^{+}	44^{+}	.6	ND	.82	26	32	37	38^{+}	.5	ND	$.69^{+}$	64^{+++}	56^{+++}	26	29	.8	ND	.40
10	32	44^{++}	44^{++}	25	21	.3	1.7^{+}	+	ND	ND	ND	ND	ND	ND	+	ND	ND	ND	ND	ND	ND	+
10	33	51^{+++}	51^{+++}	16	25	.7	3.5^{+++}	ND	ND	ND	ND	ND	ND	ND	ND	ND	ND	ND	ND	ND	ND	ND
10	34	35	59^{+++}	28	16	.4	1.3	ND	44^{++}	39	16	14	$.2^{+}$	1.1	ND	51^{++}	39	13	11	$.1^{++}$	1.3	ND
10	35	16	59^{+++}	35	25	$.2^{+}$	6.0^{+++}	.94	69^{+++}	69^{+++}	21	21	$.2^{+}$	2.0^{++}	$.60^{+++}$	69^{+++}	82^{+++}	11	8	ND	1.5^{+}	.81
10	36	29^{+}	39^{+}	16	16	.4	1.3	.70	59^{+++}	44^{++}	11	8	ND	1.1	$.01^{+++}$	69^{+++}	44	10	16	ND	1.1	$.03^{+++}$
10	37	29^{+}	28	25	28	.4	1.4	.71	39^{+}	35	17	21	ND	1.0	.85	ND	ND	ND	ND	ND	ND	ND
11	38	25	39^{+}	21	16	.3	1.0	+	35	44^{++}	15	21	$.2^{+}$	1.0	+	35	39^{+}	16	13	ND	ND	ND
11	39	31	39^{+}	21	21	.4	1.0	–	35	44^{++}	18	16	$.1^{++}$	1.0	–	31	21	11	10	ND	ND	–
11	40	44^{+++}	44^{+++}	25	11	.4	1.6^{+}	+	51^{+++}	51^{+++}	10	11	$.2^{+}$	1.1	+	44^{++}	51^{++}	6	8	.3	1.3	ND
11	41	21	25	21	21	.4	1.5^{+}	–	21	35	17	21	$.2^{+}$	1.0	–	19	31	21	25	.3	1.2	–
11	42	42^{++}	51^{+++}	10	10	ND	ND	ND	ND	ND	ND	ND	ND	ND	ND	ND	ND	ND	ND	ND	ND	ND
11	43	21	35^{+}	21	23	.3	1.0	.80	39^{+}	44^{++}	25	31	$.2^{+}$	1.2	.72	59^{+++}	69^{+++}	16	16	$.2^{+}$	1.5^{+}	.81
12	44	13	28	13	16	.6	1.5^{+}	–	39^{+}	35	11	15	ND	ND	+	59^{+++}	51^{+++}	8	11	$.2^{+}$	2.0^{++}	ND
13	45	28	35^{+}	13	12	$.2^{+}$	1.2	+	31	35	13	13	.3	1.0	+	31^{+}	39^{++}	17	9	ND	1.0	ND
13	46	31^{+}	34^{+}	16	17	.6	1.2	–	39	39	17	17	.3	1.2	+	69^{+++}	69^{+++}	11	13	ND	ND	ND
13	47	16	21	16	21	.5	1.0	–	39	39	21	21	.3	1.5	+	21	15	51	15	$.2^{+}$	1.2	–
14	48	21	51^{+++}	25	21	.6	2.0^{++}	+	16	28	19	21	$.2^{+}$	3.0^{+++}	+	ND	ND	ND	ND	ND	ND	+
14	49	39	51^{+++}	21	14	$.2^{+}$	6.3^{+++}	+	44^{+++}	39^{++}	17	25	$.1^{++}$	1.8	+	59^{+++}	35^{+}	10	11	$.1^{++}$	1.5^{+}	ND
14	50	11	8	13	11	.7	1.1	$.62^{+}$	44^{+++}	21	11	16	.3	1.2	.74	39^{++}	39^{++}	17	18	.4	1.3	.68
Total:	+	5	10	1	2	9	8	16	12	8	0	1	14	2	18	4	3	0	0	9	3	7
	++	7	6	0	0	1	2	1	2	7	0	0	4	2	3	5	8	0	0	10	2	6
	+++	6	14	0	0	0	3	4	7	3	0	0	0	1	9	20	16	0	0	0	0	2
Grand Totals:		18	30	1	2	10	13	21	21	18	0	1	18	5	30	29	27	0	0	19	5	9

[a] ART = excessive artifacts made analysis impossible.
[b] ND = no data.

neous EEG. In 6 of these children, the observed abnormalities were only significant at the .05 level. All the remainder showed abnormalities significant at the .01 or .001 level.

Examining Table 8.1, we see that the most common abnormality in the EEG was excessive slow wave in the delta frequency band. In the parieto-occipital region, only one child shows a significant excess of theta activity in the *absence* of excessive delta waves, and only 2 children show excessive theta activity which reaches the .05 level. In contrast, 35 children displayed a statistically significant excess of delta activity. In the temporal region, only one child showed excessive theta activity in the absence of excessive delta, while 27 children showed an excess of delta activity. In the fronto-temporal derivations, no child showed an excess of delta activity. *Forty-six of the children showed a significant excess of delta activity in at least one brain region.*

Ten children showed an abnormal PCC in parietooccipital derivations, 20 in temporal regions, and 19 in fronto-temporal leads. There were 13 children who showed an abnormal SR in parieto-occipital leads, 5 in temporal, and 5 in fronto-temporal derivations. A total of 34 children showed an abnormal AER symmetry in at least one head region; 21 in parieto-occipital, 30 in temporal, and 9 in frontal derivations.

Every one of the 50 children showed some abnormal features in the EEG and/or the AER. The details are discussed further below.

B. Age-Dependent Delta Quotient

Examination of the data in Table 8.1 shows clearly that these children with learning disorders commonly showed an excess of slow activity in the delta frequency band, while abnormal energy in the theta band alone was rare. Accordingly, it seemed desirable to construct a metric for the extent of the abnormal delta activity, ignoring activity in other bands. In analogy to the age-dependent quotient used so successfully by Matoušek and Petersén (1973a) in their studies on tumor localization (see Chapter 3), we defined the age-dependent delta quotient as the ratio of the percentage of delta activity normal for a given head region in a healthy individual of a specific age to the amount of delta activity actually observed in the corresponding head region of a patient with a learning disorder. Table 8.2 shows the result of such normalization.

The criterion values for an ADQ (δ) which was significantly abnormal at the .05, .01, and .001 levels were obtained by calculating the ADQ for the average value divided by the average plus 2 standard deviations (.05 level), 2.5 standard deviations (.01 level), and 3 standard deviations (.001 level). The resulting numerical criteria are shown in Table 8.2 for each age and head region. Slight discrepancies exist between Tables 8.1 and 8.2 due to round-off errors.

Examination of Table 8.2 shows that the values of ADQ (δ) significant at the .05 level vary from .67 to .52, with one exception due to an unusually large

TABLE 8.2

$$ADQ\ (\delta) = \frac{\text{Delta (Average \% for Age and Derivation)}}{\text{Delta (Patient \% for Age and Derivation)}}$$

Age	No.	P	P_3O_1 L	P_4O_2 R	P	T_3T_5 L	T_4T_6 R	F_7T_3	F_8T_4 L	R
7	1	.05=.58	.63	.98	.05=.57	.73	.65	.05=.57	.69	.51^{++}
	2	.01=.52	.63	.63	.01=.52	.58	.87	.01=.52	.44^{+++}	.51^{++}
		.001=.47			.001=.47			.001=.47		
8	3	.05=.68	1.34	.69	.05=.52	.93	.53^{+}	.05=.64	.45^{+++}	.45^{+++}
	4	.01=.63	.63^{++}	.55^{+++}	.01=.47	.63	.63	.01=.59	.71	.71
	5	.001=.58	.55^{+++}	.31^{+++}	.001=.42	.48^{+}	.48^{+}	.001=.54	–	–
	6		.86	.61^{++}		.73	.56		.71	.80
	7		.74	.86		.63	.48		.61^{+}	.71
	8		.49^{+++}	.42^{+++}		.48^{+}	.48^{++}		.53^{++}	.53^{++}
	9		.69	.86		.42^{+++}	.43^{+}		.45^{+++}	.45^{+++}
	10		.86	.69		.70	.56		–	–
	11		.86	.85		.56	.63		.38^{+++}	.45^{+++}
	12		.40^{+++}	.38^{+++}		.53	.55		–	–
9	13	.05=.58	.64	.34^{+++}	.05=.58	.51^{++}	.50^{++}	.05=.62	.47^{+++}	.47^{+++}
	14	.01=.53	.50^{++}	.45^{+++}	.01=.52	.51^{++}	.63	.01=.56	.47^{+++}	.82
	15	.001=.48	.64	.53^{+}	.001=.48	.50^{++}	.51^{++}	.001=.52	.63	.47^{+++}
	16		.86	.50^{++}		.63	.43^{+++}		.55^{++}	.55^{++}
	17		.54^{+}	.52^{++}		.64	.50^{++}		.60	.58^{+}
	18		.99	1.41		.82	1.04		.50^{+}	.70
	19		.89	.66		.83	.76		.66	.73
	20		.60	1.80		.51^{++}	.76		.46^{+++}	.46^{+++}
	21		.76	.50^{++}		.51^{++}	.51^{++}		.53^{++}	.71
	22		.58^{+}	.80		.91	1.10		.80	.46^{+++}
	23		1.10	.86		.38^{+++}	.54^{+}		.42^{+++}	.47^{+++}
	24		.66	–		.91	–		.77	.75
	25		.50^{++}	–		.61	–		.49^{+++}	.50^{+++}
	26		.45^{+++}	.45^{+++}		.79	.85		.49^{+++}	.73
	27		.50^{++}	.42		.61	.81		.44^{+++}	.46^{+++}
	28		.64	.52^{++}		.58^{+}	.66		.82	.75

contd.

TABLE 8.2 *(contd.)*

Age	No.	P	P_3O_1 L	P_4O_2 R	P	T_3T_5 L	T_4T_6 R	F_7T_3	F_8T_4 L	R
	29		.95	.58^{+}		.81	.71		.65	.94^{++}
	30		.85	.80		.67	.67		.46^{+++}	.53^{++}
	31		.74	.62		.84	.63		.43^{+++}	.50^{+++}
10	32	.05=.52	.47^{++}	.47^{++}	.05=.57	–	–	.05=.62	–	–
	33	.01=.47	.41^{+++}	.41^{+++}	.01=.52	–	–	.01=.59	–	–
	34	.001=.42	.59	.35^{+++}	.001=.47	.51^{++}	.58	.001=.52	.57^{++}	.75
	35		1.30	.35^{+++}		.33^{+++}	.33^{+++}		.43^{+++}	.36^{+++}
	36		.53	.53		.38^{+++}	.51^{++}		.43^{+++}	.67
	37		.53	.75		.58	.65		–	–
11	38	.05=.55	.79	.50^{+}	.05=.60	.66	.52	.05=.64	.81	.73
	39	.01=.49	.63	.50^{+}	.01=.54	.66	.52	.01=.59	.92	1.35
	40	.001=.45	.45^{+++}	.45^{+++}	.001=.49	.45	.45	.001=.54	.65	.56^{++}
	41		.94	.79		1.10	.66		1.50^{++}	.92
	42		.47^{++}	.39^{++}		–	–		–	–
	43		.34	.56		.59	.52		.48^{+++}	.41^{+++}
12	44	.05=.56	1.40^{++}	.65	.05=.59	.52^{++}	.63	.05=.66	.46^{+++}	.53^{+++}
		.01=.50			.01=.54			.01=.61		
		.001=.46			.001=.49			.001=.57		
13	45		.62	.50^{++}		.78	.70		.86	.60
	46	.05=.56	.56^{+}	.51^{++}	.05=.46	.62	.62	.05=.60	.39^{+++}	.39^{+++}
	47	.01=.52	1.10	.83	.01=.37	1.16	.62	.01=.55	1.26	1.78^{++}
		001=.46		.83	.001=.29			.001=.50		
14	48	.05=.57	.89	.37^{+++}	.05=.56	1.26	.72	.05=.57	–	–
	49	01=.51	.48^{++}	.37^{+++}	.01=.51	.46	.52	.01=.51	.44^{+++}	.74
	50	.001=.47	1.70^{+++}	2.35^{+++}	.001=.46	.46	.96	.001=.46	.67	.67

LEGEND: Values for ratio of average % delta (age derivation) to patient % delta (age, derivation). These values represent a normalization of the percentage of activity in the delta band and can be considered as an Age Dependent Quotient for delta activity [ADQ (δ)]. The .05, .01, and .001 values for each age and derivation were calculated by using the average values for each age and derivation, plus 2 standard deviations (.05 level), 2.5 standard deviations (.01 level) and 3 standard deviations (.001 level), as the denominators to determine the corresponding criterion values for ADQ (δ).

standard deviation (13-year-olds, temporal region). A shift of .10 in ADQ (δ) below this criterion carries the significance level to .001 in almost every case. A total of 31 children show an abnormal ADQ (δ) in parieto-occipital regions, 18 in temporal, and 34 in fronto-temporal derivations; 46 children show an abnormal ADQ (δ) in at least one head region.

A wide variety of patterns of abnormal delta activity exists in this population. Fifteen children showed abnormality in all three head regions (parieto-occipital, temporal, and fronto-temporal). Abnormalities in parieto-occipital alone, fronto-temporal alone, fronto-temporal *plus* temporal, fronto-temporal *plus* parieto-occipital, or parieto-occipital *plus* temporal are evident in 31 children, and are found with approximately equal incidence, about one-half to one-third as often as abnormalities involving all three regions. No instance was observed of an abnormality in spontaneous EEG in the temporal regions in isolation. This is particularly interesting in view of the widespread language and reading difficulties displayed by this group of children.

C. Abnormality Index

In order to provide some safeguard against erroneous conclusions which might be drawn from this pilot study because of possible inappropriateness of the normative data used to construct our criteria, a composite abnormality index was devised to reflect the overall electrophysiological abnormalities found in this population of children with learning disorders. The abnormality index reflected the results of computation of the ADQ (δ), PCC, and SR for the spontaneous EEG, as well as symmetry measures for the AER elicited by unpatterned flashes. Each item was scored for each head region as – (normal), + (significant at the .05 level), ++ (significant at the .01 level), or +++ (significant at the .001 level). The abnormality index was defined as the total number of pluses. The results of this evaluation of abnormal EEG and AER features are presented in Table 8.3.

It can be seen that the values of the abnormality index for this group of children with learning disorders range from 2 to 16, with a mean value of 7.4 and a median value of 7. Since multiple abnormal features were found in *every* child in the group, and since the different features represent separate and presumably independent measures, it seems highly unlikely that these results merely represent a concatenated series of errors arising from the inappropriateness of the reference norms. This is particularly unlikely when one realizes that 35 of these children showed at least one neurometric feature abnormal at the .001 level. It is doubtful that this high incidence of extreme deviations from norms in the literature arose because all of the norms were inapplicable, for reasons not yet clear. Although this seems improbable, these results must nonetheless be considered as tentative until norms that are directly applicable

TABLE 8.3

Sum of Abnormal EEG and AER Features in 50 Children with Learning Disorders—The 'Abnormality Index'

		EEG features					AER features						EEG features					AER features			
		Spectral analysis (δ)			Symmetry		Symmetry						Spectral analysis (δ)			Symmetry		Symmetry			
Age	Patient No.	PO	T	F	PCC	SR	PO	T	F	Abnormality index	Age	Patient No.	PO	T	F	PCC	SR	PO	T	F	Abnormality index
7	1	–	–	++	–	–	+	+	+	5	9	26	+++	–	+++	–	ND	–	+++	–	9
	2	–	+	+++	–	–	ND	ND	ND	4		27	++	–	+++	++	ND	–	–	–	7
8	3	+	+	+++	+	–	+	+	–	8		28	++	+	–	+	ND	+++	+++	–	10
	4	+++	–	–	++	–	–	–	–	5		29	+	–	++	+	ND	ND	ND	ND	4
	5	+++	+	ND	ND	–	–	–	–	4		30	–	–	+++	++	ND	–	+++	–	8
	6	++	–	–	+	–	+	+	–	5		31	–	–	+++	–	ND	–	+	–	4
	7	–	+	+	+	–	–	+	+	5		32	++	ND	ND	–	+	+	+	+	6
	8	+++	+	+++	–	+	+	+	+	11		33	+++	ND	ND	–	+++	ND	ND	ND	6
	9	+	+++	+++	–	+	–	–	–	8		34	+++	++	++	++	–	ND	ND	ND	9
	10	+	–	ND	+	–	+	ND	ND	3		35	+++	+++	+++	+	+++	–	+++	–	16
	11	–	–	+++	+	++	+++	+++	–	12		36	+	+++	+++	–	–	–	+++	+++	13
	12	+++	+	ND	–	–	–	++	ND	6		37	+	+	ND	–	–	–	–	–	2
9	13	+++	++	+++	++	–	+	–	ND	11		38	+	++	+	+	–	+	+	ND	7

14	+++	+	+++	+	++	+	+	+	13
15	+	++	+++	+	+	+	+	ND	10
16	++	+++	++	–	+	+	+	+	11
17	++	++	+	+	ND	++	–	–	8
18	–	–	+	+	ND	–	–	–	2
19	–	–	–	++	ND	+++	++	–	7
20	–	+	+++	–	ND	–	++	–	6
21	++	+	++	++	ND	–	–	–	7
22	+	–	+++	+	ND	–	–	–	5
23	–	+++	+++	+	ND	+++	+++	–	13
24	–	–	–	–	–	–	+++	+++	6
25	++	–	+++	++	ND	–	+++	–	10
+	5	9	3	11	4	9	8	5	
++	5	3	3	5	2	1	3	0	
+++	6	3	12	0	0	3	4	1	

39	+	++	–	++	–	–	–	–	5	
40	+++	+++	++	+	+	+	+	ND	12	
41	–	–	–	+	+	–	–	–	–2	
42	+++	ND	ND	ND	ND	ND	ND	ND	3	
43	+	++	+++	+	+	–	–	–	8	
44	–	+	+++	–	++	–	+	ND	7	
45	+	–	++	+	–	+	+	ND	6	
46	+	–	+++	–	–	–	+	ND	5	
47	–	–	–	+	–	–	+	–	2	
48	+++	–	ND	+	++	+	+	+	10	
49	+++	+++	+++	++	+++	+	+	ND	16	
50	–	+++	++	–	–	+	–	–	6	
+	8	3	1	10	4	7	10	2	Totals	+ 98
++	3	4	5	5	1	0	0	0		++ 40
+++	8	5	10	0	4	1	5	1		+++ 63
Totals	35	27	34	31	15	21	30	8		201 abnormal indices

NOTE: Abnormal EEG and AER features in 50 children with learning disorders. A crude abnormality index was constructed by counting a feature as *1* if it deviated from the norm at 0.05 level, *2* if it reached the .01 level, *3* if it reached the .001 level.

can be constructed. Studies with that purpose are now in progress, and preliminary results are presented later in this chapter.

D. Conclusions

These findings represent our initial effort to apply neurometric methods, already demonstrated as extremely powerful for the identification of gross neuropathology and senile deterioration, to the more subtle problems of assessment of brain dysfunction in children with learning disorders. The results justified cautious optimism and suggested that energetic development of these techniques might be extremely useful. The data revealed widespread signs of brain dysfunction in all of the 50 children in our study, representing 50 consecutive cases with learning disorders referred for examination to the Neurophysiology Clinic. At the very least, this would seem to indicate that the wide variety of individuals making these referrals, including neurologists, pediatricians, and school psychologists, were using a sufficiently shared set of criteria so that all or most of the children selected by those criteria had demonstrable brain dysfunction. At the same time, all of these children were sufficiently handicapped to be considered enough at risk for brain damage that neurophysiological evaluation was recommended. Therefore, this sample probably represents relatively extreme instances of dysfunction rather than typical cases. The data permit no estimate as to the percentage of children with learning disabilities who would display comparable dysfunctions. Since many of the children in our sample had previously been judged as normal on the basis of conventional neurological and/or EEG examinations, it seems clear that neurometric methods might provide objective signs of brain dysfunction in many of the millions of children with learning disabilities who do not seem organically impaired with conventional means of assessment.

No single pattern emerged in this sample of 50 learning disabled children. Many different profiles of dysfunction were apparent, ranging from excessive slow waves generally distributed over all head regions to slow waves in various combinations of different head regions to slow waves in only a single head region. On occasion, symmetric and reactive AERs were obtained from some head regions with asymmetric slow waves, suggesting that the dysfunction responsible for the abnormality in the spontaneous EEG was not blocking the ability of the region to respond differentially to incoming sensory information. In other cases, asymmetric or nonreactive AERs were obtained from brain regions in which no abnormalities were discerned in the spectral analyses. Much effort will be needed to relate each of this wide variety of different neurometric profiles to patterns of individual learning and behavioral performance, to learn which neurometric profiles are systematically associated with meaningful functional deficits and which are functionally of little significance, and most important, whether specific therapeutic procedures are differentially effective with particular profiles of dysfunction.

Two features were seen most commonly in this group of children and were often associated. First, a very large number (70%) of these learning disabled children showed a clear excess of posterior (parietooccipital) slow waves, confirming numerous similar findings cited in the literature reviewed early in this chapter. These children might be expected to show difficulty in processing visual information, especially in view of the studies implicating parietal regions in mediation of language-related abstractions about visual input (e.g., Grinberg & John, unpublished results, 1974). Second, many of these children (68%) showed a great excess of slow activity in frontal regions, suggesting that they might well display impairment of the ability to control impulsive behavior and to plan and perform systematically organized tasks. Certainly, both of the neurometric features most frequently found abnormal in these children have considerable face validity, corresponding to major behavioral features of the minimal brain dysfunction syndrome. It is noteworthy that not one case of excessive slow wave was found in the temporal lobes *alone*, unaccompanied by other abnormal findings.

It must be emphasized that the results presented in Section II of this chapter revealed widespread dysfunction using measures of brain activity which must be considered relatively gross, that is, the organization of spontaneous EEG rhythms and AERs to blank flashes constituting no challenge of functions involved in information processing. When children with learning dysfunctions are presented with challenges requiring the processing of information, more subtle abnormalities of brain activity that cannot be inferred from these gross measures often become glaringly evident. For example, we have seen cases in which the evoked potential from certain brain regions actually became almost isopotential when stimuli related to letters of the alphabet were presented, although apparently normal responses were elicited in the same region by geometric patterns or blank flashes!

III. COMPARISON BETWEEN PREVIOUSLY CATEGORIZED GROUPS OF "NORMAL" AND "LEARNING DISABLED" CHILDREN USING THE NEUROMETRIC TEST BATTERY (NB)[1]

During the same period in which we were evaluating the utility of the small set of selected neurometric indices used in our Neurophysiology Clinic for the assessment of children with learning problems, as described above, we were engaged in constructing the DEDAAS system, developing the neurometric test

[1]The original research reported in this section was sponsored by the RANN Program (Research Applied to National Needs) of the National Science Foundation under Grant No. GI 34946 to E. R. John. It is a pleasure to acknowledge the collaboration of Hansook Ahn, Dan Brown, Ronald Siegel, Michaela Lobel, Marion Howard, Margo Marek, Jane Kloecker, and Drs. Paul Easton, Miriam John, Herbert Kaye, and Bernard Karmel in the interdisciplinary team that carried out these studies.

battery (NB) and selecting a standardized battery of psychometric tests generally considered to reflect perceptual or cognitive disorders related to brain dysfunction. When this task was finished, we began to investigate the relationships between the full set of measures derived from the conditions and challenges of the neurometric test battery and the measures available from the psychometric instruments, and to carry out a more systematic and thorough examination of the extent to which these neurometric and psychometric indices might reveal differences within a reasonably large sample of normal children and children considered at risk for learning disability owing to the nature and amount of difficulty displayed with learning tasks. In this last section of Chapter 8, we will present the results thus far obtained in this endeavor.

A. Subjects, Behavioral Methods, and Criteria for Disability

1. Description of Subjects

In order to minimize the variance that might be introduced into our measures due to factors related to chronological age or sex, this initial study was restricted to male children between the ages of 7 years and 6 months through 9 years and 6 months. This age group was selected because of the widespread belief that the behavioral consequences of learning disability become fairly well evident in the majority of cases by the time a child reaches the third-grade level. The sample studied consisted of a total of 175 children. Of these children, 118 were preclassified as "normal," based upon the fact that neither their parents nor their teachers expressed any concern about their learning capabilities. These children were recruited as paid volunteers from the local community and schools in the metropolitan New York area, using letters to parents and school psychologists together with follow-up phone calls. A total of 57 children were preclassified as "learning disabled," based upon the fact that they were performing about two grades below level in a major component of their school activity and were considered at risk for brain damage on the basis of psychological evaluations performed by school personnel. These children were recruited on the basis of contacts provided by psychologists in schools in the New York area, the Pediatric Development Evaluation Center of Grasslands Hospital in Westchester County, and the Learning Disability Clinic of the Department of Pediatrics of New York Medical College at Metropolitan Hospital in New York.

2. Psychometric Battery and the Composite Dysfunction (3M) Score

The psychometric battery administered to these subjects included the following standardized instruments or subscales: The Wechsler Intelligence Scale for Children (WISC); the Bender Gestalt Test; the Peabody Picture Vocabulary Test (PPVT); the Wide Range Achievement Test (WRAT); the memory subtests from the McCarthy Scales; selected subtests from the Illinois Test of Perceptual Ability (ITPA); the Piaget Test; the Birch Test of Analysis and Synthesis of Forms (BT); the Gesell Incomplete Man Test; the Roswell–Chall Diagnostic Test (RCDT); and tests of eye, hand, and foot dominance (DOM). In addition, the New York City Board of Education provided access, for those of our subjects who attended the New York City public schools, to performance on the standardized Metropolitan Achievement Tests, which are regularly administered to children in the New York City school system. Socioeconomic status (SES) was assessed by a questionnaire or estimated on the basis of available demographic information. Testing was carried out in several sessions, usually extending across two different days. On one of these days, the NB measures were also obtained. Test sessions, whether psychometric or NB, could be interrupted whenever a child indicated the desire to "take a break."

Many children, especially those in the LD subgroup, became restless before the full NB could be administered. In order to minimize any possible discomfort for such children, the NB was terminated when such distress was observed. As a result, fairly large sample sizes were obtained for those NB conditions and challenges presented early in the testing situation, and this sample size diminished for the later measures. In the later stages of our study, children were subjected to the EEG conditions and challenges of the NB, a few "standard" conditions and challenges, and then received those AER conditions and challenges for which the sample size was still inadequate before those for which sufficient data had already been accumulated. This process continued, with continuous reordering of the sequence of AER conditions in the NB as a function of the sample size already obtained for each condition, until data had been obtained for every AER condition and challenge from at least 20 normal and 20 LD children.

A composite dysfunction score, which will be referred to subsequently as the *3M score*, was derived by assigning relative weights to selected subscales of the instruments listed above. These weights were considered to possess construct validity, particularly in defining learning disability on the basis of the psychological literature. The 3M score spanned a range from 0 to 16 and was defined as in Table 8.4.

One of our goals in this study was to examine the relationship between the various instruments in the psychological test battery, generally considered to be sensitive to learning disability, and school performance, on the one hand, and neurometric assessment of brain dysfunction on the other hand. By combining those subscales of the instruments that showed a reliable relation to observed abnormalities of brain function as well as to academic performance, and by discarding those items that failed to display any such relation, it might be possible to construct a new composite test which would be much more rapid to administer and score while at the same time providing a more accurate indication of brain dysfunction than the original psychometric test battery.

TABLE 8.4
Definitions of the 3M Scale of Brain Dysfunction

Scores on instrument	Value Assigned on Composite Score 0	1	2
Bender-Gestalt:[a]			
Koppitz scores	0–2	3 (+ 1 pt for 1 or 2 signs)	4 and up (+ 2 pts for more than 2 signs)
WRAT:			
Spelling & reading		Low in 1 area[b]	Low in 2 areas[b]
WISC			
Block design			
Scaled scores	8 or above	7	6
Object assembly			
scaled scores	8 or above	7	6
Coding			
Scaled scores	8 or above	7	6
Digit span			
Scaled scores	8 or above	7	6 (+ 1 pt for 4 pt discrepancy between F & B)
McCarthy:[c]			
Memory tests	5,5,14,8	Low in 1 area	Low in 2 areas
PPVT			
Expressive scores	18–22 (1–4 errors)	16–17 (5 or 6 errors)	15 and below (7 or more errors)
Receptive scores		80 or below	70 and below
ITPA			
Auditory Association		2 years below	
Visual association		2 years below	

[a]On Bender, in addition to points given on the basis of Koppitz scores, additional points are added if signs specifically correlated with brain dysfunction are found.

[b]Low means 1.6 years below grade level.

[c]On McCarthy, scores refer to Pictorial Memory, Tapping Sequence, Verbal Memory I, and Verbal Memory II in this order. Perfect scores would be 6, 9, 15, and 11.

3. *The Criterion Problem*

It must be readily apparent to the reader, as it became painfully obvious to us, that perhaps the most serious difficulty in the attempt to identify neurometric correlates of brain function related to learning disability is the absence of a reliable independent indication that any particular child with a learning problem is actually learning disabled. Since the functional significance of NB measures has not yet been established, it would not be legitimate to assert that any child with some NB measure significantly deviant from the average value has thereby been identified as learning disabled. The NB measure cannot be assumed a priori to be relevant to brain processes involved in learning. Establishing which of the NB measures constitute valid indices of brain dysfunction with functional implications for learning capability is precisely the problem. Tautological answers can only be avoided by correlating NB measures with an independent and valid measure of learning impairment of organic origin. If such an unequivocal independent measure existed, there would be no need for direct assessment of brain dysfunction except insofar as such assessment might provide useful insights into possible individualized prescriptive therapies.

Unfortunately, not even the most ardent proponents of psychological testing would contend that incontrovertible evidence demonstrates that any presently available psychological instrument unfailingly provides an unequivocal index of brain dysfunction responsible for learning disability. While strong indication of such brain dysfunction is plausibly inferred in cases with severe impairment, more subtle damage such as that assumed to exist in minimal brain dysfunction is only tentatively identifiable. The incidence of false positive and false negative findings is not objectively ascertainable. Thus, psychometric measures, except in extreme instances, cannot be relied upon as indicators of an organic basis for learning difficulties. Many psychometric instruments merely substantiate the presence of a deficiency in prior learning.

"Real-life" behavior, as manifested by performance in the school, is even less reliable. Whether a child is considered to have learning difficulty, and whether that difficulty results in referral to a school psychologist for psychometric evaluation or chastisement for unruly behavior or inadequate performance depends upon the sensitivity of the teacher, the academic level of the school, and a variety of cultural and socioeconomic factors. Learning difficulty itself is hard to define. Quite aside from the distinction between learning difficulty and lack of motivation or emotional distress, whether or not a child is learning at a rate and level appropriate for that age depends heavily upon the standards by which the child is judged. For example, examination of the norms for different schools in the New York City public school system, based upon average performance on the standardized Metropolitan Achievement Test, reveals that entire schools might be considered "learning disordered" because they fall more than one grade level below the norms for the school system as a whole. The standard deviation describing interschool variability on these tests is large. Thus, a child considered learning disordered or disabled in one school might be considered perfectly normal in another or might actually be learning disabled.

We found that the correlation in our sample of children between academic performance as assessed in the school by the Metropolitan Achievement Test and as assessed by our psychological team using the Wide Range Achievement Test was .7, casting grave doubt on whether preclassification of children as normal or learning disabled based upon the opinion of the referral source from which the child entered our study could be accepted. Some children preclassified as normal might well be learning disabled and some children preclassified as learning disabled might well be normal!

For these reasons, we explored the results of defining groups to be compared within our total population in a number of different ways, as we sought for reliable relationships between the NB measures and learning impairment or learning disability. We separated children on the following bases: preclassification as normal or LD, scores on the psychometric test battery, the 3M composite dysfunction score alone, preclassification plus cultural membership, preclassification as normal plus no psychometric or 3M indication (3M score $\leqslant 2$) of abnormality (clear normal), preclassification as LD plus clear psychometric or 3M confirmation (3M score $\geqslant 6$) of abnormality (clear LD), disagreement between preclassification and psychometric or 3M evaluation (doubtful normal or doubtful LD), and preclassification plus psychometric or 3M score plus cultural membership. For some purposes, groups defined as above were further subdivided into 8-year-olds (7 years and 6 months to 8 years and 6 months) and 9-year-olds (8 years and 6 months to 9 years and 6 months).

As we examined the results of comparing groups defined in these different ways, it became perfectly clear that although statistically significant differences could be demonstrated between these various subgroups, the *nature* of the differences depended upon the a priori categorization. Thus, a priori categorization biases the results of any attempt at accurate or objective classification of children according to characteristic patterns of brain function, superimposing upon the natural patterns of covariance of features within the population an overriding distortion that depends on the criteria chosen for the initial categorization of potentially different subgroups. Since the "correct" criteria are unknown, a definitive classification of the distinctive patterns of brain dysfunction which characterize significantly different subgroups within the learning disabled population, or for that matter within any of the heterogeneous populations in which evaluation of brain dysfunction is essential for diagnosis and proper treatment, can only be accomplished by the use of data analysis methods that reveal the natural clusters of functional features inherent in the data without biases imposed by any a priori categorization. Such methods, in our opinion, are most likely to arise from the field of numerical taxonomy, in which highly sophisticated techniques have been and are being developed for this purpose. An excellent overview of this field is provided by the recent volume on numerical taxonomy presented by Sneath and Sokal (1973). The numerical indices of brain function yielded by the NB constitute the necessary prerequisite for application of these multivariate analytic methods in this area of inquiry.

Multiple analyses of variance have been used to compare groups of children according to their prelabeled classification, with respect to the waveshapes elicited in every EP condition and challenge of the NB. Multiple discriminant analyses have been performed which separated normal and LD subgroups, defined in various ways, using the set of psychometric measures, indices derived from the eyes open and eyes closed EEG conditions and the eyes open minus eyes closed challenge of the NB, a set of indices derived from some AER conditions and challenges of the NB, and indices derived from the combination of certain EEG and AER conditions and challenges. Using indices derived from EEG conditions and challenges, additional multiple discriminant analyses have been performed for classifications based upon age or WISC IQ. Canonical correlations have been computed between various psychometric and EEG measures.

The results show that (1) in spite of the shortcomings of a priori categorization, NB measures are extremely sensitive indices of brain dysfunction in children with learning disabilities, and (2) clear differences exist between carefully defined subgroups of learning disabled children.

4. Need to Compress Measure Sets

The statistically oriented reader will have realized long since that we suffer from a superabundance of riches; the number of psychometric and neurometric variables gathered in this study was orders of magnitude greater than the number of subjects. Before meaningful analyses of variance or multiple discriminant functions could be computed, it was necessary to compress this data set drastically so as to reduce the variables entering into such multivariate analyses to a number somewhat smaller than the number of subjects. Not only was there substantial redundancy in the measure set, but many variables were expected to be quite useless for discriminating between normal and LD subgroups. Unless these measures were removed, they would contribute so much random variance to the overall data space that detection of significant effects might well become impossible. The following sections detail the tedious but essential procedures followed to compress the data to a meaningful and manageable residue of worthwhile measures.

B. Factor Analysis of Psychometric Measures

As the initial step in data analysis and reduction, an analysis of variance was performed on the scores derived from the various subtests comprising the instruments in our psychometric test battery. Twenty-four of these variables showed a significant relationship to learning disability as assessed by the preclassification of our subjects. The following scores were found to discriminate between the normal and learning disabled subgroups:

1. *Weschler Intelligence Scale for Children*: verbal IQ (IQV), performance IQ (IQP), full scale IQ (IQFS), information (INFO), comprehension (COMP), arithmetic (ARITH), similarities (SIM), vocabulary (VOCAB), digit span (DIG), coding (COD), digit span backward (DIGBACK), picture completion (PICOM).
2. *Wide Range Achievement Test*: spelling (SP), arithmetic (WRARITH), reading (WORD).
3. *McCarthy Scales*: tapping sequence (TAP), verbal memory I (VMEM I), verbal memory II (VMEM II).
4. *Peabody Picture Vocabulary Test*: expressive vocabulary (EXPVOC), receptive vocabulary (RECVOC).
5. *Illinois Test of Perceptual Ability*: auditory association (AUDASS), visual association (VISASS).
6. *Bender Gestalt Test*: score by Koppitz method.
7. *Composite Dysfunction Scale*: total score (3M).

The scores on these variables for our full sample were then subjected to principal components factor analysis, followed by a Varimax rotation. The remaining psychometric variables showed no significant relevance for the discrimination between normal and learning disabled children, and were therefore not included in the factor analysis. Four factors were found to account for the variance of this set of measures. The loading coefficients describing the contribution of each factor to each of the discriminating psychometric variables is shown in Table 8.5.

Examination of the factor loadings in Table 8.5 shows that Factor 1 loads especially heavily on verbal and full scale IQ scores from the WISC, and on the Information, Comprehension, Similarities and Vocabulary subtests of the WISC. Thus, Factor 1 seems to be related to general IQ and language-related abstractions. The F ratio obtained for this factor in an analysis of variance with respect to preclassification as "normal" or "learning disabled" was .70, which was not significant. Factor 4 loads heavily upon the expressive and receptive vocabulary scores of the PPVT and on the auditory and visual association scores of the ITPA. This factor seems to reflect visual and auditory memory. The F ratio between normal and LD yielded by an analysis of variance was 1.32 for this factor, which was not significant.

Factor 3 loads heavily on achievement and numerical memory scores from the WRAT and WISC. This factor relates significantly at the 0.01 level to the difference between normal and LD, with an F ratio of 4.11. Factor 2 loads heavily upon performance and full-scale IQ scores from the WISC, on the coding score from the WISC, on the Bender-Gestalt score and on the composite dysfunction or 3M score. This factor is by far the most significant with respect

TABLE 8.5
Loading Coefficients of Factors on Discriminating Psychometric Tests

Test[a]	Factor 1	Factor 2	Factor 3	Factor 4
WISC–IQV	.86	.32	.37	.15
IQP	.41	.80	.03	.20
IQFS	.72	.63	.23	.20
INFO	.71	.24	.24	.25
COMP	.72	.18	.05	.17
ARITH	.45	.47	.35	.12
SIM	.68	.18	.26	.13
VOCAB	.81	.09	.13	.26
DIG	.25	.32	.60	–.04
COD	.07	.54	.11	–.01
DIGBACK	.10	.26	.54	–.08
PICOM	.42	.43	.07	.24
WRAT–SP	.25	.06	.73	.28
WRARITH	.10	.36	.51	.23
WORD	.35	.06	.73	.36
McCARTHY–TAP	–.08	.27	.33	.23
VMEM I	.16	–.05	.35	.25
VMEM II	.26	–.05	.13	.22
PPVT EXPVOC	.37	.10	.19	.52
RECVOC	.42	.15	.11	.62
ITPA AUDASS	.46	.31	.16	.63
VISASS	.09	.29	.12	.49
BENDER-GESTALT	–.11	–.52	–.26	–.24
COMPOSITE 3M	–.33	–.59	–.35	–.38

[a] Abbreviations are defined in the text.

to the differentiation between normal and learning disabled, with an F ratio of 12.

Once this factor structure and loading coefficients were ascertained, the four corresponding factor scores were calculated for each subject and added to the initial list of 24 variables. Further analysis of the psychometric data, such as the conputation of discriminant functions, were based upon this total set of 28 variables.

C. Factor Analysis of Neurometric Indices Extracted from the EEG Conditions of the NB

For each derivation of the 10/20 system, the indices extracted from every EEG condition of the NB included the absolute power and its standard deviation and the relative power and its standard deviation in each frequency band. For all homologous pairs, the amplitude ratio and coherence in each band were also calculated. Since there were several EEG conditions and challenges, these measures represented many thousands of numbers. In order to evaluate this enormous volume of data, several steps of data analysis and comparison were essential.

1. Factor Analysis of the Frequency Spectrum

As the first step toward reduction of this mass of data into a manageable distillate, we submitted the results of frequency analysis of the resting EEG from all leads of the 10/20 system to factor analysis. The results of this factor analysis revealed that two factors accounted for most of the variance in the whole body of frequency data. The first factor, referred to as the "slow" factor (S), displayed its major loadings on the low delta, high delta and theta bands. The second factor, referred to as the "fast" factor (F), displayed loadings on the alpha, low beta and high beta bands.

2. Construction of a Compressed Set of Spectral Values

The next step in data reduction was to construct a compressed set of spectral values. For each derivation, 8 variables were constructed: S_1 (eyes open, absolute power), S_2 (eyes closed, absolute power), S_3 (eyes open, relative power), S_4 (eyes closed, relative power), F_1 (eyes open, absolute power), F_2 (eyes closed, absolute power), F_3 (eyes open, relative power), F_4 (eyes closed, relative power). Each of these variables was defined by multiplying the power in the appropriate bands by the corresponding loading coefficient for that factor.

3. Factor Analyses of the 10/20 System

Four separate factor analyses were now carried out, using the principal component method followed by a Varimax rotation. Thirty-eight variables were included in each of these analyses. The first set of variables consisted of the *absolute power* accounted for in each of the 19 electrodes of the 10/20 system by the "slow" factor in the "eyes-open" and "eyes-closed" conditions of the NB (S_1 and S_2). The second set of variables consisted of the comparable *relative power* indices (S_3 and S_4), the third set consisted of the *absolute power* accounted for across the full electrode set by the "fast" factor in the same EEG conditions (F_1 and F_2), while the fourth set consisted of the comparable *relative power* indices (F_3 and F_4).

In all four of these factor analyses, six factors accounted for somewhat over 82% of the communality. Careful examination of the factor loading coefficients in the regression equations for each derivation in every analysis revealed a remarkable and quite unexpected fact: whether the variables used in the factor analysis of the 10/20 system were absolute slow power, relative slow power, absolute fast power, or relative fast power, the factor structure was essentially invariant! In other words, the resting EEG activity of the set of brain regions monitored by the 10/20 system displays a relatively constant pattern of topographical covariance that is the same for both slow and fast frequencies, whether these are quantified in absolute or relative terms. It remains to be seen whether comparable and equally stable factor structures exist in an adult population, or whether these patterns of covariance are only found in children. The actual factor loadings obtained for every factor in each of the four analyses are shown for each derivation of the 10/20 system in Tables 8.6, 8.7, and 8.8.

Note that in these tables, derivations are presented bilaterally in a sequence which moves in three chains: (1) from anterior to posterior positions along the middle portion of each hemisphere (F_{p_1}/F_{p_2}, F_3/F_4, C_3/C_4, P_3/P_4, O_1/O_2), followed by (2) derivations arranged from anterior to posterior along the lateral portion of each hemisphere (F_7/F_8, T_3/T_4, T_5/T_6), followed by (3) derivations arranged from anterior to posterior along the midline (F_z, C_z, P_z). Bearing this in mind, it can be seen that in general the factor loadings show a smooth increase and decrease as positions change along these chains of electrodes in an anterior–posterior direction. In almost every case, the high factor loadings (i.e., >.50) are all located in either one or another of these chains, *but not in more than one chain*. When high loadings are found in two chains, as for Factors 1 and 2, the coefficients for the two chains have opposite signs.

a. Topography of Factors 1–6. The major weightings of Factor 1 are on C_4, P_3/P_4, O_1/O_2, and F_7, eyes open, *minus* T_5/T_6, F_z, C_z, and P_z, eyes closed.

The major weightings of Factor 2 are on C_3/C_4, P_3/P_4, O_1/O_2, eyes closed. F_7 is questionably also a member of this set.

The major weightings of Factor 3 are on F_z, F_7/F_8, T_3/T_4, T_5/T_6, eyes open.

The major weightings of Factor 4 are on F_{p_1}/F_{p_2}, F_3/F_4, C_3/C_4, and possibly P_3, eyes open.

The major weightings of Factor 5 are on F_{p_1}/F_{p_2}, F_3/F_4, and possibly C_3, eyes closed, *minus* F_z, C_z, and P_z, eyes open.

The major weightings of Factor 6 are on F_7/F_8, T_3/T_4, T_5/T_6, and possibly F_z, eyes closed.

b. Similar factor structure, eyes open or closed. Examining the topographical structure of the regions which share high loadings on each of Factors 1–6, listed in the preceding section, it becomes evident that although the details of the actual frequency spectrum change drastically as a function of whether the eyes are open or closed, *the patterns of covariance between cortical regions are essentially the same in either condition*.

This examination reveals the same three basic anatomical patterns of covariance, or factors, in both the eyes open and the eyes closed condition. Both Factors 1 and 2 contribute major influences to frontal, central, parietal, and occipital leads. Factor 1 loads both on the eyes-open and eyes-closed spectrum; Factor 2 loads only on the eyes-closed spectrum. The difference between Factor

TABLE 8.6
Loading Coefficients: EEG Factors 1 and 2

	Factor 1				Factor 2			
Analysis:	Slow		Fast		Slow		Fast	
Derivation	Absolute	Relative	Absolute	Relative	Absolute	Relative	Absolute	Relative
Eyes open								
F_{p1}	.12	.10	.10	.13	−.47	−.39	−.36	−.34
F_{p2}	.17	.11	.12	.18	−.37	−.28	−.27	−.26
F_3	.19	.20	.17	.23	−.21	−.14	−.11	−.09
F_4	.24	.29	.22	.34	−.06	−.00	.05	.04
C_3	.33	.41	.34	.43	.07	.10	.16	.17
C_4	.49	.56	.53	.55	.15	.16	.19	.25
P_3	.64	.69	.66	.68	.21	.24	.22	.25
P_4	.75	.80	.78	.75	.22	.18	.20	.21
O_1	.80	.80	.79	.77	.22	.17	.23	.13
O_2	.75	.76	.75	.65	.19	.18	.23	.09
F_7	.63	.61	.62	.51	.17	.12	.19	.09
F_8	.44	.44	.45	.33	.16	.14	.19	.15
T_3	.27	.27	.30	.15	.16	.21	.17	.20
T_4	.15	.13	.19	.02	.23	.26	.21	.27
T_5	−.02	−.01	−.06	−.09	.28	.34	.26	.27
T_6	−.10	−.10	−.07	−.16	.33	.35	.28	.27
F_Z	−.20	−.17	−.16	−.23	.33	.29	.26	.20

C_Z	−.23	−.16	−.15	−.24	.27	.21	.18	.09
P_Z	−.21	−.15	−.16	−.21	.17	.05	.06	−.05
Eyes closed								
F_{p1}	.45	.38	.38	.38	−.03	−.01	.07	.03
F_{p2}	.40	.32	.32	.32	.07	.14	.21	.13
F_3	.29	.24	.22	.23	.18	.30	.33	.32
F_4	.15	.17	.11	.16	.33	.45	.47	.51
C_3	.11	.14	.12	.14	.52	.62	.64	.72
C_4	.14	.12	.14	.18	.67	.76	.74	.82
P_3	.16	.13	.16	.19	.77	.84	.80	.81
P_4	.19	.21	.20	.19	.82	.85	.82	.74
O_1	.19	.20	.22	.16	.84	.79	.82	.63
O_2	.19	.19	.23	.10	.75	.66	.72	.46
F_7	.12	.11	.16	−.00	.61	.48	.55	.29
F_8	.00	−.03	.05	−.18	.44	.29	.37	.16
T_3	−.13	−.18	−.09	−.33	.26	.14	.15	.06
T_4	−.27	−.36	−.21	−.53	.12	.05	.06	−.05
T_5	−.41	−.55	−.36	−.66	.06	.01	.00	−.07
T_6	−.57	−.66	−.51	−.77	.02	−.02	−.04	−.05
F_Z	−.72	−.76	−.67	−.82	−.00	−.03	−.04	−.07
C_Z	−.76	−.76	−.71	−.76	−.05	−.09	−.10	−.09
P_Z	−.72	−.67	−.65	−.65	−.10	−.13	−.16	−.12

Note for Tables 8.6, 8.7, 8.8: Factor loading coefficients obtained from four separate factor analyses for each of the six different factors accounting for the power in the EEG spectrum. Note the extremely high concordance between the coefficients yielded by the 4 different factor analyses. Each

contd.

TABLE 8.6 (*contd.*)

factor analysis utilized 38 variables, obtained from the 19 monopolar electrodes of the 10/20 system recorded relative to linked earlobes under eyes open and eyes closed conditions. The four separate factor analyses evaluated 4 different sets of data: (1) Absolute slow power (low delta plus high delta plus theta); (2) Relative slow power (% low delta plus % high delta plus % theta); (3) Absolute fast power (alpha plus low beta plus high beta); (4) Relative fast power (% alpha plus % low beta plus % high beta).

In each table, coefficients underlined with a solid line (XY) relate to those derivations which receive the major contributions from that factor; coefficients underlined with a dotted line (XY) represent derivations expected to receive major contributions based upon the overall pattern of results, but which *failed* to meet that expectation; coefficients underlined with an interrupted bar (XY) represent derivations receiving major contributions but not belonging to a clearly discernible pattern.

There are 152 loading coefficients for each factor. For factor 1, 42 of the 152 are quite high (.50) and fall in a clear pattern. Two coefficients expected to correspond to that pattern fail to do so, while one coefficient which is large belongs to no discernible pattern. For factor 2, 24 of the coefficients are high and fit a clear pattern. Three other coefficients are high but fail to fit a pattern. For factor 3, 28 high coefficients fit a clear pattern while one other seems aberrant. For factor 4, 23 high coefficients fit a clear pattern, one coefficient in the pattern is unexpectedly low, and two high coefficients belonged to no pattern. For factor 5, 26 high coefficients fit a clear pattern, while 2 coefficients belonging to that pattern are slightly lower than expected. For factor 6, 21 high coefficients fit a clear pattern, while 3 coefficients in the pattern are slightly lower than expected. Two high coefficients seem to belong to no pattern.

Selection of .50 as the size of a loading coefficient reflecting a 'major' contribution is of course quite arbitrary and has no valid statistical basis. It does serve, however, as a crude criterion drawing attention to salient features of the pattern of factor loadings and further provides a crude index of concordance between the results of the various analyses. Thus, of a total of 912 factor loadings in the four factor analyses, 164 coefficients were greater than .50 and fell into clear and consistent patterns. Eight coefficients expected to be high on the basis of those patterns were lower than expected. Nine coefficients which were high failed to fit into any discernible patterns.

TABLE 8.7

Loading Coefficients: EEG Factors 3 and 4[a]

	Factor 3				Factor 4			
Analysis:	Slow		Fast		Slow		Fast	
Derivation	Absolute	Relative	Absolute	Relative	Absolute	Relative	Absolute	Relative
Eyes open								
F_{p1}	–.30	–.21	–.23	–.23	.54	.54	.63	.41
F_{p2}	–.21	–.14	–.14	–.16	.68	.65	.74	.56
F_3	–.13	–.09	–.08	–.09	.78	.72	.79	.67
F_4	–.10	–.09	–.07	–.08	.76	.71	.78	.67
C_3	–.04	–.03	–.02	–.04	.72	.69	.74	.62
C_4	.02	.04	.04	.03	.63	.64	.61	.59
P_3	.06	.05	.09	.05	.50	.51	.48	.48
P_4	.11	.17	.15	.14	.38	.38	.36	.40
O_1	.21	.29	.22	.29	.24	.26	.25	.22
O_2	.36	.41	.36	.49	.10	.09	.14	.07
F_7	.54	.59	.54	.66	–.00	.03	.06	–.01
F_8	.73	.76	.71	.81	–.07	–.04	.00	–.07
T_3	.84	.84	.82	.87	–.12	–.03	–.05	–.09
T_4	.83	.86	.83	.83	–.11	–.06	–.09	–.06
T_5	.81	.81	.81	.77	–.09	–.06	–.09	–.01
T_6	.73	.71	.72	.66	–.09	–.12	–.13	.00

contd.

TABLE 8.7 (*contd.*)

Analysis:	Factor 3				Factor 4			
	Slow		Fast		Slow		Fast	
Derivation	Absolute	Relative	Absolute	Relative	Absolute	Relative	Absolute	Relative
F_Z	.63	.57	.61	.52	–.16	–.20	–.19	–.01
C_Z	.50	.40	.46	.38	–.23	–.23	–.22	–.11
P_Z	.31	.23	.27	.25	–.39	–.36	–.39	–.19
Eyes closed								
F_{p1}	–.28	–.18	–.21	–.17	.25	.21	.27	.28
F_{p2}	–.13	–.02	–.05	–.03	.23	.17	.25	.23
F_3	.06	.13	.11	.14	.25	.19	.27	.20
F_4	.19	.21	.21	.24	.21	.17	.28	.14
C_3	.28	.29	.27	.29	.20	.17	.26	.07
C_4	.28	.31	.28	.25	.13	.15	.17	.05
P_3	.29	.26	.29	.24	.05	.07	.06	.04
P_4	.25	.22	.23	.19	–.05	–.09	–.08	–.00
O_1	.20	.18	.15	.18	–.16	–.16	–.19	–.10
O_2	.20	.17	.13	.24	–.28	–.23	–.29	–.15
F_7	.21	.19	.15	.26	–.42	–.33	–.42	–.20
F_8	.25	.22	.20	.29	–.45	–.33	–.39	–.28
T_3	.27	.28	.26	.30	–.47	–.31	–.41	–.28
T_4	.26	.31	.27	.26	–.37	–.27	–.32	–.25
T_5	.24	.29	.25	.24	–.29	–.20	–.25	–.17
T_6	.21	.22	.22	.13	–.18	–.10	–.13	–.12
F_Z	.14	.11	.11	.05	–.10	–.07	–.06	–.02
C_Z	.00	–.03	–.01	–.12	–.09	–.03	–.00	–.06
P_Z	–.12	–.23	–.13	–.26	–.12	–.04	–.04	–.10

[a]See note to Table 8.6 for explanation of the classification of factor loading coefficients.

TABLE 8.8
Loading Coefficients: EEG Factors 5 and 6[a]

Analysis:	Factor 5				Factor 6			
	Slow		Fast		Slow		Fast	
Derivation	Absolute	Relative	Absolute	Relative	Absolute	Relative	Absolute	Relative
Eyes open								
F_{p1}	.33	.45	.41	.40	–.06	–.27	–.09	–.36
F_{p2}	.34	.44	.40	.41	–.06	–.31	–.13	–.35
F_3	.32	.41	.31	.40	–.17	–.32	–.22	–.39
F_4	.30	.39	.26	.34	–.29	–.34	–.31	–.38
C_3	.23	.27	.16	.25	–.35	–.33	–.34	–.37
C_4	.15	.16	.10	.16	–.34	–.24	–.29	–.27
P_3	.13	.11	.11	.12	–.26	–.14	–.22	–.10
P_4	.07	.05	.05	.10	–.17	–.07	–.16	.05
O_1	.12	.11	.12	.15	–.06	.03	–.04	.16
O_2	.20	.20	.20	.20	.05	.13	.03	.24
F_7	.22	.20	.21	.22	.09	.22	.11	.23
F_8	.19	.15	.14	.11	.07	.24	.14	.16
T_3	.09	.06	.02	–.00	.06	.22	.12	.16
T_4	–.01	–.06	–.06	–.15	.13	.18	.15	.20
T_5	–.12	–.22	–.14	–.30	.22	.18	.24	.26
T_6	–.27	–.39	–.30	–.44	.20	.14	.19	.31
F_Z	–.43	–.58	–.49	–.57	.17	.12	.18	.34
C_Z	–.58	–.68	–.66	–.65	.13	.15	.15	.35
P_Z	–.64	–.70	–.68	–.64	.11	.15	.14	.36

contd.

TABLE 8.8 *(contd.)*

	Factor 5				Factor 6			
Analysis:	Slow		Fast		Slow		Fast	
Derivation	Absolute	Relative	Absolute	Relative	Absolute	Relative	Absolute	Relative
Eyes closed								
F_{p1}	.60	.69	.65	.66	–.07	–.16	–.14	–.01
F_{p2}	.74	.79	.75	.77	–.02	–.11	–.10	.01
F_3	.78	.80	.74	.76	–.07	–.10	–.13	–.04
F_4	.72	.73	.61	.65	–.20	–.13	–.19	–.08
C_3	.61	.56	.44	.48	–.23	–.08	–.15	–.11
C_4	.44	.38	.31	.27	–.20	–.03	–.11	.04
P_3	.27	.19	.19	.06	–.07	.02	–.04	.24
P_4	.12	.01	.06	–.07	.05	.14	.07	.44
O_1	–.01	–.11	–.09	–.13	.16	.34	.19	.58
O_2	–.09	–.17	–.14	–.16	.31	.51	.36	.69
F_7	–.13	–.18	–.16	–.14	.40	.62	.47	.76
F_8	–.13	–.16	–.14	–.13	.52	.74	.51	.76
T_3	–.13	–.17	–.16	–.09	.60	.77	.66	.74
T_4	–.13	–.18	–.16	–.07	.70	.73	.74	.64
T_5	–.12	–.16	–.12	–.13	.74	.63	.77	.51
T_6	–.17	–.19	–.19	–.19	.63	.55	.67	.38
F_Z	–.23	–.31	–.29	–.26	.46	.34	.51	.25
C_Z	–.32	–.41	–.41	–.34	.25	.18	.33	.13
P_Z	–.38	–.46	–.45	–.36	.12	.07	.21	.03

[a]See note to Table 8.6 for explanation of the factor loading coefficients.

1 and Factor 2 consists of the negative loadings on posterior temporal and midline loci, eyes closed, which are found only in Factor 1.

Factor 3 and Factor 6 both contribute major influence to frontal and temporal regions. Factor 3 loads only on the eyes-open spectrum; Factor 6 loads only on the eyes-closed spectrum.

Factor 4 and factor 5 both contribute major influences to the frontal pole, frontal, and central regions. Factor 4 contributes major influences to the eyes-open spectrum; Factor 5 contributes both to the eyes-open and eyes-closed spectrum. Reminiscent of Factor 1, Factor 5 displays negative loadings on the midline loci, but in the eyes-open spectrum.

These topographies can be seen in Fig. 8.2.

4. Construction of Set of Multivariate EEG Indices

The factor analysis results presented in the preceding section (III.C.3) showed that 24 factors accounted for the absolute and relative slow and fast power

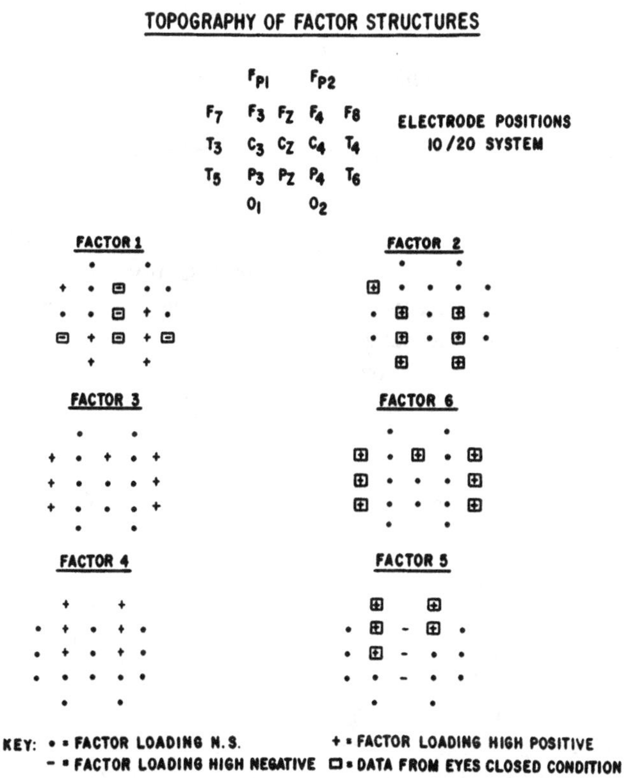

FIG. 8.2 Topography of factor structure in the EEG. For details, see text. Each set of patterns refers to positions of electrodes on the head, as illustrated at the top of the figure.

across the whole 10/20 system in the eyes open and eyes closed EEG conditions of the NB. The next step in the analysis of these data was to compute the 24 factor scores for each individual. This operation consisted of multiplying the absolute or relative slow or fast power in each derivation by the appropriate loading coefficient and summing these values for each factor in each individual.

These 24 factor scores were then used to compute a discriminant function between the normal and LD subgroups. Eleven of these variables entered the discriminant function at a significant level. They were: Factor 1, absolute slow power and relative fast power; Factor 2, absolute slow power and absolute fast power; Factor 3, absolute and relative fast power; Factor 4, absolute and relative slow power and relative fast power; Factor 5, absolute fast power; Factor 6, relative fast power.

The reader may at first be surprised that, in spite of the striking similarities between the factor structures resulting from different analyses, discussed in detail in the preceding section, these variables contributed to the discriminant function in substantially different ways. This apparent paradox merely reflects the fact that the different estimates of each factor structure, although closely similar in their most salient features, nonetheless differ in many small details. The cumulative effects of these small differences can make a substantial difference in discriminative power when the full multivariate factor score is computed.

D. Psychometric Multiple Discriminant Function Separating Normal from Learning Disabled Children

In Section III.B above, we defined a set of 28 psychometric variables: Based upon univariate discriminant tests, 24 scores from the psychometric instruments in our test battery were selected as sensitive to the differences between the normal and learning disabled subgroups in our sample; four additional factor scores representing multivariate combinations of these variables were computed for each subject, yielding a total set of 28 variables related to psychometric measures. Using these 28 variables, a multiple discriminant function was computed which optimally separated the normal and learning disabled subgroups.

The variables making significant contributions to the multiple discriminant function were:

1. WRAT arithmetic score.
2. Composite dysfunction score.
3. PPVT picture recognition scale.
4. WISC similarities subscale.
5. WISC digit span subscale.
6. WISC digit span backward subscale.
7. McCarthy Scale tapping sequence.
8. "Achievement" Factor (Factor 3).
9. "Disability" Factor (Factor 2).

The overall accuracy of separation in accordance with the prelabeled classification as normal or LD was 82%. A leave-one-out replication of this psychometric discriminant function was carried out. (For a description of this method, see Chapter 7, Section III.D.7.b). The concordance performance achieved on this replication is shown in the upper portion of Table 8.9. The replicated discriminant accuracy was 71%.

E. Neurometric Multiple Discriminant Function Separating Normal from Learning Disabled Children

1. Further Reduction of EEG Measure Set

In addition to the neurometric factor scores defined in Section III.C.4 above, 18 different EEG measure sets were available on each subject, providing a total of well over 1000 measures. Table 8.10 lists these measure sets. The first column in Table 8.10 indicates the total number of variables in each measure set. The ability of each separate variable in each measurement set to distinguish disability was evaluated using an *F*-test criterion from a simple analysis-of-variance procedure. The second column in the table shows the number of variables in each set which showed a significant univariate *F*-test difference between the normal and LD subgroups. Since it seemed reasonable to expect that many of these variables would be sensitive to age, IQ, socioeconomic status, or cultural variables, and since we desired to select variables which would be maximally sensitive to neural processes related to learning disability, uncontaminated by such secondary

TABLE 8.9a
A. Concordance Performance of Leave-One-Out Replication of Psychometric Discriminant

	Predicted "Normal"	Predicted "LD"	Total
Prediagnosed "normal"	101 (61%)	9 (5%)	110
Prediagnosed "LD"	39 (23%)	17 (10%)	56
Total	140	26	166

Accuracy = 71%

TABLE 8.9b
B. Concordance Performance of Leave-One-Out Replication of Neurometric Discriminant

	Predicted "Normal"	Predicted "LD"	Total
Prediagnosed "normal"	92 (58%)	11 (7%)	102
Prediagnosed "LD"	25 (16%)	29 (19%)	54
Total	116	40	156

Accuracy = 77%

TABLE 8.10
Summary of Number of Variables from Eyes Open and Eyes Closed EEG Used in Neurometric Discriminant Function

	Measure set	Total possible variables in set	Number variables significant	Number variables improved	Number variables worsened
EEG	Absolute power, eyes open	133[a]	27	13	14
EEG	Absolute power, eyes closed	133	43	8	35
EEG	Relative power, eyes open	114[b]	24	18	6
EEG	Relative power, eyes closed	114	39	13	26
EEG	Real pair correl. coef., eyes open	56[c]	0	–	–
EEG	Real pair correl. coef., eyes closed	56	26	9	17
EEG	Imag. pair correl. coef., eyes open	56	20	16	4
EEG	Imag. pair correl. coef., eyes closed	56		not included yet	
EEG	Absolute power (eyes closed–eyes open)	133	110	50	60
EEG	Relative power (eyes closed–eyes open)	114	75	40	35
EEG	Real pair correl. coef. (eyes closed–eyes open)	56	0	–	–
EEG	Imag. pair correl. coef. (eyes closed–eyes open)	56		not included yet	
EEG	Absolute power slow frequencies ($D_1 + D_2 + \theta$)				

	(eyes open + eyes closed)	38[d]	11	3	8
EEG	Relative power slow frequencies ($D_1 + D_2 + \theta$) (eyes open + eyes closed)	38	11	4	7
EEG	Absolute power fast frequencies ($\alpha + \beta_1 + \beta_2$) (eyes open + eyes closed)	38	10	3	7
EEG	Relative power fast frequencies ($\alpha + \beta_1 + \beta_2$) (eyes open + eyes closed)	38	14	7	7
EEG	Ratio slow/fast frequencies (absolute power) (eyes open + eyes closed)	38	4	4	0
EEG	Ratio slow/fast frequencies (relative power) (eyes open + eyes closed)	38	7	5	2
		1193 (100%)	421 (35%)	193 (16%)	228 (19%)

[a]133 variables represent the values of the *absolute power* in each of the 6 frequency bands as well as the total power values for the 19 leads in the 10/20 system (7×19=133).

[b]114 variables represent the *relative power* calculated for each of 6 frequency bands for the 19 leads (6×19=114).

[c]56 variables represent the information for each of 6 frequency bands and the total band for the 8 pairs of homologous bilateral pairs in this 19 lead derivation set (e.g., $O_1 - O_2$ pair, etc.; 8×7=56).

[d]38 variables represent the values for each of the 19 leads for two conditions, eyes open and eyes closed (19×2=38).

influences, the "discriminating" variables of Column 2 were subjected to further analysis in which each of these influences was regressed out. This was accomplished by a regression analysis of variance on each measure set. (SPSS procedures and programs were used here.) The third and fourth columns of the table indicate the number of variables in which ability to discriminate between normal and learning disabled children improved or worsened as a result of such regression. Obviously, the variables entered in Column 3 were satisfactorily robust, representing indices that were not sensitive to contaminating influences not directly related to learning disability, while the variables in Column 4 were likely to yield misleading results. Thus, the 193 variables of column 3 were selected as the first compressed measure set.

A stepwise discriminant function between the normal and LD subgroups was next computed for the "robust" variables in each measure set. The 30 variables contributing most significantly to the discrimination between the two groups are listed in Table 8.11.

It is perhaps advisable to emphasize that these variables can only be evaluated properly as a multivariate set. Many variables were eliminated from the 193, indicated in Table 8.10 as possessing significant univariate discriminative power, because other variables in the set were more sensitive. It should be apparent to the reader that the particular variables which emerge as most contributory to a stepwise discriminant depend upon the features of the measure set as a whole.

2. *Neurometric Multiple Discrminant Function*

These 30 variables listed in Table 8.11, together with the 11 discriminating factor scores itemized above in Section III.C.4, constituted our final EEG neurometric set. The full set of 41 variables was then used to compute the multiple discriminant function which would optimally separate normal from LD subgroups in our sample. The initial accuracy of discrimination was 92%. A leave-one-out replication of the neurometric discriminant was next carried out. The concordance of the replicated discriminant with the prelabeled classifications is shown in the bottom portion of Table 8.9. The accuracy of discrimination in this replication was 77%.

F. Correction of Misclassified Subjects

Using the discriminant scores obtained by each subject on the psychometric and neurometric discriminant functions, it is possible to begin to come to grips with the problem of unreliability of preclassification of "normality" or "learning disability," discussed previously in Section III.A.3. Three types of classification data are now available for each child: prelabeling (PRE), psychometric discriminant score (PSYCH), and neurometric discriminant score (NEURO). Denoting classification as normal by any method as *N* and learning disabled as *LD*, 8 classes of subjects can be identified, as shown in Table 8.12.

An analysis of the concordance of accurate and inaccurate diagnosis among the possible descriptors of handicap, that is, prediagnosis based on psychometric or

TABLE 8.11
Most Significant EEG Variables in Stepwise Discriminant

Absolute power, eyes open, T_6, high beta
Absolute power, eyes open, F_Z, low delta
Difference in absolute power, eyes closed minus eyes open, C_4, low delta
Relative power, eyes open, P_4, high delta
Relative power, eyes closed, O_2, low delta
Relative power, eyes closed, T_4, theta
Relative power, eyes closed, P_4, low beta
Relative power, eyes closed, F_7, low beta
Relative power, eyes closed, O_1, high beta
Differences in relative power, eyes closed minus eyes open, C_4, low beta
Differences in relative power, eyes closed minus eyes open, C_4, high beta
Real correlation coefficient, eyes closed, F_7/F_8, low beta
Imaginary correlation coefficient, eyes open, F_7/F_8, low beta
Imaginary correlation coefficient, eyes open, T_3/T_4, alpha
Imaginary correlation coefficient, eyes open, T_3/T_4, low beta
Imaginary correlation coefficient, eyes open, T_3/T_4, high beta
Slow frequencies, eyes open, F_Z, absolute power
Slow frequencies, eyes closed, O_1, absolute power
Slow frequencies, eyes open, P_4, relative power
Slow frequencies, eyes open, T_6, relative power
Slow frequencies, eyes closed, T_3, relative power
Fast frequencies, eyes open, P_4, relative power
Fast frequencies, eyes closed, O_2, relative power
Fast frequencies, eyes closed, F_8, relative power
Fast frequencies, eyes closed, T_4, relative power
Fast frequencies, eyes closed, P_Z, relative power
Slow/fast ratio, eyes closed, T_3, absolute power
Slow/fast ratio, eyes open, T_3, relative power
Slow/fast ratio, eyes open, T_5, relative power
Slow/fast ratio, eyes open, T_6, relative power

Note: "slow refers to sum of low delta, high delta, and theta; "fast" refers to sum of alpha, low beta, and high beta.

TABLE 8.12
Classes of Subjects and Numbers in Each Class

Type	Prediagnosis	Psychometrics	Neurometrics	Most probable class	*N*
1	N	N	N	N	85
2	N	N	LD	?	10
3	N	LD	N	N	8
4	N	LD	LD	LD	0
5	LD	N	N	N	20
6	LD	N	LD	LD	15
7	LD	LD	N	?	3
8	LD	LD	LD	LD	12
				Total:	153

EEG neurometrics, revealed interesting results. Whereas no child prediagnosed as "normal" was subsequently "diagnosed" as dysfunctional by both psychometrics and EEG neurometrics, almost all children prelabeled as handicapped, who were considered normal by EEG neurometrics, were also considered normal by psychometrics. An approximately equal number of normal children were classified as dysfunctional by one metric but not the other. However, only a small proportion of children ($N = 3$) prediagnosed as dysfunctional were considered dysfunctional by psychometrics and not neurometrics. A much larger proportion of children ($N = 15$) prediagnosed as dysfunctional were considered dysfunctional by neurometrics and not psychometrics. These data reinforce our conclusion that the two types of measurements as presently constituted reflect different aspects of dysfunction. It is now possible to reclassify some of our subjects. Subjects of Type 1 can be considered unequivocally normal, since all three methods of evaluation agree. Subjects of Type 2 display abnormal brain function, but are behaviorally normal. These children constitute an uncertain category, possibly representing overachievers who have successfully compensated for brain dysfunction or whose brain dysfunction has no relevance for behavior functions reflected in school or psychometric test performance. Subjects of Type 3 are considered to be normal, with the psychometric indication of abnormality probably attributable to cultural factors. Subjects of Type 4 are considered to be LD, representing either overachievers or classification errors reflecting the high variability of academic criteria. Subjects of Type 5 are considered to be organically normal, their learning problems probably reflecting emotional disorders or behavioral problems in the school setting. Subjects of Type 6 are considered to be LD, their normal performance on psychometric tests reflecting the relatively narrow scope of functions presently assessed by such measures. Subjects of Type 7 are considered to be an uncertain category. While the prelabeling and psychometric classifications are in agreement, no evidence of organic dysfunction can be found. Children in this group may have emotional, behavioral, or environmentally derived learning disorders. Subjects of Type 8 can be considered unequivocally learning disabled, since all three evaluation methods agree.

If we adopt these new criteria, some of our subjects can be more rigorously classified. The result of this reexamination of each individual is to revise the accuracy of the psychometric and neurometric discriminant functions presented in Table 8.9. The revised concordance performance is shown in revised Table 8.9a and b.

G. Comparison Between Psychometric and Neurometric Discriminant Functions

Sufficient data to compute psychometric discriminant scores were available for 153 children. It is informative to compare the two sets of results. If we ignore the arguments in the preceding section supporting the reclassification of some of

TABLE 8.9 (Revised)

Revised Concordance Performance of Leave-One-Out Replication of Discriminant Functions, If

A. Reclassification upon assumption that EEG neurometrics give a more valid insight into brain function than psychometrics.

	Neurometrics			Psychometrics		
	Predicted "normal"	Predicted "LD"	Total	Predicted "Normal"	Predicted "LD"	Total
Reclassified "normal"	A1 61%	A2 0%	A3 61%	a1 55.7%	a2 5.3%	a3 61%
Reclassified "LD"	A4 0%	A5 17.6%	A6 17.6%	a4 9.8%	a5 7.8%	a6 17.6%
Reclassified ?	A7 15.2%	A8 6.2%	A9 21.4%	a7 19.6%	a8 1.8%	a9 21.4%
Total	A10 76.2%	A11 23.8%	A12 100% $n = 153$	a10 85.1%	a11 14.9%	a12 100% $n = 153$%

B. Reclassification based upon assumption that psychometrics give a more valid insight into brain function than EEG neurometrics.

	Neurometrics			Psychometrics		
	Predicted "normal"	Predicted "LD"	Total	Predicted "normal"	Predicted "LD"	Total
Reclassified "normal"	B1 55.8%	B2 6.2%	B3 62%	b1 62%	b2 0%	b3 62%
Reclassified "LD"	B4 2.0%	B5 7.8%	B6 9.8%	b4 0%	b5 9.8%	b6 9.8%
Reclassified ?	B7 18.4%	B8 9.8%	B9 28.2%	b7 23.1%	b8 5.1%	b9 28.2%
Total	B10 76.2%	B11 23.8%	B12 100% $n = 153$	b10 85.1%	b11 14.9%	b12 100% $n = 153$

our prelabeled subjects, the neurometric score correlates at about .6 with the prediagnosis, whereas the psychometric score correlates at about .3. Both of these correlations are significant. Yet, it is essential to emphasize that the neurometric measures entering into the discriminant function are derived from two minutes of data which can be obtained efficiently and economically by a technician with a few days of training, from a subject who need only stay awake and remain moderately motionless. In contrast, the psychometric measures are derived from instruments that take over 4 hrs to be administered by a person with a substantial amount of training, require further hours of scoring by a highly sophisticated psychometrician, and can only be obtained from a cooperative and verbally competent subject. It is, of course, possible that the psychometric battery could be made more robust by the inclusion of additional tests. Even were this so, the psychometric evaluation would become more demanding of time and cooperation and it would be less probable that any subject, but most of all any learning disabled subject, would tolerate the effort required to provide the additional data.

1. Canonical Correlations between Psychometric and Neurometric Measures

Canonical correlations were computed between every psychometric measure and all neurometric indices entering into the discriminant functions. No significant canonical correlations could be found. Thus, whatever it may be that the psychometric indices measure about learning disability, it appears to be orthogonal to what is measured by those indices derived from the EEG included in the discriminant function.

The absence of canonical correlations between the psychometric and neurometric EEG measures suggests four conclusions:

1. Psychometric measures may or may not be significantly correlated with measures derived from the more dynamic EP conditions of the NB. The data available in this study are inadequate to answer this question.

2. The EEG measures provide more direct access to brain processes than the psychometric measures, which merely reflect the products of such processes. EEG measures are not only more precise and numerous, but far more efficient to obtain and evaluate. Thus, assessment of children for learning disability by neurometric methods utilizing the EEG is of greater utility than assessment by conventional psychometric methods.

3. Since the EEG and psychometric measures are so poorly correlated, it will necessarily be difficult, if not impossible, to find reliable EEG signs of learning disability if the dysfunction is defined in terms of psychometrics. Psychometric measures reflect age, socioeconomic status, IQ, and cultural factors to an extent that obscures the identification of abnormal brain function. If learning dysfunction is defined in terms of clinical evaluations by teachers, parents, and physi-

cians observing real-life performance, a substantial amount of consensual validation is usually likely to contribute to the decision that a particular child is learning disabled. Thus, one might expect neurometric measures to correlate better with clinical than with psychometric assessments.

4. The cost–benefit ratio is far higher for neurometric evaluation of learning disabilities. The feasibility of neurometric assessment is high, since essentially 100% of the children in our sample gave adequate data for this evaluation. The feasibility of psychometric assessment is far lower. Most learning-disabled children in our sample did not complete the psychometric test battery. Perhaps of greatest importance, the EEG measures used in the neurometric discriminant function can readily be obtained (and have been obtained by our research team) down to newborn infants. Even using different metrics which are extremely difficult to standardize and equate, reliable psychometric measures are simply not available below the age of 3.

All of these considerations support the contention that efficient mass screening for brain dysfunction related to learning disability can only be carried out by neurometric methods. This conclusion holds even more strongly if the goal is early identification of the learning handicapped child at the preschool level. As will be demonstrated later, neurometric methods provide the additional overwhelming advantage that they permit separation of the heterogeneous learning disabled population into homogeneous subgroups sharing common profiles of brain dysfunction.

2. *Regression Analysis of Covariance for Psychometric and Neurometric Discriminant Scores*

Correlation coefficients were computed between the psychometric and neurometric discriminant score for each child in this study. The average correlation coefficient was .2. This calculation revealed that the factors contributing to the psychometric and neurometric discrimination between children preclassified as normal or as LD were almost completely independent. These results provide further confirmation of the absence of any significant relation between these two sets of measures, as indicated by the failure to find significant canonical correlations.

In order to explore this finding further, a regression analysis of covariance was carried out for each of the two kinds of discriminant scores. The results are presented in Table 8.13. The top half of Table 8.13 shows the results of regression analysis of covariance on the EEG discriminant scores. Note that the prelabeled diagnosis accounts for 33% of the total variance, significant far beyond the $p = .001$ level. Age contributed about 1% to the total variance, while culture and SES were not significant sources of variance. Most important was the fact that the psychometric discriminant score did not make a significant contribution to the total variance.

TABLE 8.13
Regression Analysis of Covariance for Two Types of Discriminant Scores

Source of variation	Sums of squares	*df*	*F*	Total variance (%)
EEG discriminant score				
Prediagnosis	204.50	1	87.63***	33.3
Culture	1.02	1	1.02	–
SES	.19	1	.19	–
Age	6.54	1	2.80	1.1
Psychometric discriminant score	.07	1	.03	–
Residual	338.38	145		
Total	613.75	150		

Source of variation	Sums of squares	*df*	*F*	Total variance (%)
Psychometric discriminant score				
Prediagnosis	25.98	1	15.24***	8.9
Culture	1.70	1	1.00	–
SES	.01	1	.01	–
Age	1.71	1	1.00	.6%
EEG discriminant score	0.05	1	.03	–
Residual	247.25	145		
Total	292.56	150		

***$p < .001$.

The comparable data for the psychometric discriminant scores are presented in the lower half of Table 8.13. The prediagnosis only accounts for about 9% of the total variance. Age, SES, and culture are insignificant as sources of variance in these scores. Confirming the results reported above, the EEG discriminant score did not make a significant contribution to the total variance.

Thus, a more sensitive and sophisticated technique of evaluation leads to the same conclusion: the psychometric and neurometric indices sensitive to the difference between normal and learning disabled children may present two essentially orthogonal or mutually independent sets of measures. Presumably the greatest overall sensitivity would be afforded by a composite psychometric and neurometric assessment, since the two techniques appear to be sensitive to radically different kinds of processes.

3. Density Distribution of Psychometric and Neurometric Discriminant Scores

In order to help visualize the results obtained by the psychometric and neurometric discriminant functions, and the relationship between these measures, a visual representation was devised. For each subject in our sample, the psychometric discriminant was plotted versus the neurometric discriminant score. A surface was then constructed such that the greater the density of subjects at any point, the higher the surface. Computer graphic methods were then utilized to depict the resulting three-dimensional distribution, as illustrated in Fig. 8.3.

Figure 8.3 illustrates the density distribution of psychometric versus neurometric discriminant scores. The eight pictures around the periphery of this figure provide views of the constructed density surface from different positions. These pictures can be likened to what one would see as one walked around such a surface in a clockwise direction, looking at it from different viewpoints.

Examination of the figure shows that the distribution of psychometric discriminant scores is essentially Gaussian. No unequivocal indication of heterogeneity in the population emerges from inspection of these data, although a suggestive bulge does appear toward the learning handicapped (LD) pole. By way of contrast, the distribution of neurometric discriminant scores can be seen to be clearly multimodal.

The neurometric discriminant distribution suggests that the learning disabled subgroup may be heterogeneous. In the remainder of this chapter, data are presented not only to demonstrate that the EP conditions and challenges of the neurometric battery are powerfully and differentially sensitive measures of learning disability, but that those neurometric indices provide the basis for applying the methods of numerical taxonomy to fractionate the learning disabled population into homogeneous subgroups with different neurometric profiles.

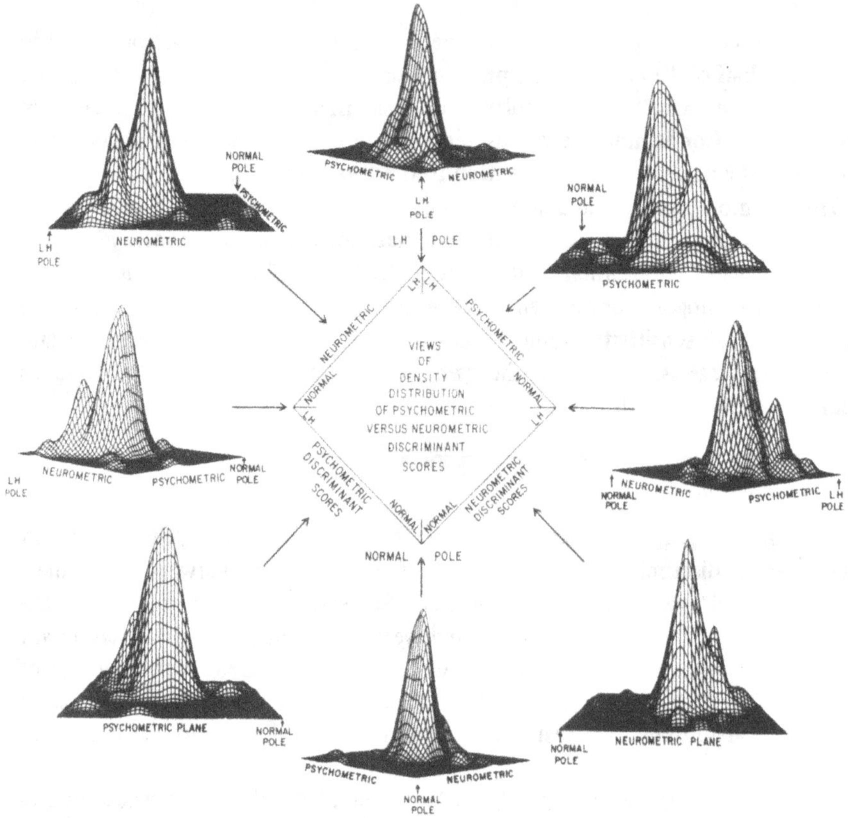

FIG. 8.3 Density distribution of psychometric versus neurometric discriminant scores. The two scores for each subject have been plotted with respect to a psychometric and neurometric axis. The surfaces illustrated in the figure represent the number of density of subjects whose scores fell at the corresponding points. These surfaces are presented as if viewed from different vantage points, to permit visualization of the distributions (LH = LD).

H. Analysis of Variance of EP Conditions of the NB

The detailed analysis of the EEG indices yielded by the neurometric battery, discussed in the previous sections, undoubtedly made the reader more concretely cognizant of the substantial technical problems involved in reducing such a mass of numerical data to manageable and meaningful proportions. None of the previous studies nor the conventional techniques encountered in neuroscience or in psychology served as an adequate model to provide guidelines for the proper treatment of these huge quantities of data.

The full set of EEG challenges and conditions of the NB provide on the order of 10,000 numerical indices per subject. Each EP challenge or condition of the

NB provides about 450 derived numerical indices. Since there are a total of 92 EP test items, about 40,000 indices are extracted from the EP portions of the NB for each subject. If one realizes further that each of the 100 latency points in the analysis epoch can be treated as a quantitative variable, it becomes apparent that well over half a million additional data points are available from the EP test items, or about 50 times more data than is provided by the EEG portions of the NB. Since these measures were devised to provide far more direct insight into sensory, perceptual, and cognitive processes than those afforded by the less dynamic spontaneous EEG measures, they offer far greater potential utility for the identification of meaningful individual profiles of information processing than the EEG indices. At the same time, they pose an even more difficult task of data compression.

One approach to achieving a substantial reduction in the amount of data requiring high-resolution analysis is to ascertain for each EP condition and challenge those portions of the analysis epoch in which processes sensitive to the differences between normal and learning disabled children are located. We have utilized multivariate factor analysis and multiple analysis of variance (MANOVA) for this purpose, and found both methods to be extremely useful. Only the results obtained from MANOVA techniques are presented here. The factor analysis results will be presented in a more detailed forthcoming publication (John, Easton, Karmel, Prichep, Brown, & Ahn, in preparation).

The reader may recall the results obtained initially by Valdes (1973) and later confirmed by us, cited in Chapter 3. Those studies revealed that the EPs recorded from many placements in the 10/20 system could be represented as linear combinations of two or three factors. The implication of those results is that the latency and polarity of EP components in many different regions are highly correlated, although their relative amplitude may vary from region to region. In view of the fundamental coherence of the EP morphology across the topography of the 10/20 system which was revealed by those observations, we decided to evaluate the variance of the EPs obtained in every EP condition and challenge of the NB, summing across the full 10/20 system and comparing the normal and LD subgroups.

On the one hand, this might be a hazardous approach, since fine differences between anatomical regions might well be obscured by such a gross treatment, leading to Type II errors. On the other hand, this approach was certainly maximally conservative, practically ensuring no Type I errors. Sources of variance sufficiently robust to survive such treatment, were any such to be found, would surely reflect pervasive and consistent differences between the normal and LD subgroups which appeared at many different locations in the 10/20 system. Once such major sources of variance were identified, subsequent analyses of the contributions of individual leads to the overall variance should provide much higher anatomical resolution and the elimination of many of the possible Type II errors inherent in this procedure.

The approach outlined above was implemented in the Computer Center of the University of Connecticut at Storrs, in analyses carried out by Dr. Bernard Z. Karmel and Mr. Jack Davis. The initial results were so encouraging that all of the EP conditions and challenges of the NB were subjected to this treatment. Only a small fraction of the results will be presented here, to illustrate the utility of the method and the salient features of our findings. Full details will be presented subsequently (John, Karmel, Prichep, Davis, & Ahn, in preparation).

There are two essential steps in this procedure. First, a multiple analysis of variance is performed on the set of EPs obtained from all leads of the 10/20 system in a mixed sample of normal and LD children. In this analysis, the main variables are the 100 successive time points or latencies along the EP analysis epoch. The variance at each time point contributed by every lead in the 10/20 system is determined and then summed across the full set of electrodes. At each latency point, the *F* ratio between the normal and LD subgroups is computed. The result is a statement about the magnitude of the *F* ratio as a function of latency, summed across the 10/20 system.

Obviously, successive points along the EP waveshape are not necessarily independent. EP components are relatively slow processes that occupy substantial time intervals. For this reason, a criterion is required to establish the points along the analysis epoch at which essentially independent processes appear. In order to accomplish this, a second step is required. A step-down discriminant analysis was utilized for this purpose. In this procedure, the predictability of the *F* ratio at each latency point is estimated as a function of the values of the *F* ratio at all of the preceding points in the epoch. A criterion of $p = .01$ was selected as the index that an independent process had emerged.

Figure 8.4 shows the results of the MANOVA evaluation of the first 11 EP conditions of the NB. Each of the 11 graphs displays the values of the F ratio between the normal and LD subgroups as a function of latency, based upon data collapsed across the 19 electrodes of the 10/20 system.

The data are organized into 3 groups. The first group consists of the EPs elicited by presentation of 3 spatial grids, each providing 50% transmission. The top graph in the left column shows the *F* ratios obtained from EPs elicited by presentation of a grid of 65 lines/inch, perceived as a blank, uniformly grey visual field. Independent processes, significantly different between the normal and LD subgroups, occurred at 100, 250, and 425 msec. The middle graph in the left column shows the *F* ratios for a grid of 7 lines/inch, perceived as a checkerboard by subjects with 20/200 visual acuity or better. Again, the salient discriminating processes were located at 100, 250, and 425 msec, but a marked and highly significant process appeared at about 575 msec. The bottom graph in this column shows the *F* ratios for a grid of 27 lines/inch, perceived as a checkerboard by subjects with visual acuity of 20/20. It is known that when visual structure appears in a previously blank visual field, a new EP component or a change in polarity appears at about 150 msec. Note that the *F* ratios elicited by this stimulus show an initial peak somewhat later in latency than the upper graphs, at about 150 msec. This suggests that some of the population has visual acuity poorer than 20/20. Note further that a huge process now appears at 425 msec, dwarfing the other processes seen in this condition.

The stimuli generating the EPs analyzed in the second column of graphs, from top to bottom were a large square, a small square, a large diamond, and a small diamond. Note that these stimuli show a common pattern of highly significant *F* ratios, quite different from those observed with spatial grids. The early process at 100 msec has vanished. Two major discriminative processes are seen, one at 250 msec and the other at 450–550 msec.

The right column represents data elicited, top to bottom, by presentation of the letters b, d, p, and q. While variable early processes are observed, possibly representing entrainment of long latency components from the previous presentation of stimuli in the train, the major processes of interest are found at 250, 350, and 425 msec.

These data suggest that each of these 3 subclasses of visual stimuli elicit EPs which discriminate between normal and LD subgroups in somewhat different latency domains. Note that the baseline of the *F* ratio curves in these graphs is at the .01 significance level for differences between the normal and LD subgroups. Thus, the peaks of these *F* ratio curves represent the time points of the independent processes in the EPs elicited by these stimuli, which discriminate between the normal and LD subgroups. By constraining the EP data obtained in these various conditions only to the values of EP voltages at these latency points, and discarding all the data from the remainder of the epoch, we succeed in compressing a total of 1100 latency points in each of 19 derivations (almost 21,000 data points) to less than 30 points selected for optimal sensitivity. Assuming that these 30 points must ultimately be examined for each of the 19 derivations, the data have been compressed from 21,000 to 600 points, or by a factor of 35. Assume a comparable reduction for each of the 92 EP conditions and challenges of the NB. This gives an estimated 275 critical variables for the collapsed or 5,225 for the full 10/20 system, across the total set of EP conditions and challenges of the full NB. Further compression can be achieved by factor analysis of the 10/20 system and of the EP test items, both of which are substantially redundant. At the moment, our strategy is to deal with the collapsed 10/20 system in the first stage of data analysis and to identify the leads contributing the major sources of variance in each condition in a subsequent stage.

I. Visual Display as a Method of Data Compression: The Density-Coded *Z* Transform

Once the critical latency points in the analysis epoch for each EP test item have been identified, it becomes possible to construct visual displays which provide a

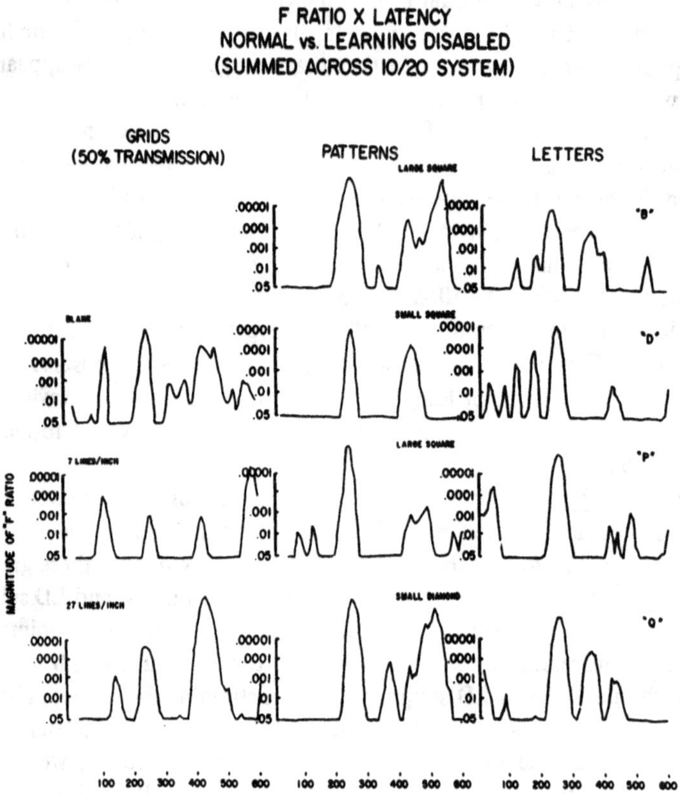

FIG. 8.4 Graphs describing the F ratio obtained by computing MANOVA's between samples of normal and learning disabled children, combining AER data from the full 10/20 system in each of a variety of conditions. Each graph shows the statistical significance of the difference between the two groups as a function of the latency of the AER components.

(a) F ratios obtained from 11 different conditions of the NB. The 3 graphs in the left column show results obtained using stimuli consisting of spatial grids of different frequency; those in the middle column show results obtained with different letters of the alphabet. Note that the latencies at which peaks in the F ratio curve occur are very similar for stimuli belonging to the same general subgroups arranged within any column, but are substantially altered when the *nature* of the information in the stimuli changes from column to column. This suggests that critical steps in neural processing of different types of information occur at different latencies.

Opposite (b) As above, but data for random (upper graphs) and regular (lower graphs) flash, click and tap stimuli.

(c) As above, but data for flash, click and tap stimuli presented while the subjects watched a slient video cartoon (upper graphs) or listened to music in dim light (lower graphs).

Note the striking changes in the latencies of the salient differences between these two groups of children elicited by visual stimuli of comparable energy as the information processing mechanisms of the subject are engaged in different activities.

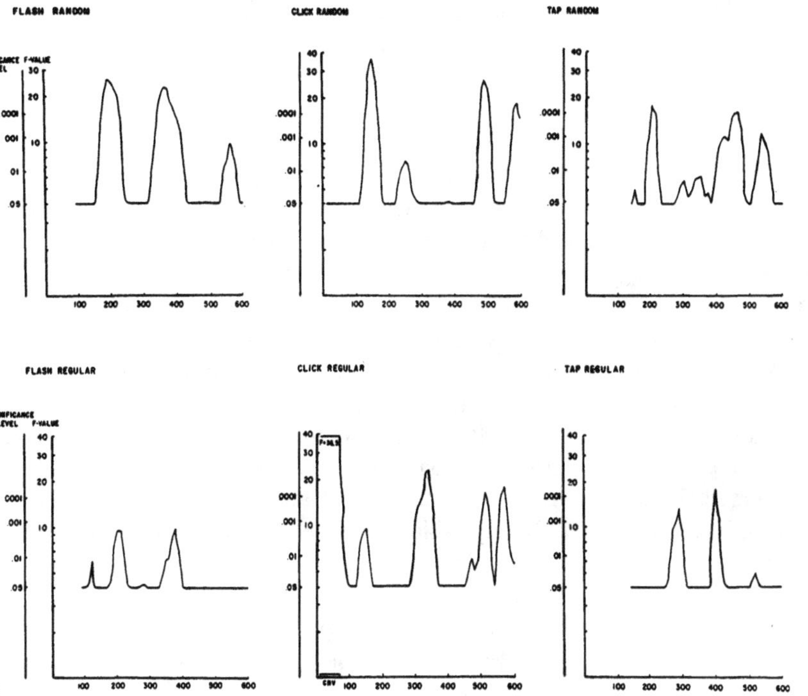
FLASH RANDOM
CLICK RANDOM
TAP RANDOM
SIGNIFICANCE LEVEL
F-VALUE
FLASH REGULAR
CLICK REGULAR
TAP REGULAR
CSV

(b)

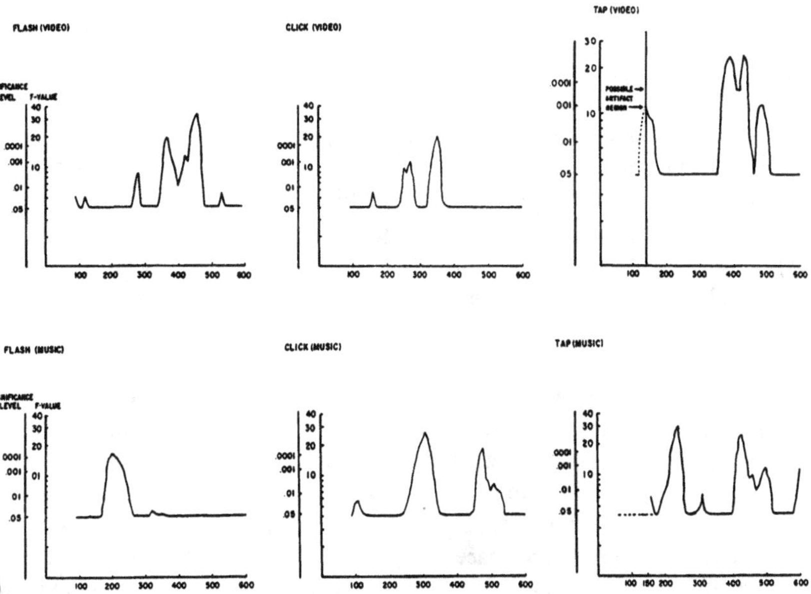
FLASH (VIDEO)
CLICK (VIDEO)
TAP (VIDEO)
SIGNIFICANCE LEVEL
F-VALUE
POSSIBLE ARTIFACT REGION
FLASH (MUSIC)
CLICK (MUSIC)
TAP (MUSIC)

(c)

topographically understandable representation of deviation from normality for each individual on every discriminating index. In order to do this, the mean value and standard deviation of each index is calculated for the whole population of subjects of a given age.

For every electrode location, the value of the index obtained from each individual is subjected to a z transform. The resulting z-transformed values are then translated into a density code: if the value was less than $\pm z = 2$ ($p = 0.01$), a pair of small spots (· .) is placed on the position corresponding to that electrode location on a diagram of a head. If the value deviates from the population average at the .01 level, a small plus is entered at the appropriate position on the head diagram if the value was *above* the average, whereas a small minus is entered if the value was below the mean. As the z-transformed value deviates more and more from the population average, the plus or minus sign becomes larger and more dense. The result of this is to produce a series of head diagrams for each subject on which deviations from average values of indices which discriminate between normal and LD subjects are represented as densities of entries. Such diagrams are comparable to an electrophysiological "brain scan" showing the anatomical distribution of processes related to abnormal sensory, perceptual or cognitive processes.

An example of this method of data presentation is illustrated in Fig. 8.5. Note that such depictions can describe the values of univariate measures, or of multivariate derived measures representing combinations of sensitive indices. This example was constructed for heuristic purposes and represents "prototypic" rather than real individuals.

J. Cluster Analysis

Once all data from a group of individuals have been subjected to z transformation, it becomes possible to apply powerful methods of numerical taxonomy for the objective classification of individual subjects. While such methods can be applied to the full set of data without prior z transformation, there is a great danger, especially with as enormous a measure set as that provided by the NB, that significant effects will be overwhelmed by the random variance contributed by irrelevant processes. Further, different metrics are hard to combine.

Imagine a measure space in which one dimension exists for every index found to be discriminating between normal and LD, by analysis of variance procedures like those just described. Imagine further that for each of these indices, the data from every subject have undergone z transformation. Every subject can now be represented as a z vector in this discriminating measure space. The z vector defines a point in the space reached by translating the subject's representational point from the origin of the space along the complex trajectory constructed by vectorial summation of that individual's z scores on each of the discriminating indices or dimensions. Data from different indices can be readily combined.

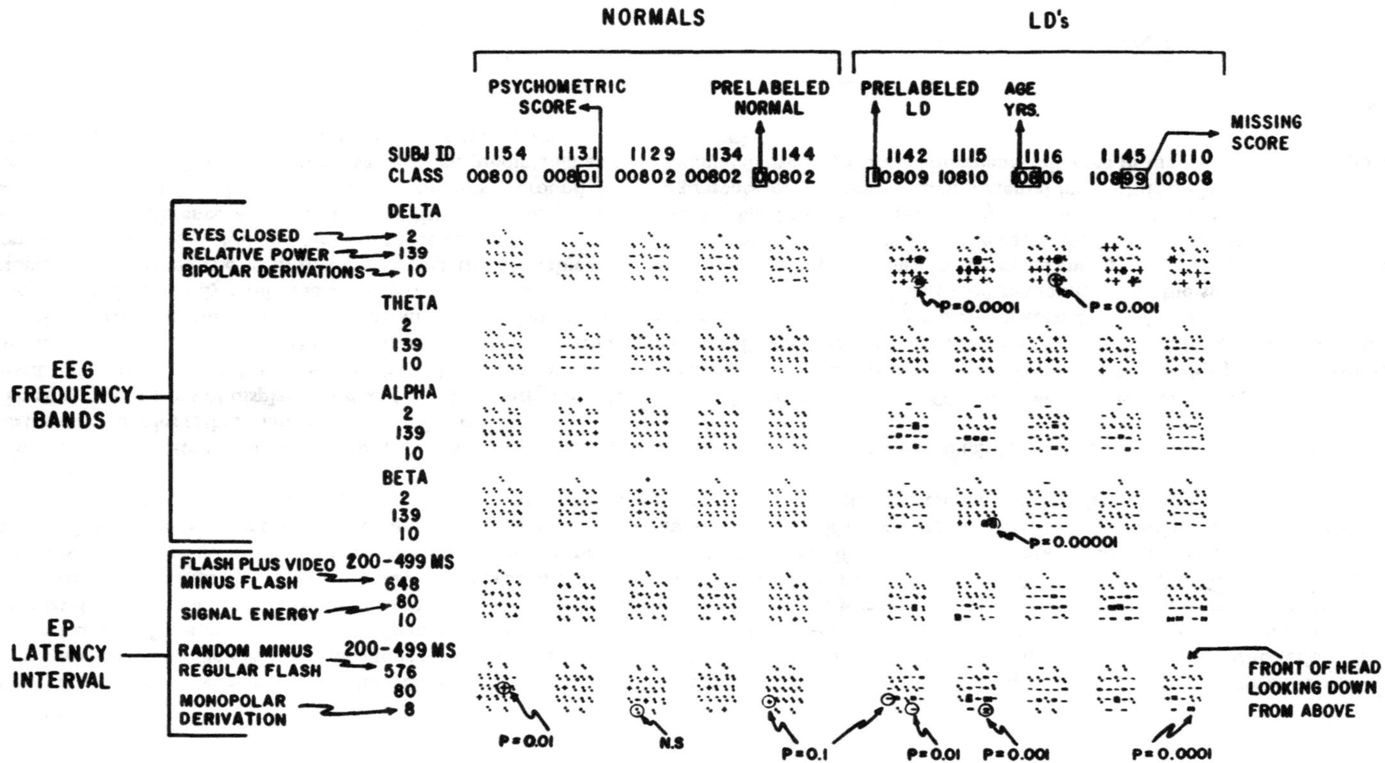

FIG. 8.5 Density coded z-transformed displays of neurometric indices extracted from various EEG and EP conditions of the NB. Each column of displays represents data obtained from one subject, while each row represents one univariate or multivariate index. Each display represents an array

contd.

FIG. 8.5 (*contd.*)

of entries: each entry corresponds to the value of the index measured at that point on the subject's head, while the position of the entries in the array corresponds to the electrode locations of the 10/20 system. For each index, the entry at any location has been density coded to reflect the z transformation of the measure obtained from that subject referred to the mean of the whole population. If the z-transformed value was such that the p level of obtaining that value by chance was not less than .1, two small spots (˙ .) were entered to convey that the measure was assessed and found within the normal range. If the value of z was such that the p level was between .1 and .01, a small plus was entered to show that the index was unusually large and a small minus was entered if the index was unusually small; if the p level was between .01 and .001, a large plus was entered if the value was abnormally *high* and a large minus if it was abnormally *low;* if the p level was between .001 and .0001, a double ++ or = was entered; between .0001 and .00001, a triple +++ or ≡ was entered; p levels below 0.00001 were indicated by large solid shapes in the form of plus or minus signs.

The prototypic data illustrated in the figure show 5 normal and 5 LD children selected from a much larger sample and are to be considered as illustrative examples rather than as invariable findings.

The upper four rows of displays represent the distribution of relative power in the spontaneous EEG, recorded from bipolar derivations with eyes closed, respectively from top down in the delta, theta, alpha, and beta bands. Note the typical excess of slow delta activity predominantly in posterior head regions of the LD subjects, usually coupled with a deficiency of alpha and sometimes of beta activity. The fifth row shows that the LD subjects show significantly less change in the signal energy of the bipolar EP in the latency region between 200 and 500 milliseconds when a flash delivered randomly while the subject is watching a TV cartoon is compared with a flash delivered randomly while the subject looks at the defocused TV screen. The sixth row shows that the LD subjects display significantly less change in the signal energy of the monopolar EP in the latency region between 200 and 500 msec when a random is compared with a regular flash. Although the particular head region displaying this less than expected difference when the two conditions are compared vary from subject to subject, such findings show that LD children tend to display less suppression of P-300 to an irrelevant stimulus ("ground") in the presence of meaningful environmental input ("figure") and, analogously, display less of a tendency to distinguish between predictable and unpredictable events in the environment, reflected in the similarity of late positive components in EPs elicited by these two different kinds of events.

It should be clear to the reader that any individual, all of whose indices lie within the "normal" or average range, will be represented by a vector which lies at the origin of this z deviate space. The more "abnormal" an individual, that is, the further his indices lie from the population mean, the *longer* will be the vector representing that individual in the z space. Thus, z-vector length represents degree of abnormality. z-vector orientation in the z space defines the type of abnormality.

If we adopt the conventions defined above, the basis for numerical taxonomic classification of types of abnormality has been achieved. It now becomes possible to use methods of numerical taxonomy, such as cluster analysis, to identify groups or "clusters" of z vectors in the z space that are nonrandomly associated. Individuals whose z vectors lie significantly more closely together than the average distance between points in the space represent members of the same numerically defined taxonomic class. (For further discussions of numerical taxonomic and cluster analytic methods, see Chapters 3 and 7.)

Figure 8.6 shows the results of applying the same cluster analysis method used for the classification of senile patients in Chapter 7 to the categorization of learning handicapped (presumably learning disabled) children. These results were obtained by clustering vectors representing the whole EP wave, without reducing the number of data points by analysis of variance to determine the optimally discriminating latencies and without z transform to optimize the signal to noise ratio in the data vectors. These results are presented to illustrate the power of the cluster analytic procedure even without the stratagems described to maximize the sensitivity of the technique.

The cluster analyses were carried out on data from 3 EP conditions: blank flash, 27 lines/inch, and large square. The data vector for each individual consisted of the full set of 100 data points for the EP obtained from leads P_4 and C_3 in each of the three conditions.

Five significantly distinct clusters of LD children were obtained from this analysis. For each cluster, the average cluster EP waveshape is illustrated for both leads. The top row of EP waveshapes illustrate the average response of the first cluster in each condition, the second row illustrates the average response of the second cluster, and so on. The bottom waveshape represents the average response of the whole population, also overlaid as the background waveshape accompanying each prototypic cluster average.

Examination of those data shows that the five different clusters of LD children display five distinctly different EP waveshapes in each derivation for each EP condition. Furthermore, the distinctive features displayed by a given cluster under one stimulus condition are manifested in their essential aspects in the other EP conditions. For example, the second cluster (Row 2) displays a waveshape in lead P_4 under each of the 3 conditions which is distinctively different from the waveshapes obtained from the other four clusters in that it lacks the large negative–positive complex N_2–P_3 evident in the other clusters.

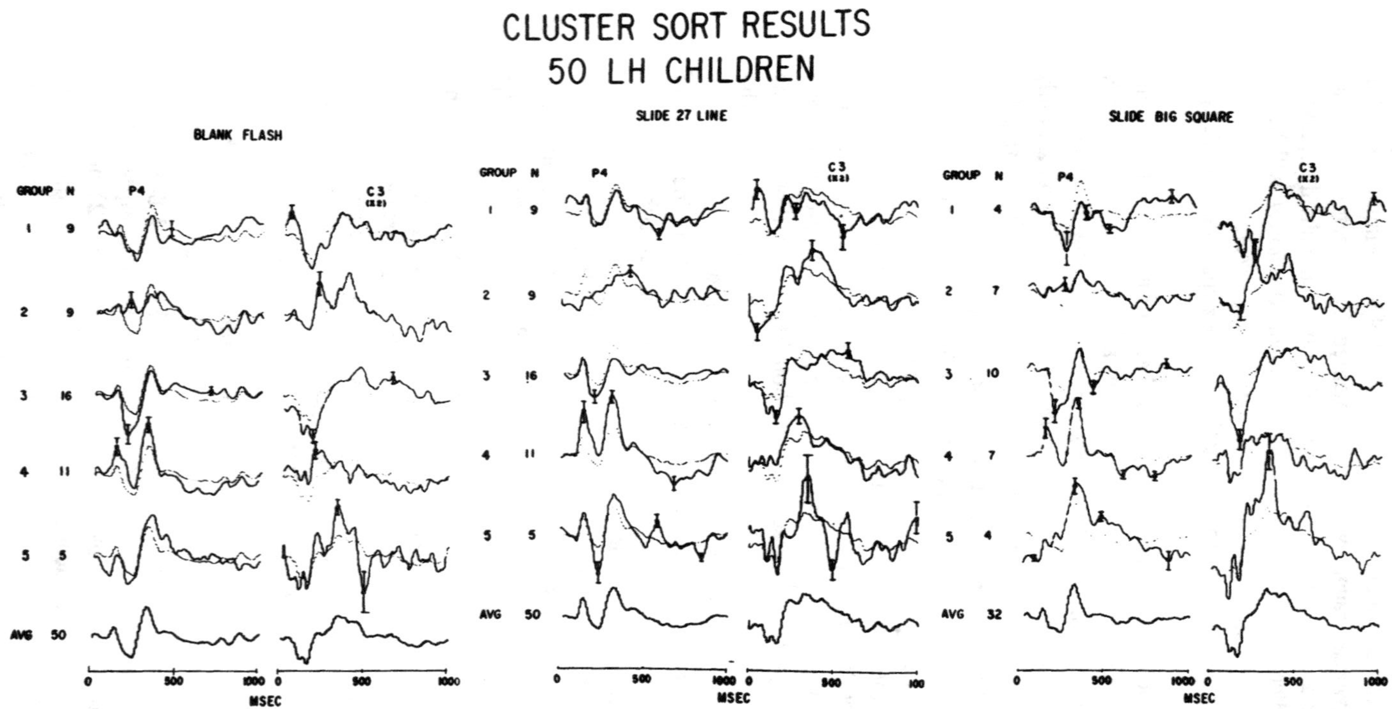

FIG. 8.6 Average evoked responses characteristic of leads P_4 and C_3 in 5 clusters of learning disabled children under 3 different EP conditions of the NB. Each row depicts the average response of a different cluster. The bottom row of waveshapes represents the average response of the whole population, also depicted as the light waveshapes over which the individual cluster averages are superimposed. The bracket in the EP curves indicates the mean ±1 *SD* at each time point where the cluster average deviates significantly (<.01) from the population average. Note the relatively minor differences between the 'population averager' waveshapes obtained in the three different conditions, although each of the five clusters displays clear differences between conditions. This illustrates how combination of heterogeneous subgroups into an ostensibly homogeneous pool can obscure important features of electrophysiological data. Note also the relatively consistent features which characterize salient aspects of the EP waveshape displayed by any particular cluster across the 3 different conditions; for example, the large positive–negative wave between 350 and 500 msec shown in electrode C_3 by Cluster 5. (LH = LD).

As another example, Cluster 5 shows a marked positive–negative process in the interval 350–500 msec in lead C_3, which is absent in EPs obtained from the other clusters.

Thus, cluster analysis of EP waveforms, even without the enhancement of discriminative features available by combining analysis of variance and *z* transforms, reveals a clear heterogeneity within the population of learning disabled children. Potentially, the distinctions between children, classified as a homogeneous group called "learning disabled" by behavioral and psychometric techniques, which are made possible by applying numerical taxonomic methods to neurometric data, will eventually lead to the identification of distinct etiologies for different types of learning disabilities and, more important, to the development of individualized prescriptive therapies optimal for the differential remediation of members of these different subgroups, subsumed within the present indiscriminate label of "learning disability."

Obviously, as additional conditions of the NB are included in the numerical taxonomic procedure, the subtlety of the differential diagnosis and the number of the distinctive subgroups will increase. Given sufficient effort by enough investigators so that a truly representative sample of learning disabled children can be studied, and sufficient time to explore the functional implications of membership in any subgroup, we can look forward to a definitive classification of the distinct types of learning disability, an understanding of the critical factors contributing to their etiology, and development of the optimal individual prescriptive therapies that will ameliorate many of these conditions.

TABLE 8A.1

Absolute Energy Distribution in Different Frequency Bands of the EEG: Means and Standard Deviations of Amplitude of the EEG in Various Frequency Bands as a Function of Electrode Position and Age[a]

Age group	1-	2-	3-	4-	5-	6-	7-	8-	9-	10-	11-	12-	13-	14-	15-	16-	17-	18-	19-	20-	21-22
n =	19	18	18	22	26	18	26	41	25	31	24	29	29	49	26	26	32	21	29	25	27
F_7-T_3 & F_8-T_4																					
δ	23.6	23.6	19.2	18.9	19.9	16.6	15.0	14.7	14.2	14.6	12.6	11.2	12.0	10.2	8.7	7.8	7.5	7.1	7.5	7.4	6.8
	5.3	7.3	3.9	5.2	7.3	3.0	5.6	4.1	4.4	4.5	3.5	2.9	4.0	3.9	2.5	2.3	2.2	1.9	2.1	2.8	2.3
θ	15.4	16.9	13.8	15.0	15.1	12.5	11.1	10.2	11.2	10.8	9.9	8.5	8.8	7.4	6.2	4.9	5.2	4.4	4.7	4.3	4.2
	5.1	5.2	3.5	4.7	5.2	3.3	3.6	2.3	3.8	4.3	2.8	2.6	3.3	3.5	2.0	1.6	2.8	1.3	1.6	1.5	2.0
α1	6.3	7.1	6.8	7.7	8.4	7.3	6.5	6.0	7.4	7.1	6.3	6.0	6.8	6.3	5.1	4.5	4.6	4.1	4.7	5.0	4.2
	1.9	2.0	2.5	2.3	3.1	2.4	1.8	1.7	3.6	3.8	2.6	2.6	3.6	5.0	2.4	2.4	2.8	2.0	3.2	3.0	2.2
α2	4.8	5.1	4.8	5.3	6.0	5.4	5.2	4.9	5.3	5.4	5.1	5.3	5.5	5.4	5.4	3.8	3.9	4.1	4.0	4.6	4.3
	1.8	1.3	1.2	1.5	2.4	1.9	2.3	1.6	1.8	3.0	2.8	2.7	2.5	3.3	2.3	2.3	1.7	1.9	2.3	2.4	2.3
β1	8.4	7.4	6.5	6.7	6.9	5.9	5.5	5.3	6.3	5.5	5.2	4.9	5.5	4.8	4.6	3.9	4.0	4.5	3.7	4.1	3.4
	3.5	2.3	1.6	2.2	3.2	1.5	1.4	1.4	1.8	2.4	1.4	1.8	3.3	2.6	1.4	1.4	1.3	1.8	1.0	2.3	1.2
β2	11.7	8.6	7.5	7.7	7.9	6.2	5.9	5.6	6.4	6.4	5.2	5.0	6.4	4.9	4.6	4.7	4.4	5.6	4.2	4.9	4.5
	5.9	3.4	2.5	3.5	3.9	1.4	1.9	2.0	2.1	3.4	1.8	2.1	4.7	2.9	1.4	2.6	1.6	3.2	1.1	2.9	2.8
C_0-C_3																					
δ	24.4	24.0	17.7	17.1	17.7	15.9	14.1	13.5	13.6	11.9	10.7	11.1	10.5	8.2	6.7	6.0	6.1	4.7	4.8	4.2	4.8
	7.5	8.8	4.8	3.2	6.8	3.9	7.3	3.7	4.1	3.7	2.5	3.7	3.0	2.8	2.0	1.5	3.5	1.3	1.1	1.3	2.0
θ	18.2	19.4	16.8	17.5	18.0	16.6	14.0	13.6	14.8	12.2	12.4	11.1	11.3	8.5	7.7	5.8	6.0	4.8	5.1	4.4	4.9
	6.4	7.4	4.0	4.5	7.7	5.7	4.5	3.0	5.5	4.1	3.4	3.5	3.1	2.9	2.7	2.1	2.8	1.4	2.1	1.8	2.3
α1	9.0	10.8	11.1	11.8	12.5	12.2	10.2	8.9	10.4	9.9	9.3	8.8	8.7	6.8	7.8	4.7	5.5	3.8	5.1	4.7	5.2
	3.6	6.4	5.0	4.6	5.1	6.5	4.0	3.1	5.2	6.4	4.0	5.3	4.7	4.1	4.5	2.5	4.2	1.6	4.0	2.8	3.9
α2	5.7	6.0	6.7	6.5	8.8	8.4	6.9	7.7	7.8	7.9	7.3	9.2	9.2	6.7	9.0	4.6	4.6	4.6	5.3	5.0	5.3
	2.4	2.2	2.7	1.9	4.1	4.2	2.8	3.1	3.6	4.5	3.4	5.3	5.3	4.0	5.5	2.8	2.6	2.2	3.4	3.0	2.8
β1	7.9	6.5	5.8	5.8	6.2	5.8	5.6	5.4	6.2	5.1	5.3	5.8	5.5	4.4	4.7	3.7	3.7	3.5	3.3	3.4	3.7
	4.7	2.5	1.4	1.6	2.9	2.0	1.9	1.4	2.5	2.2	1.3	2.7	2.4	1.9	1.9	1.2	1.4	1.8	1.2	1.8	1.9
β2	8.8	5.8	5.5	5.2	6.1	4.9	4.8	4.7	5.4	4.9	4.7	5.3	6.1	4.4	4.9	3.8	3.8	3.8	3.8	4.3	4.5
	5.8	1.8	2.0	1.6	4.8	1.4	2.1	1.4	2.4	2.7	1.4	2.3	4.4	2.0	2.1	1.2	1.2	1.5	1.4	2.6	1.7

T_3–T_5 & T_4–T_6																					
δ	23.7	22.9	18.7	20.0	19.8	16.9	16.1	16.1	15.5	14.8	13.9	12.7	14.1	9.8	8.5	7.2	7.6	6.9	6.5	6.3	5.9
	6.8	5.3	4.7	6.5	6.7	4.6	6.0	7.2	5.7	5.6	4.7	4.4	11.7	3.8	3.0	2.4	3.4	2.0	2.2	2.0	2.3
θ	17.6	17.8	17.9	21.2	19.4	16.8	15.1	13.9	15.2	14.0	13.3	10.4	10.5	9.8	7.8	5.4	6.4	5.4	5.4	5.5	5.0
	7.7	5.0	7.7	8.1	7.0	7.5	5.6	6.3	6.1	6.1	6.5	4.2	3.7	5.0	3.6	2.2	4.6	1.8	2.7	2.8	2.7
α1	6.7	7.4	8.2	10.5	13.4	11.1	11.1	12.5	15.0	13.2	12.4	11.6	10.1	9.3	9.6	7.2	8.9	7.9	7.6	9.0	8.0
	2.0	2.1	4.1	3.6	5.4	5.8	4.2	7.0	10.8	8.2	8.5	7.4	6.0	6.7	6.5	6.8	8.6	5.4	5.8	7.4	7.3
α2	5.0	5.1	5.1	6.5	7.6	7.5	8.3	10.1	9.8	10.3	9.5	11.4	10.7	8.9	11.0	6.7	8.1	8.8	7.7	9.4	8.3
	1.5	1.4	1.5	2.6	3.0	3.4	4.7	5.8	4.2	8.0	5.5	7.4	5.1	4.5	5.5	4.8	5.0	5.3	5.9	5.0	4.8
β1	8.4	7.4	6.6	7.3	7.7	6.8	6.6	6.9	8.3	6.8	6.4	6.3	6.5	5.7	6.0	4.7	4.9	5.5	4.6	5.1	4.5
	2.8	2.7	2.0	2.7	3.5	2.5	1.6	3.1	2.3	2.8	2.0	2.7	2.9	2.2	1.8	1.7	1.9	3.2	1.3	2.4	1.6
β2	12.5	8.3	7.0	7.5	7.8	5.7	5.7	5.7	6.6	6.0	4.9	4.9	5.7	4.8	5.2	4.6	4.7	5.7	4.5	5.1	4.5
	6.7	3.4	2.4	3.7	4.3	1.5	1.7	2.7	2.2	2.7	1.5	2.0	3.6	2.1	1.5	2.0	1.8	3.2	1.5	1.8	1.6
P_3–O_1 + P_4–O_2																					
δ	28.3	28.3	20.1	21.7	23.4	18.0	19.2	16.7	16.8	16.8	16.4	13.9	13.4	11.1	9.3	7.5	7.4	7.0	6.3	6.9	6.1
	9.3	8.0	3.8	5.3	9.6	5.6	7.2	4.0	6.0	7.7	6.8	5.6	5.2	4.3	3.6	2.8	3.6	1.9	2.2	3.9	2.1
θ	19.0	22.5	21.1	22.4	25.6	21.7	17.7	16.5	18.0	16.7	17.0	12.6	12.6	10.1	9.6	6.6	7.0	6.5	6.2	6.3	5.8
	7.0	7.7	7.3	6.6	11.2	8.7	6.2	6.1	7.9	8.0	8.2	4.9	5.0	4.5	3.8	3.0	4.0	2.3	3.1	3.5	3.0
α1	8.7	9.4	12.0	15.3	23.4	18.0	17.0	18.2	20.5	19.2	20.0	16.6	17.0	12.5	12.8	10.7	11.7	8.9	9.5	10.1	10.6
	3.3	3.3	6.1	5.3	12.2	6.7	7.0	7.3	10.3	9.6	11.7	9.2	9.0	8.3	7.4	10.0	9.2	5.8	6.5	7.6	8.5
α2	5.7	6.0	6.8	7.9	11.0	11.1	11.6	14.5	14.0	14.4	15.1	17.8	18.8	13.5	16.6	12.4	12.8	11.8	11.4	13.1	13.1
	2.0	1.8	2.0	2.1	5.1	4.3	6.6	6.5	5.7	11.0	7.4	10.5	10.2	8.4	8.3	9.0	9.0	6.6	8.8	7.1	7.6
β1	7.3	6.0	7.1	7.3	9.0	8.0	7.5	7.7	8.9	7.7	8.5	8.4	8.0	6.6	7.2	5.8	5.3	6.1	5.1	5.4	4.8
	3.8	1.5	1.6	2.1	4.5	2.6	2.0	2.1	3.2	3.4	2.6	4.0	3.0	2.7	2.5	2.2	2.3	4.6	1.7	2.0	1.9
β2	9.1	5.9	5.7	5.5	6.2	5.2	5.3	5.5	5.9	5.6	6.2	7.0	6.8	5.2	6.5	4.9	4.5	4.9	4.6	5.5	5.1
	5.5	1.3	1.7	2.5	2.9	1.3	1.9	1.7	1.9	2.8	2.4	3.3	2.9	2.0	2.2	1.7	1.9	2.0	2.1	2.1	2.4

[a]From Matoušek and Petersén (1973b).

TABLE 8A.2

Relative Frequency Distribution in Different Frequency Bands in the EEG (Means and Standard Deviations of Percentage Energy of the EEG and Various Frequency Bands as a Function of Electrode Position and Age)

,65,c63,c63,,	Age																			
	1		2		3		4		5		6		7 8		9 10					
%	M	SD	M	SD	M	SD	M	SD	M	SD	M	SD	M	SD	M	SD	M	SD	M	SD
F–T																				
δ	52.3	2.6	53.2	5.0	50.7	2.1	46.2	3.5	47.0	6.3	46.9	1.5	46.6	6.5	49.2	3.8	41.1	4.0	44.4	4.2
θ	22.3	2.4	27.3	2.6	26.2	1.7	29.1	2.9	27.1	3.2	26.6	1.8	25.5	2.7	23.7	1.2	25.6	2.9	24.3	3.8
α	5.9	.6	1.3	.5	9.5	1.1	11.3	1.0	12.7	1.8	14.0	1.6	14.4	1.8	13.7	1.3	16.9	3.3	16.6	4.9
β	19.5	4.4	12.3	1.6	13.6	1.2	13.5	2.2	13.1	3.0	12.5	.7	13.5	1.2	13.5	1.4	16.4	1.5	14.8	3.6
C_o–C_3																				
δ	50.5	4.7	48.7	6.5	37.9	2.8	34.8	1.2	33.1	4.7	31.4	1.9	26.2	8.9	32.7	2.5	28.9	2.6	28.3	2.7
θ	28.1	3.5	31.8	4.6	34.1	1.9	36.4	2.4	34.2	6.3	34.2	4.0	25.8	3.4	33.2	1.6	34.2	4.7	29.7	3.4
α	9.6	1.6	12.9	3.9	20.3	3.9	21.6	2.9	24.7	4.5	27.2	7.4	40.8	4.0	24.9	3.4	26.4	6.2	32.0	12.2
β	11.9	4.7	6.5	.8	1.1	.7	1.2	.6	8.0	3.3	7.2	.7	7.2	1.3	9.2	.7	10.6	1.9	10.0	2.4
T–T																				
δ	48.1	4.0	50.2	2.7	40.9	2.6	35.1	3.8	34.8	4.0	34.6	2.5	34.3	4.8	32.8	6.6	26.6	3.6	28.2	4.0
θ	26.5	5.1	30.3	2.4	31.4	6.9	40.2	5.9	33.4	4.4	34.2	6.8	30.2	4.2	24.4	5.0	25.5	4.1	25.2	4.8
α	6.0	.5	1.1	.6	10.9	2.2	13.8	1.8	21.1	3.4	21.7	5.4	25.4	8.3	32.7	10.4	35.5	14.8	36.1	16.9
β	19.4	4.5	11.8	1.8	10.8	1.1	9.9	1.9	10.7	2.7	9.5	1.0	10.1	.7	10.7	2.1	12.4	1.1	10.6	1.9
O–P																				
δ	57.0	6.2	53.3	4.3	36.0	1.3	34.8	2.1	27.5	4.6	24.3	2.5	31.0	4.4	23.6	1.4	21.1	2.7	23.0	4.8
θ	25.7	3.5	33.7	3.9	39.7	4.7	31.1	3.2	32.9	6.3	35.3	6.1	26.3	3.2	23.0	3.1	24.2	4.7	22.7	5.2
α	1.7	1.0	8.3	.9	11.0	3.7	21.9	2.4	33.6	8.8	33.5	5.1	35.6	7.8	45.2	8.0	46.1	10.4	46.9	10.0
β	9.3	3.2	4.7	.3	7.4	.5	6.2	.8	6.0	1.4	6.8	.7	7.1	.6	7.6	.6	8.5	1.0	7.4	1.6

	Age																					
	11		12		13		14		15		16		17		18		19		20		21	
%	M	SD	M	SD	M	SD	M	SD	M	SD	M	SD	M	SD	M	SD	M	SD	M	SD	M	SD
F–T																						
δ	42.2	3.3	40.4	2.7	39.0	4.3	37.9	5.5	35.8	3.0	38.8	3.4	36.3	3.1	32.5	2.4	38.1	3.0	34.2	4.9	35.1	4.0
θ	26.0	2.1	23.2	2.2	21.0	3.0	19.9	4.5	18.2	1.9	15.3	1.7	17.4	5.0	12.5	1.1	14.9	1.8	11.5	1.4	13.4	3.0
α	17.4	3.8	20.6	4.5	20.7	5.2	25.1	13.1	26.1	5.2	22.1	7.9	23.5	6.9	21.7	5.0	25.8	10.5	28.8	9.2	27.4	7.7
β	14.4	7.1	16.8	2.4	19.3	8.9	17.1	5.5	20.0	1.9	23.8	5.6	22.8	2.8	33.3	8.8	21.2	1.5	25.5	8.5	24.1	7.0
C_0–C_3																						
δ	25.0	1.4	26.2	2.9	23.7	1.9	25.0	2.9	15.8	1.4	25.5	1.6	24.4	8.0	20.6	1.6	17.9	.9	15.5	1.5	16.9	2.9
θ	23.6	2.5	26.2	2.6	21.4	2.1	26.8	3.1	30.3	2.5	23.9	3.1	23.6	5.1	21.4	1.9	20.2	3.4	17.0	2.8	17.3	3.9
α	30.5	6.0	34.5	12.0	34.4	10.8	33.8	12.2	48.6	17.3	30.7	10.0	33.7	16.0	33.1	6.9	42.1	2.5	41.3	14.7	40.5	16.9
β	11.0	.8	13.1	2.7	14.5	5.4	14.4	2.8	15.8	2.2	19.9	2.0	18.4	2.2	24.8	5.1	19.7	2.6	26.3	8.8	24.9	4.8
T–T																						
δ	28.5	4.6	27.0	3.2	33.1	2.3	23.2	3.5	17.7	2.2	23.5	2.6	19.9	4.0	17.0	1.4	18.4	2.1	13.6	1.4	14.9	2.3
θ	26.0	8.8	18.1	2.9	18.4	2.3	23.2	6.1	14.9	3.2	13.2	2.2	14.1	7.3	10.4	1.1	12.7	3.2	10.4	2.7	10.7	3.1
α	35.9	21.4	44.3	18.3	36.1	10.3	40.1	15.8	52.1	17.7	43.8	31.3	50.0	34.2	50.1	20.5	50.9	29.7	58.1	27.4	57.0	32.1
β	9.6	1.3	10.7	1.9	12.5	3.6	13.4	2.2	15.4	1.3	19.6	3.1	15.9	2.3	22.5	7.3	18.0	1.7	17.9	3.1	17.4	2.2
O–P																						
δ	20.7	3.6	18.2	2.9	16.5	2.5	19.4	2.9	12.1	1.8	13.2	1.8	12.1	2.9	13.2	1.0	11.5	1.4	11.3	3.6	9.2	1.1
θ	23.3	5.2	14.9	2.3	14.6	2.3	17.1	3.2	12.9	2.0	10.2	2.1	10.8	3.5	11.4	1.4	11.1	2.8	9.4	2.9	8.3	2.2
α	42.4	14.8	55.7	18.3	58.9	17.0	53.4	22.0	61.7	17.8	63.0	42.5	66.4	36.6	58.9	20.7	63.7	34.6	65.1	25.7	70.3	32.2
β	8.5	1.0	11.2	2.5	10.1	1.6	11.1	1.8	13.2	1.8	13.5	1.8	10.7	2.0	16.5	6.8	14.1	2.0	14.1	2.0	12.1	2.3

TABLE 8A.3
Deviations from Normal Frequency Distribution Significant at the .05 and .01 Level in Each Frequency Band as a Function of Electrode Position and Age

	Age																			
	1		2		3		4		5		6		7		8		9		10	
	.05	.01	.05	.01	.05	.01	.05	.01	.05	.01	.05	.01	.05	.01	.05	.01	.05	.01	.05	.01
F–T																				
δ	57.4	58.8	63.0	65.7	54.8	55.9	53.1	55.0	59.3	62.8	49.8	50.8	49.3	62.8	56.6	58.7	48.9	51.1	52.0	54.9
θ	27.0	28.3	32.4	33.8	29.5	30.4	34.8	38.9	33.4	35.1	30.4	31.1	30.8	32.3	26.1	26.7	31.3	32.8	31.7	33.8
α	4.7	4.4	0.3	0.1	7.3	6.8	9.3	8.8	10.9	10.0	10.9	10.0	10.9	9.9	11.2	10.5	10.5	8.7	7.0	4.4
β	28.1	30.5	15.4	16.3	16.0	16.6	17.8	19.0	19.0	20.6	13.7	14.3	15.9	18.5	16.2	17.0	19.3	20.2	21.9	23.8
C_0–C_3																				
δ	59.7	62.3	61.4	64.9	43.4	44.9	37.2	37.8	42.7	45.3	35.1	36.2	43.6	48.4	37.6	35.2	34.0	35.4	33.6	35.1
θ	35.0	36.9	35.8	43.3	37.8	38.9	41.1	42.4	46.5	50.0	42.0	44.2	32.5	34.3	36.3	37.2	43.4	46.0	36.4	38.2
α	6.5	5.6	5.3	3.2	12.7	10.6	16.0	14.4	15.9	13.5	12.7	8.7	33.0	30.8	11.3	16.4	14.3	10.9	8.1	1.5
β	21.1	23.7	8.1	8.5	2.5	2.9	2.4	2.7	14.5	16.2	8.6	9.0	9.8	10.4	10.6	11.0	14.3	15.4	14.7	16.0
T–T																				
δ	55.9	58.1	55.5	57.0	46.0	47.4	43.5	45.6	42.6	44.8	39.5	40.8	43.7	46.3	45.7	49.3	33.6	35.6	36.0	38.2
θ	36.5	32.3	35.0	36.3	44.9	48.6	51.8	55.0	42.0	44.8	47.5	51.2	38.4	40.7	34.2	36.9	33.5	35.7	34.6	37.2
α	5.0	4.8	5.9	5.6	6.6	5.4	10.3	9.3	14.5	12.6	11.2	3.2	15.0	12.2	12.3	6.7	6.5	1.5	3.0	–
β	28.2	30.6	15.3	16.3	13.0	18.6	13.6	14.7	16.0	17.5	11.5	12.0	11.5	11.9	14.8	15.9	14.6	15.7	14.3	15.4
O–P																				
δ	69.2	72.5	61.7	64.1	38.5	39.2	38.9	40.0	36.5	39.0	29.2	30.5	39.6	42.0	26.3	27.1	26.4	27.9	32.4	35.0
θ	32.6	34.5	41.3	43.5	48.9	51.5	37.4	39.1	45.2	48.7	47.3	50.5	32.6	34.3	29.1	30.8	33.4	36.0	32.9	35.7
α	–	–	6.6	6.1	3.8	1.8	17.2	15.9	16.4	11.6	23.6	20.8	20.4	16.1	30.2	25.8	25.7	20.1	27.3	21.9
β	16.0	17.7	5.3	5.5	8.4	8.6	7.8	8.2	8.7	9.5	8.2	8.6	8.3	8.6	8.8	9.1	10.5	11.0	10.5	11.4

	Age																					
	11		12		13		14		15		16		17		18		19		20		21	
	.05	.01	.05	.01	.05	.01	.05	.01	.05	.01	.05	.01	.05	.01	.05	.01	.05	.01	.05	.01	.05	.01
F–T																						
δ	48.7	50.4	45.7	47.2	47.4	49.8	48.7	51.7	41.7	43.3	45.5	47.3	42.4	44.1	37.2	38.5	43.9	45.6	43.8	46.4	42.9	45.1
θ	30.1	31.2	27.5	28.8	26.9	28.5	28.7	31.1	21.9	23.0	18.6	19.5	27.2	29.9	14.7	15.3	18.4	19.4	14.2	15.0	19.3	20.9
α	10.0	7.9	11.8	9.4	10.5	7.7	–	–	15.9	13.1	8.2	4.4	10.0	6.3	11.9	9.2	5.2	0.6	10.8	5.8	12.3	8.2
β	28.3	32.2	20.5	21.8	36.7	41.5	27.9	30.9	23.7	24.8	34.8	27.8	28.3	29.8	50.5	55.	24.1	25.0	42.1	46.7	37.8	41.6
C_0–C_3																						
δ	27.7	28.5	31.9	33.4	27.4	28.4	30.7	32.2	18.1	18.9	28.6	29.5	40.1	44.4	23.7	24.6	19.7	20.1	18.4	19.3	22.6	24.1
θ	28.5	29.8	31.3	32.7	25.5	26.6	32.9	34.6	25.2	26.5	30.0	31.7	33.6	36.3	25.1	26.2	26.8	28.7	22.5	24.0	24.9	27.1
α	18.7	15.5	11.0	4.5	13.2	7.4	9.9	3.3	14.7	5.4	11.1	5.7	2.3	–	19.6	15.9	–	–	12.5	4.5	7.4	–
β	12.6	13.0	18.4	19.9	25.1	28.0	19.9	21.4	20.1	21.3	23.8	24.9	22.7	23.9	34.8	37.6	24.8	26.2	43.5	48.3	34.3	36.5
T–T																						
δ	37.5	40.0	33.3	35.0	37.6	38.9	30.1	32.0	22.0	23.2	28.6	30.0	27.7	29.9	19.5	20.5	22.5	23.6	16.3	17.1	19.4	20.3
θ	43.2	48.0	23.8	25.3	22.9	24.2	35.2	38.4	21.2	22.9	17.5	18.7	28.4	32.3	12.6	13.2	19.0	20.7	15.7	17.2	16.3	18.5
α	–	–	8.5	–	15.9	10.4	9.2	0.6	17.4	7.9	–	–	–	–	9.9	–	–	–	4.4	–	–	–
β	12.1	12.8	14.4	15.5	19.6	21.5	17.7	18.9	17.9	18.6	25.7	27.4	20.4	21.7	36.8	40.7	21.3	22.2	23.9	25.7	21.7	22.9
O–P																						
δ	27.7	29.7	23.9	25.4	21.4	22.7	25.1	26.6	15.6	16.6	16.7	17.7	17.8	19.3	15.2	15.7	14.2	15.0	13.3	20.3	11.3	12.0
θ	32.5	35.3	19.4	20.7	19.1	20.4	22.4	24.1	16.8	17.9	14.3	15.4	17.7	14.6	14.1	14.3	16.6	18.1	15.1	16.6	12.6	13.8
α	19.4	11.4	19.8	10.0	25.6	16.4	10.3	–	27.6	–	–	–	–	–	18.2	6.9	–	–	14.7	0.9	7.9	–
β	10.5	11.0	16.1	17.4	13.2	14.1	14.6	15.6	16.7	17.7	17.0	18.0	14.6	15.7	29.8	33.5	17.8	18.9	18.0	19.1	16.6	17.9

NOTE: δ, θ, β: M + 1.96σ 0.05 level. M + 2.50σ 0.01 level. α: M – 1.96σ 0.05 level. M – 2.50σ 0.01 level.

9

The Perspective for Neurometrics

It is my hope that this volume may constitute the cornerstone of a structure that will become a new field of neuroscience, applied to pressing clinical problems. I feel no personal ostentation in this hope. The extensive bibliography of this volume constitutes concrete evidence of the labors of countless dedicated scientists who have devoted untold effort to the identification of functional correlates of the electrical activity of the brain and to the quantitative evaluation of multivariate data. My colleagues and I are fortunate in arriving upon the scene at a time when the body of accumulated facts and technological capability reached the "critical mass" necessary to systematize their knowledge into a standardized, clinically useful form.

In this volume, I have reviewed the state of our present knowledge about the functional implications of various features of the spontaneous and evoked electrical activity of the brain. Methods for supplanting visual examination of such features by multivariate statistical pattern recognition techniques have been surveyed. A technology has been described which permits the rapid, automatic acquisition of precise data about a spectrum of functionally significant electrophysiological measures and their reduction to numerical form. Techniques for the numerical taxonomic classification of such data have been surveyed. These several endeavors clearly established the theoretical practicality of quantitative and objective evaluation of the structural and functional integrity of the brain and the evaluation of sensory, perceptual and cognitive processes by the combination of neurometric and numerical taxonomic methods.

Three bodies of evidence have been adduced to establish that these methods are not only of theoretical interest but have demonstrable practical utility. Concrete results have been presented to establish that numerical taxonomic methods can be applied to neurometric data to achieve highly accurate discrimination between normal individuals and those suffering from neurological diseases and to achieve significant separation between individuals afflicted with different diseases, to discriminate between normal and senile aged individuals and to identify distinctly different subgroups within the senile population, to discriminate between normal and learning disabled children and to identify distinctly different subgroups within the learning disabled population.

The experimental results in these three clinical domains are important for two reasons. First, they demonstrate that essentially automatic, objective neurometric and numerical taxonomic methods surpass the accuracy of the conventional EEG in the identification of brain dysfunction arising from neuropathology and from far more subtle disorders in sensory, perceptual and cognitive processes. I believe that this evidence establishes incontrovertibly that clinical electrophysiology can be extended beyond its traditional domain of gross neuropathology, into the new domains of sensory, perceptual and cognitive dysfunctions from which many more people suffer than those afflicted with neurological diseases. The objective, automatic nature of this technology not only provides a powerful adjunct to the neurologist, but makes the equivalent of high-caliber neurological evaluation routinely accessible to the clinical practitioner, physician or psychologist, in the most remote rural area.

Second, our results demonstrate that dysfunctional conditions conventionally treated as homogeneous functional entities, such as "senility" or "learning disability," can, in fact, be decomposed into homogeneous subgroups. Presumably, each of these subgroups potentially reflects a unique etiology and will be optimally responsive to a particular treatment. Hitherto, attempts to develop pharmacotherapeutic or other approaches to the amelioration of these conditions have foundered upon the rocks of heterogeneity of variance. Evaluation of the efficacy of any treatment has been impeded by the presence of disparate subgroups within the "target" population. Hopefully, once distinct subgroups can be identified, it will be possible to evaluate the *differential* efficacy of treatments for any single subgroup. Similarly, it should become feasible to search for the factors contributing decisively to the etiology of each distinct manifestation of disordered cerebral process. Finally, it must be borne in mind that this technology may well reveal the existence of distinctly different subgroups within the normal population. Differences in cognitive style or other functional attributes of membership in a particular normal subgroup may have important pedagogical or social implications.

Not only can early identification of abnormal brain processes lead to a substantial improvement in prevention of debilitating disease or remediation of dysfunction, but identification of a particular normal mode of brain function may lead to selection of an optimal type of education or social environment.

It should be clear from the foregoing remarks, as well as from the text of this volume, that I believe that neurometrics has enormous contributions to make to the welfare of human beings. It has been a pleasure for my colleagues and me to serve as midwives to the birth of what I firmly believe is an important new field of medicine.

References

Akiyama, Y., Schulte, F. J., Schultz, M. A., & Parmelee, A. H. Acoustically evoked responses in premature and full-term newborn infants. *Electroencephalography & Clinical Neurophysiology*, 1969, *26*, 371–380.

Arnal, D., Gerin, P., Salmon, D., Ravault, M. P., Nakache, P., & Peronnet, F. *Electroencephalography & Clinical Neurophysiology*, 1972, *32*, 499–511.

Ball, G. H. Data analysis in the social sciences: What about the details? *Proceedings of the Fall Joint Computer Conference*, 1965, 533–599.

Barlow, J. S. Rhythmic activity induced by photic stimulation in relation to instrinsic alpha activity of the brain in man. *Electroencephalography & Clinical Neurophysiology*, 1960, *9*, 340–343.

Barlow, J. S., Morrell, L., & Morrell, F. Some observations on evoked responses in relation to temporal conditioning to paired stimuli in man. *Proceedings of International Colloquium on Mechanisms of Orienting Reactions in Man*, 1967, (Bratislava-Smolenice, Czechoslovakia). (Published by Slovak Academy of Science, 1967.)

Barnes, R. H., Busse, E. W., Friedman, E. L. The psychological functioning of aged individuals with normal and abnormal EEG, II. *Journal of Neurology and Mental Disease*, 1956, *124*, 585–593.

Barnet, A. Sensory evoked response recording. In A. Rémond (Ed.), *Handbook of electrophysiology and clinical electrophysiology* (Vol. 15B). Amsterdam: Elsevier, 1972. Pp. 30–33.

Barnet, A., & Goodwin, R. S. Averaged evoked electroencephalographic responses to clicks in the human newborn. *Electroencephalography & Clinical Neurophysiology*, 1965, *18*, 441–450.

Barnet, A., & Lodge, A. Diagnosis of deafness in infants with the use of computer-averaged electroencephalographic responses to sound. *Journal of Pediatrics* (St. Louis), 1966, *69*, 753–758.

Barnet, A., & Lodge, A. Click evoked EEG responses in normal and developmentally retarded infants. *Nature*, 1967, *214*, 252–255.

Begleiter, H., & Platz, A. Cortical evoked potentials to semantic stimuli. *Psychophysiology*, 1969, *6*, 91–100.

Bendat, J. S., & Piersol, A. G. *Random data: Analysis and procedures.* New York: Wiley, 1971.

Bender, L. *A visual–motor gestalt test and its clinical use.* New York: American Orthopsychiatric Assoc., 1938.

Bennett, J. F., MacDonald, J. S., Drance, S. M., & Venoyama, K. Some statistical properties of the visual evoked response in man and their application as a criterion of normality. *Institute of Electrical & Electronics Engineers Transactions on Bio-Medical Engineering,* 1971, *18,* 23–24.

Benton, A. L., & Bird, G. W. The EEG and reading disability. *American Journal of Orthopsychiatry,* 1963, *33,* 520–531.

Bergamini, L., & Bergamasco, B. *Cortical evoked potentials in man.* Springfield, Ill.: Charles C Thomas, 1967.

Berger, H. Über das Elektrenkephalogramm des Menschen. *Archives Psychiatric Nervenkrankhen,* 1929, *87,* 527–570.

Bickford, R. G. Scope and limitations of frequency analysis. *Electroencephalography and Clinical Neurophysiology,* 1961, *20,* 9–13.

Bickford, R. G., Brimm, J., Berger, L., & Aung, M. Application of compressed spectral array in clinical EEG. In P. Kellaway & I. Petersén (Eds.), *Automation of Clinical Electroencephalography.* New York: Raven Press, 1973. Pp. 55–64.

Bickford, R. G., Fleming, N. I., & Billinger, T. W. Compression of EEG data by isometric power spectral plots. *Electroencephalography & Clinical Neurophysiology,* 1971, *31,* 632.

Bigum, H. B., Dustman, R. E., & Beck, E. C. Visual and somatosensory evoked responses from mongoloid and normal children. *Electroencephalography & Clinical Neurophysiology,* 1970, *28,* 576–585.

Birch, H. G. *Brain damage in children.* Baltimore: Williams & Wilkins, 1964.

Birchfield, R. I., Wilson, W. P., & Heyman, A. An evaluation of electroencephalography in cerebral infarction and ischemia due to arteriosclerosis. *Neurology* (Minneapolis), 1959, *9,* 859–870.

Brazier, M. A. B. Physiologic mechanisms underlying the electrical activity of the brain. *Journal of Neurology, Neurosurgery & Psychiatry,* 1948, *11,* 118–133.

Brazier, M. A. B., & Barlow, J. S. Some applications of correlation analysis to clinical problems in electroencephalography. *Electroencephalography & Clinical Neurophysiology,* 1956, *8,* 325–331.

Brazier, M. A. B., & Casby, J. U. Cross-correlation and autocorrelation studies of electroencephalographic potentials. *Electroencephalography & Clinical Neurophysiology,* 1952, *4,* 201–211.

Buchsbaum, M., & Fedio, P. Visual information and evoked responses from the left and right hemispheres. *Electroencephalography & Clinical Neurophysiology,* 1969, *26,* 266–272.

Buchsbaum, M., & Fedio, P. Hemispheric differences in evoked potentials to verbal and nonverbal stimuli in the left and right visual fields. *Physiology and Behavior,* 1970, *5,* 207–210.

Buchsbaum, M., & Wender, P. Average evoked responses in normal and minimally brain dysfunctioned children treated with amphetamine. *Archives of General Psychiatry,* 1973, *29,* 764.

Burkhardt, D. A., & Riggs, L. A. Modification of the human visual evoked potential by monochromatic backgrounds. *Vision Research,* 1967, *7,* 453–459.

Burks, H. F. Diagnostic implications of the electroencephalogram for behavior problem children. *Journal of School Psychology,* 1968, *6,* 284–291.

Buser, P., & Borenstein, P. Réponses corticales (secondaires) à la stimulation sensorielle chez le chat curarisé non anesthésié. *Electroencephalography & Clinical Neurophysiology,* 1956, Supplement 6, 89–108.

Butler, B. V., & Engel, R. Mental and motor scores at 8 months in relation to neonatal photic responses. *Developmental Medicine and Child Neurology*, 1969, *11*, 77–82.

Cael, W. W., Nash, A., & Singer, J. J. The late positive components of the human EEG in a signal detection task. *Neuropsychologia*, 1974, *12*, 385–387.

Callaway, E. Correlations between averaged evoked potentials and measures of intelligence. *Archives of General Psychiatry*, 1973, *29*, 553–558.

Callaway, E., Jones, R. T., & Layne, R. S. Evoked responses and segmental set in schizophrenia. *Archives of General Psychiatry*, 1965, *12*, 83–89.

Capute, A. J., Niedermeyer, E. F. L., & Richardson, F. The electroencephalogram in children with minimal cerebral dysfunction. *Pediatrics*, 1968, *41*, 1104–1114.

Carmon, A., Lavy, S., & Schwartz, A. Correlation between electroencephalography and angiography in cerebrovascular accidents. *Electroencephalography & Clinical Neurophysiology*, 1966, *21*, 71–76.

Carter, S. Of priorities, promise and the path ahead. *Neurology*, 1971, *21*, 877–888.

Chen, C. H. *Statistical pattern recognition*. Rochelle Park, N.J.: Hayden, 1973.

Christian, C., & April, R. EEG aspects of different methods of obstetrical delivery in newborn infants. Paper presented at the 28th Annual Meeting of American EEG Society, Seattle, July, 1974.

Ciganek, L. The EEG response (evoked potential) to light stimulus in man. *Electroencephalography & Clinical Neurophysiology*, 1961, *13*, 165–172.

Ciurea, E., & Crighel, E. Neurogenic mechanisms in the evolution of cerebral circulatory insufficiency. Investigations on the role of the brain stem reticular formation. *Confinia Neurologica*, 1967, *29*, 351–359.

Clark, G. M. A summary of the literature in brain damaged children. In C. Haywood (Ed.), *Brain damage in school children*. Washington, D.C.: Council for Exceptional Children, 1968.

Clynes, M. Brain space analysis of evoked potential components applied to chromaticity waves. *Proceedings 6th International Conference Medical Electronics & Biological Engineering*, Tokyo, 1965.

Clynes, M., & Kohn, M. Spatial VEPs as physiologic language elements for color and field structure. *Electroencephalographic Journal* (Suppl. 1). 1967, *26*, 82–96.

Clynes, M., Kohn, M., & Gradijan, J. Computer recognition of the brain's visual perception through learning the brain's physiologic language. *Institute of Electrical and Electronics Engineers, International Conference Record*, Pt. 9, 1967, 125–142.

Cobb, W. A. The normal adult EEG. In D. Hill & G. Parr (Eds.), *Electroencephalography: A symposium on its various aspects*. New York: Macmillan, 1963. Pp. 232–239. (a)

Cobb, W. A. The EEG of lesions in general. In D. Hill and G. Parr (Eds.), *Electroencephalography*. London: Macdonald, 1963. Pp. 295–316. (b)

Cody, D. T. R., & Bickford, R. G. Cortical audiometry; an objective method of evaluating auditory acuity in man. *Proceedings of the Mayo Clinic*, 1965, *40*, 273–287.

Cohn, R. Rhythmic after-activity in visual evoked responses. *Annals of the New York Academy of Sciences*, 1964, *112*, 281–291.

Cohn, R., & Nardini, J. The correlation of bilateral occipital slow activity in the human EEG with certain disorders of behavior. *American Journal of Psychiatry*, 1958, *115*, 44–54.

Cohn, R., Raines, G. N., Mulder, D. W., & Neumann, M. A. Cerebral vascular lesions: electroencephalographic and neuropathologic correlations. *Archives of Neurology & Psychiatry*, 1948, *60*, 165–181.

Conners, C. K. A teacher rating scale for use in drug studies with children. *American Journal for Psychiatry*, 1969, *126*, 152–156.

Conners, C. K. Symptom patterns in hyperkinetic neurotic and normal children. *Child Development*, 1970, *41*, 667–673.

Conners, C. K. Cortical visual evoked response in children with learning disorders. *American Journal of Psychiatry*, 1971, 152–156.

Conners, C. K. Psychological assessment of children with minimal brain dysfunction. *Annals of the New York Academy of Sciences*, 1973, *205*, 283–302.

Cooley, J. W., & Tukey, J. W. An algorithm for the machine calculation of complex Fourier series. *Mathematics of Computation*, 1965, *19*, 297–301.

Copenhaver, R. M., & Perry, N. W., Jr. Factors affecting visually evoked cortical potentials such as impaired vision of varying etiology. *Investigative Ophthalmology*, 1964, *3*, 665–675.

Courjon, J. Traumatic disorders. In O. Magnus (Ed.), *Handbook of electroencephalography and clinical neurophysiology* (Part B, Vol. 14). Amsterdam: Elsevier, 1972.

Cress, C. M., & Gibbs, E. L. Electroencephalographic asymmetry during sleep in patients with vascular and traumatic hemiplegia. *Diseases of The Nervous System*, 1948, *9*, 327–329.

Creutzfeldt, O. D., & Kuhnt, U. The visual evoked potential: physiological, developmental and clinical aspects. *Electroencephalography & Clinical Neurophysiology*, 1967, Supplement *26*, 29–41.

Crighel, E., & Poilici, I. Photic evoked responses in patients with thalamic and brain stem lesions. *Confinia Neurologica*, 1968, *30*, 301–312.

Davis, H. (Ed.), The young deaf child: identification and management. *Acta Otolaryngologia* (Stockholm), Supplement 206, 1965.

Davis, H. Validation of evoked-response audiometry (ERA) in deaf children. *International Audiology*, 1966, *5*, 77–81.

Davis, H. Averaged evoked response EEG audiometry in North America. *Acta Otolaryngologia* (Stockholm), 1968, *65*, 79–85.

Davis, H., Hirsch, S. K., Shelnutt, J., & Bowers, C. Further validation of evoked response audiometry (ERA). *Journal of Speech & Hearing Research*, 1967, *10*, 717–732.

Davis, P. A., Davis, H., & Thompson, J. W. Progressive changes in human EEG under low oxygen tension. *American Journal of Physiology*, 1938, *123*, 51–52.

Davis, R. E., & Agranoff, B. W. Stages for memory formation in goldfish: Evidence for an environmental trigger. *Proceedings of National Academy of Science, United States*, 1966, *55*, 555–559.

De la Cruz, F. F., Fox, B. H., & Roberts, R. H. (Eds.), *Minimal brain dysfunction*, (Vol. 205). New York: New York Academy of Sciences, 1973.

Deisenhammer, E., Hofer, R., & Jellinger, K. Korrelation der Szintigraphie und Morphologie von Hirngeschwulsten. *Aertzl Forschrift*, 1968, *22*, 349–355.

Deisenhammer, E., & Jellinger, K. Morphological and EEG correlations in demented old people. *Electroencephalography & Clinical Neurophysiology*, 1971, *31*, 529.

Desmedt, J. E., & Manil, J. Somatosensory evoked potentials of the normal human neonate in REM sleep, in slow wave sleep and in waking. *Electroencephalography & Clinical Neurophysiology*, 1970, *29*, 113–126.

Donald, M. W., Jr., & Goff, W. R. Attention-related increases in cortical responsivity dissociated from the contingent negative variation. *Science*, 1971, *172*, 1163–1166.

Donchin, E. A multivariate approach to the analysis of averaged evoked potentials. *Institute of Electrical and Electronic Engineers Transactions on Bio-Medical Engineering*, 1966, *BME-13*, 131–139.

Donchin, E., Callaway, E., & Jones, R. T. Auditory evoked potential variability in Schizophrenia II. The application of discriminant analysis. *Electroencephalography and Clinical Neurophysiology*, 1970, *29*, 429–440.

Donchin, E., & Cohen, L. Average evoked potentials and intramodality selective attention. *Electroencephalography & Clinical Neurophysiology*, 1967, *22*, 537–546.

Donchin, E., Kubovy, M., Kutas, M., Johnson, R., & Herning, R. I. Graded changes in evoked response (P-300) amplitude as a function of cognitive activity. *Perception & Psychophysiology,* 1974, *14,* 319–324.

Donchin, E., & Smith, B. D. The CNV and late positive wave of the average evoked potential. *Electroencephalography & Clinical Neurophysiology,* 1970, *29,* 201–203.

Donchin, E., Tueting, P., Ritter, W., Kutas, M., & Heffley, E. On the independence of the CNV and the P300 components of the human averaged evoked potential. *Electroencephalography & Clinical Neurophysiology,* 1975, *38,* 449–461.

Doyle, J. C., Ornstein, R., Galin, D. Lateral specialization of cognitive mode: II. EEG frequency analysis. *Psychophysiology,* 1974, *11,* 567–578.

Drachman, D. A., & Hughes, J. R. Aging, memory and the hippocampal complex. *Neurology,* 1968, *18,* 288.

Dreyfus-Brisac, C., & Ellingson, R. J. Hereditary, congenital and perinatal diseases. In A. Rémond (Ed.), *Handbook of electroencephalography and clinical neurophysiology* (Part B. Vol. 15). Amsterdam: Elsevier, 1972.

Dreyfus-Brisac, C., & Monod, N. Neonatal status epilepticus. In A. Rémond (Ed.), *Handbook of electroencephalography and clinical neurophysiology* (Vol. 15B). Amsterdam: Elsevier, 1972. Section VI, pp. 38–52.

Drohocki, Z. Electrospektrographie des Gehirns. *Klinische Wochenschrift,* 1939, *18,* 536–538.

Duda, R. O., & Hart, P. E. *Pattern classification and scene analysis.* New York: Wiley, 1973.

Dumermuth, D., & Keller, E. EEG spectral analysis by means of fast Fourier Transform. In P. Kellaway & I. Petersén (Eds.), *Automation of clinical electroencephalography.* New York: Raven Press, 1973. Pp. 145–159.

Dunlop, C. W., McLachlan, E. M., Webster, W. R., & Day, R. H. Auditory habituation in cats as a function of stimulus intensity. *Nature,* 1964, *203,* 874–875.

Dustman, R. E., & Beck, E. C. Phase of alpha brain waves, reaction time and visually evoked potentials. *Electroencephalography & Clinical Neurophysiology,* 1965, *18,* 433–440.

Dustman, R. E., & Beck, E. C. The effects of maturation and aging on the waveform of visually evoked potentials. *Electroencephalography & Clinical Neurophysiology,* 1969, *26,* 2–11.

Eason, R. G., Harter, M. R., & White, C. T. Effects of attention and arousal on visually evoked cortical potentials and reaction time in man. *Physiology & Behavior,* 1969, *4,* 281–289.

Eason, R. G., White, C. T., & Bartlett, N. Effects of checkerboard pattern stimulations on evoked cortical responses in relation to check size and visual field. *Psychonomic Science,* 1970, *2,* 113–115.

Eisenberg, L. The epidemiology of reading retardation and a program of preventive intervention. In J. Money (Ed.), *The disabled reader: Education for the dyslexic child.* Baltimore: Johns Hopkins Press, 1966. Pp. 3–19.

Ellingson, R. J. Cerebral electrical responses to auditory and visual stimuli in the infant (human and subhuman). In P. Kellaway & I. Petersén (Eds.), *Neurologic and electroencephalographic correlative studies in infancy.* New York: Grune & Stratton, 1964. Pp. 78–116.

Ellingson, R. J. Methods of recording cortical evoked responses in the human infant. In A. Minkowski (Ed.), *Regional development of the brain in early life.* Oxford: Blackwell, 1967. Pp. 413–435. (a)

Ellingson, R. J. The study of brain electrical activity in infants. In L. P. Lipsitt & C. C. Spiker (Eds.), *Advances in child development and behavior* (Vol. 3). London: Academic Press, 1967. Pp. 53–97. (b)

Elmgren, J., & Lowenbard, P. *A factor analysis of the human EEG.* Report, Psychological Laboratory of the University of Göteborg, 1969, No. 2.

Endroczi, E., Fekete, T., Lissak, K., & Osvath, D. Electroencephalographic studies of the habituation in humans. *Acta Physiologica Academy of Science Hungaricae Tomus*, 1968, *34*, 311–318.

Engel, R. Neonatal EEG responses, developmental landmarks and IQ. Paper presented at the 13th International Congress of Pediatrics, Vienna, 1971.

Engel, R., & Benson, R. C. Estimate of conceptional age by evoked response activity. *Biology & Neonatology* (Basel), 1968, *12*, 201–213.

Engel, R., & Fay, W. Are electroencephalographic evoked response latencies in neonates predictors of language performance at 1 and 3 years of age? *Electroencephalography & Clinical Neurophysiology*, 1971, *30*, 159–160.

Engel, R., & Young, N. B. Calibrated pure tone audiograms in normal neonates based on evoked electroencephalographic responses. *Neuropadiätrie*, 1969, *1*, 149–160.

Epstein, J. A., & Lennox, M. A. Electroencephalographic study of experimental cerebrovascular occlusion. *Electroencephalography and clinical neurophysiology*, 1949, *1*, 491–502.

Ewalt, J. R., & Ruskin, A. The EEG in patients with heart disease. *Texas Reports on Biologic Medicine*, 1944, *2*, 161, 174.

Fernandez-Guardiola, A., Roldan, E., Fanjul, L., & Castells, C. Role of the pupillary mechanism in the process of habituation of the visual pathways. *Electroencephalography & Clinical Neurophysiology*, 1961, *13*, 564–576.

Ferris, G. S., Davis, G. D., Dorsen, M. M., & Hackett, E. R. Changes in latency and form of the photically induced averaged evoked responses in human infants. *Electroencephalography & Clinical Neurophysiology*, 1967, *22*, 305–312.

Fields, C. Visual stimuli and evoked responses in the rat. *Science*, 1969, *165*, 1377–1379.

Floris, V., Morocutti, C., Amabile, G., Bernardi, G., & Rizzo, P. A. Recovery cycle of visual evoked potentials in normal, schizophrenic and neurotic patients. In N. S. Kline & E. Laska (Eds.), *Computers and electronic devices in psychiatry*. New York: Grune & Stratton, 1968. Pp. 194–205.

Fogarty, T. P., & Reuben, R. N. Light-evoked cortical and retinal responses in premature infants. *Archives Ophthalmology*, 1969, *81*, 454–459.

Frantzen, E., Harvald, B., & Haugsted, H. The arteriographic and electroencephalographic findings in cerebral apoplexy. *Danish Medical Bulletin*, 1959, *6*, 12–19.

Friedman, D., Simson, R., Ritter, W., & Rapin, I. Cortical evoked potentials elicited by real speech words and human sounds. *Electroencephalography & Clinical Neurophysiology*, 1975, *38*, 13–19. (a)

Friedman, D., Simson, R., Ritter, W., and Rapin, I. The late positive component [P-3(00)] and information processing in sentences. *Electroencephalography & Clinical Neurophysiology*, 1975, *38*, 255–262. (b)

Fruhstorfer, H., Soveri, P., & Jarvilehto, T. Short-term habituation of the auditory evoked response in man. *Electroencephalography & Clinical Neurophysiology*, 1970, *28*, 153–161.

Fukunaga, K. *Introduction to statistical pattern recognition*. New York: Academic Press, 1972.

Gallagher, J. J. New educational treatment models for children with minimal brain dysfunction. In F. de la Cruz, B. H. Fox, & R. H. Roberts (Eds.), *Minimal brain dysfunction* (Vol. 205). New York: New York Academy of Sciences, 1973. Pp. 383–389.

Gastaut, H., Frank, G., Krolikowska, W., Naquet, R., & Roger, J. Study of visual evoked potentials in hemianopics presenting visual epileptic attacks in their blind fields. *Electroencephalography & Clinical Neurophysiology*, 1964, *16*, 627.

Gastaut, H., Gastaut, Y., Roger, A., Corriol, J., & Naquet, R. Étude électrographique du cycle d'excitabilité cortical. *Electroencephalography and Clinical Neurophysiology*, 1951, *3*, 401–428.

Gellis, S. S., & Kagan, B. M. *Current pediatric therapy* (Vol. 4). Philadelphia: Saunders, 1970.

Gergen, J., Conant, L., Hill, D., & Caudle, J. Personal communication to J. R. Hughes, 1965. Cited in H. R. Myklebust (Ed.), *Progress in learning disabilities* (Vol. 1). New York: Grune & Stratton, 1968. Pp. 113–146.

Gerson, I. M., John, E. R., Koenig, V., & Bartlett, F. Average evoked response (AER) in the electroencephalographic diagnosis of the normally aging brain: a practical application. Clinical Electroencephalography, 1976, *7*, 77–91.

Giannitrapani, D. Developing concepts of lateralization of cerebral functions. *Cortex*, 1967, *3*, 353–370.

Giannitrapani, D. EEG frequency and intelligence. *Electroencephalography & Clinical Neurophysiology*, 1969, *27*, 480–486.

Gibbs, F. A., & Gibbs, E. L. *Atlas of encephalography*. Vol. 1: *Methodology and controls*. Cambridge, Mass.: Addison-Wesley, 1950.

Gibbs, F. A., & Gibbs, E. L. *Atlas of encephalography*. Vol. 2: *Epilepsy*. Cambridge, Mass.: Addison-Wesley, 1952.

Gibbs, F. A., & Gibbs, E. L. *Atlas of encephalography*. Vol. 3: *Neurological and psychological disorders*. Cambridge, Mass.: Addison-Wesley, 1964.

Goff, W. R. Evoked potential correlates of perceptual organisation in man. In C. R. Evans & T. B. Mulholland (Eds.), *Attention in neurophysiology*. London: Butterworths, 1969. Pp. 169–193.

Goldfarb, A. I., Hochstadt, J. J., & Jacobson, J. H. Hyperbaric oxygen treatment of organic mental syndrome in aged persons. *Journal of Gerontology*, 1972, *27*, 212–217.

Gotman, J., Skuce, D. R., Thompson, C. S., Gloor, P., Ives, J. R., & Ray, W. F. Clinical applications of spectral analysis and extraction of features from electroencephalograms with slow waves in adult patients. *Electroencephalography & Clinical Neurophysiology*, 1973, *35*, 225–235.

Graziani, L. J., & Weitzman, E. D. Sensory evoked responses in the neonatal period and their application. In A. Rémond (Ed.), *Handbook of electroencephalography and clinical neurophysiology* (Vol. 15B). Amsterdam: Elsevier, 1972. Section IX, pp. 73–88.

Gross, M. B., & Wilson, W. C. (Eds.), *Minimal brain dysfunction*. New York: Brunner/Mazel, 1974.

Groth, H., Weled, B., & Batkin, S. A comparison of monocular visually evoked potentials in human neonates and adults. *Electroencephalography and Clinical Neurophysiology*, 1970, *28*, 278–287.

Hagne, I., Persson, J., Magnusson, R., & Petersén, I. Spectral analysis via fast Fourier transform of waking EEG in normal infants. In P. Kellaway & I. Petersén (Eds.), *Automation of clinical electroencephalography*. New York: Raven Press, 1973. Pp. 103–143.

Halas, E. S., & Beardsley, J. V. A factor analysis of neuronal responses during habituation in cats. *Psychology Records*, 1969, *19*, 47–52.

Halliday, R., Rosenthal, J. H., Naylor, H., & Callaway, E. Average evoked potentials: predictor of clinical improvement in hyperactive children treated with methylphenidate. *Psychophysiology*, 1976, *13*, 429–440.

Harmon, H. H. *Modern factor analysis*. Chicago: University of Chicago Press. 1960.

Harmony, T., Otero, G., Ricardo, J., & Fernández, G. Polarity coincidence correlation coefficient and signal energy ratio of the ongoing EEG activity. *Brain Research*, 1973, *61*, 133–140. (a)

Harmony, T., Ricardo, J., Otero, G., Fernández, G., Llorente, S., & Valdés, P. Symmetry of the visual evoked potential in normal subjects. *Electroencephalography & Clinical Neurophysiology*, 1973, *35*, 237–240. (b)

Harmony, T., Otero, G., Ricardo, J., Valdés, P., & Fernández, G. Paper presented at the symposium on Applications of Computation in the Study of the Nervous System. Havana, Cuba, October, 1975.

Harter, M. R. Evoked cortical responses to checkerboard patterns: effect of check-size as a function of retinal eccentricity. *Vision Research,* 1971, *10,* 1365–1376.

Harter, M. R., & Salmon, L. E. Evoked cortical responses to patterned light flashes: Effects of ocular convergence and accommodation. *Electroencephalography & Clinical Neurophysiology,* 1971, *31,* 287–305.

Harter, M. R., & Salmon, L. E. Intra-modality selective attention and evoked cortical potentials to randomly presented patterns. *Electroencephalography & Clinical Neurophysiology,* 1972, *21,* 605–613.

Harter, M. R., & Suitt, C. D. Visually evoked cortical responses and pattern vision in the infant: a longitudinal study. *Psychonomic Science,* 1970, *18,* 235–237.

Harter, M. R., & White, C. T. Effects of contour sharpness and check-size on visually evoked cortical potentials. *Vision Research,* 1968, *8,* 701–711.

Harter, M. R., & White, C. T. Evoked cortical response to checkerboard patterns: effect of check-size as a function of visual acuity. *Electroencephalography & Clinical Neurophysiology,* 1970, *28,* 48–54.

Hartley, L. R. The effect of stimulus relevance on the cortical evoked potentials. *Quarterly Journal of Experimental Psychology,* 1970, *22,* 531, 546.

Harvold, B., & Skinhoj, E. EEG in cerebral apoplexy. *Acta Psychiatrica & Neurologica Scandinavica,* 1956, *31,* 181–185.

Heninger, G. R. Lithium effects on cerebral cortical function in manic depressive patients. *Electroencephalography & Clinical Neurophysiology,* 1969, *27,* 670.

Heninger, G. R., & Speck, L. Visual evoked responses and mental status of schizophrenics. *Archives General Psychiatry,* 1966, *15,* 419–426.

Hernández-Peón, R., Scherrer, H., & Jouvet, M. Modification of electrical activity in cochlear nucleus during "attention" in unanesthetized cats. *Science,* 1956, *123,* 331–332. (a)

Hernández-Peón, R., Scherrer, R. H., & Velasco, M. Central influences on afferent conduction in the somatic and visual pathways. *Acta Neurologica Latinoamerica,* 1956, *2,* 8–22. (b)

Herrington, R. N., & Schneidau, P. The effect of imagery of the visual evoked response. *Experientia,* 1968, *24,* 1136–1137.

HEW National Advisory Committee on Dyslexia and Related Reading Disorders. *Reading disorders in the United States.* Washington, D.C.: Department of Health, Education & Welfare, 1969.

Hill, D., & Parr, G. *Electroencephalography.* New York: Macmillan, 1963.

Hillyard, S. A., Krausz, H. I., & Picton, T. W. Scalp distributions of the late positive wave (P_3) in six different decision tasks. *Electroencephalography & Clinical Neurophysiology,* 1974, *36,* 211.

Hillyard, S. A., Squires, K. C., Bauer, J. W., & Lindsay, P. H. Evoked potential correlates of auditory signal detection. *Science,* 1971, *172,* 1357–1360.

Hjorth, B. EEG analysis based on time domain properties. *Electroencephalography & Clinical Neurophysiology,* 1970, *29,* 306–310.

Hrbek, A., Hrbkova, M., & Lenard, H. G. Somato-sensory evoked responses in newborn infants. *Electroencephalography & Clinical Neurophysiology,* 1968, *25,* 443–448.

Hrbek, A., Hrbkova, M., & Lenard, H. G. Somatosensory, auditory and visual evoked responses in newborn infants during sleep and wakefulness. *Electroencephalography & Clinical Neurophysiology,* 1969, *26,* 597–603.

Hrbek, A., & Mares, P. Cortical evoked responses to visual stimulation in full-term and premature newborns. *Electroencephalography & Clinical Neurophysiology*, 1964, *16*, 575–581.

Hughes, J. R. Electroencephalography and learning. In H. R. Myklebust (Ed.), *Progress in learning disabilities* (Vol. 1). New York: Grune & Stratton, 1968. Pp. 113–146.

Hughes, J. R., & Park, G. EEG in dyslexic children. Unpublished results, 1966.

Irwin, D. A., Knott, J. R., McAdam, D. W., & Rebert, C. S. Motivational determinants of the "contingent negative variation." *Electroencephalography & Clinical Neurophysiology*, 1966, *21*, 538–543.

Ishikawa, K. Studies on the visual evoked responses to paired light flashes in schizophrenics. *Kurume Medical Journal*, 1968, *15*, 153–167.

Jarvik, L. F., & Milne, J. F. Gerovital H-3 – a review of the literature. In S. Gershon & A. Raskin (Eds.), *Genesis and treatment of psychiatric disorders in the elderly*. New York: Raven Press, 1976. Pp. 203–227.

Jasper, H. H. The ten-twenty electrode system of the International Federation. *Electroencephalography and Clinical Neurophysiology*, 1958, *10*, 371–375.

Jeffreys, D. A. Evoked potentials as indicators of sensory information processing. *Neurosciences Research Progress Bulletin*, 1969, *7*.

Jennings, W. G. An ergot alkaloid preparation (hydergine) vs. placebo for treatment of cerebrovascular insufficiency: A double-blind study. *Journal of the American Geriatric Society*, 1972, *10*, 407–412.

Jensen, D. R., & Engel, R. Statistical procedures for relating dichotomous responses to maturation and EEG measurements. *Electroencephalography and Clinical Neurophysiology*, 1971, *30*, 437–443.

John, E. R. Higher nervous functions: brain functions and learning. *Annual Review of Physiology*, 1961, *23*, 451.

John, E. R. Neural mechanisms of decision making. In W. S. Fields & W. Abbot (Eds.), *Information storage and neural control.* Springfield, Ill.: Charles C Thomas, 1963.

John, E. R. *Mechanisms of memory*. New York: Academic Press, 1967. (a)

John, E. R. Electrophysiological studies of conditioning. In G. C. Quarton, T. Melnechuk, & F. O. Schmitt (Eds.), *The neurosciences: A study program.* New York: Rockefeller University Press, 1967. (b)

John, E. R. Switchboard versus statistical theories of learning and memory. *Science*, 1972, *177*, 850–864.

John, E. R. Brain evoked potentials: Acquisition and analysis. In R. F. Thompson & M. M. Patterson (Eds.), *Bioelectric recording techniques.* Part A: *Cellular processes and brain potentials.* New York: Academic Press, 1973. (a)

John, E. R. Memory mechanisms in instrumental responding. *Science*, 1973, *181*, 685–686. (b)

John, E. R. Abstracts, *Proceedings 25th International Congress Physiological Sciences*, 1974, New Delhi, India. (a)

John, E. R. Assessment of acuity, color vision and shape perception by statistical evaluation of evoked potentials. *Annals of Ophthalmology*, 1974, *6*, 55–66. (b)

John, E. R., Bartlett, F., Shimokochi, M., & Kleinman, D. Neural readout from memory. *Journal of Neurophysiology*, 1973, *36*, 893–924.

John, E. R., Herrington, R. N., & Sutton, S. Effects of visual form on the evoked response. *Science*, 1967, *155*, 1439–1442.

John, E. R., & Laupheimer, R. A Method for the Analysis of Symmetry of Brain Wave Activity. U.S. Patent No. 3696808, 1972.

John, E. R., Ruchkin, D. S., & Villegas, J. Signal analysis and behavioral correlates of evoked potential configurations in cats. *Annals of New York Academy of Sciences*, 1964, *112*, 362–420.

John, E. R., Shimokochi, M., & Bartlett, F. Neural readout from memory during generalization. *Science,* 1969, *164,* 1519–1521.

John, E. R., Walker, P., Cawood, D., Rush, M. & Gehrmann, J. Mathematical identification of brain states applied to classification of drugs. *International Review of Neurobiology,* 1972, *15,* 273–347.

Jones, E. V., & Bagchi, B. EEG in verified thrombosis of major cerebral arteries. *Electroencephalography & Clinical Neurophysiology,* 1951, *31,* 374–379.

Jones, R. T., Blacker, K. H., & Callaway, E. Perceptual dysfunction in schizophrenia: clinical and auditory evoked response findings. *American Journal of Psychiatry,* 1966, *123,* 639–645.

Jones, R. T., & Callaway, E. Auditory evoked responses in schizophrenia. *Biological Psychiatry,* 1970, *2,* 291–298.

Jonkman, E. J. *The average cortical response to photic stimulation.* Doctoral dissertation, University of Amsterdam, 1967.

Kaiser, E., Petersén, I., & Magnusson, R. A method in automatic pattern recognition in EEG. In P. Kellaway & I. Petersén (Eds.), *Automation of clinical electroencephalography.* New York: Raven Press, 1973. Pp. 235–242.

Kaiser, E., Petersén, I., Selldén, U., & Kagawa, N. EEG data representation in broad-band frequency analysis. *Electroencephalography & Clinical Neurophysiology,* 1964, *17,* 76–80.

Kaiser, E., & Sem-Jacobsen, C. W. "Yes–no" data reduction in EEG automatic pattern recognition. *Electroencephalography & Clinical Neurophysiology,* 1962, *14,* 955.

Kaiser, H. F. The varimax criterion for analytic rotation in factor analysis. *Psychometrika,* 1958, *23,* 187–200.

Kahn, R. L., Goldfarb, A. I., Pollack, M., & Peck, A. Brief objective measures for the determination of mental status in the aged. *American Journal of Psychiatry,* 1960, *117,* 326–328.

Karlin, L. Cognition, preparation, and sensory-evoked potentials. *Psychological Bulletin,* 1970, *73,* 122–136.

Karmel, B. Z., Hoffmann, R. F., & Fegy, M. J. Processing of contour information by human infants evidenced by pattern dependent evoked potentials. *Child Development,* 1974, *45,* 39–48.

Karmel, B. Z., & Maisel, E. A neuronal activity model for infant visual attention. In L. B. Cohen & P. Salapatek (Eds.), *Infant perception: from sensation to cognition.* Part 1: *Basic processes.* New York: Academic Press, 1975. Pp. 78–125.

Kay, D. W. K. Epidemiological aspects of organic brain disease in the aged. In C. M. Gaitz (Ed.), *Aging in the brain.* New York: Plenum Press, 1972. Pp. 15–26.

Keidel, W. D., & Spreng, M. Audiometric aspects and multisensory power functions of electronically averaged evoked cortical responses in man. *Acta Otolaryngology,* Stockholm, 1965, *59,* 201–208.

Keidel, W. D., & Spreng, M. Recent status results and problems of objective audiometry in man. Parts I and II. *Journal Française d' Oto-rhinolaryngologie, 19,* 1970.

Kellaway, P. Automation of clinical electroencephalography: the nature and scope of the problem. In P. Kellaway & I. Petersén (Eds.), *Automation of clinical electroencephalography.* New York: Raven Press, 1973. Pp. 1–29.

Kellaway, P., & Petersén, I. (Eds.), *Neurological and electroencephalographic correlative studies in infancy.* New York: Grune & Stratton, 1964.

Kellaway, P., & Petersén, I. *Automation of clinical electroencephalography.* New York: Raven Press, 1973.

Keogh, B. Hyperactive and learning disorders: Review and speculation. *Exceptional Child,* 1971, *39,* 101–110.

Key, B. J. Correlation of behavior with changes in amplitude of cortical potentials evoked during habituation by auditory stimuli. *Nature,* 1965, *207,* 441–442.

Kirk, S. A. Perceptual defect and role handicap: a theory of the etiology of schizophrenia. In D. R. Hawkings & L. C. Pawling (Eds.), *Orthomolecular psychiatry: treatment of schizophrenia.* San Francisco: Freeman, 1972.

Klahr, D. A Monte Carlo investigation of Kruskal's non-metric scaling procedure. *Psychometrika,* 1969, *34,* 319–330.

Klinke, R., Fruhstorfer, H., & Finkenzeller, P. Evoked responses as a function of external and stored information. *Electroencephalography & Clinical Neurophysiology,* 1968, *26,* 216–219.

Klinkerfuss, G. H., Lange, P. H., Weinberg, W. A., & O'Leary, J. L. Electroencephalographic abnormalities of children with hyperkinetic behavior. *Neurology* (Minneapolis), 1965, *115,* 883–891.

Klove, H. Relationship of differential electroencephalographic patterns to distribution of Wechsler–Bellevue scores. *Neurology,* 1959, *9,* 871–876.

Kooi, K. A., & Bagchi, B. K. Visual evoked responses in man: Normative data. *Annals of the New York Academy of Science,* 1964, *112,* 254–269.

Kooi, K. A., Guevener, A. M., & Bagchi, B. K. Visual evoked responses in lesions of the higher optic pathways. *Neurology,* 1965, *15,* 841.

Kozhevnikov, V. A. Some methods of automatic measurement of the electroencephalogram. *Electroencephalography & Clinical Neurophysiology,* 1958, *10,* 269–278.

Kruskal, J. B. Multidimensional scaling by optimizing goodness of fit to a non-metric hypothesis. *Psychometrika,* 1964, *29,* 1–27.

Kullback, S. *Information theory and statistics.* New York: Wiley, 1959.

Larsen, L. E., Ruspini, E. H., McNew, J. J., Walter, D. O., & Adey, W. R. Classification and discrimination of the EEG during sleep. In P. Kellaway & I. Petersén, (Eds.), *Automation of clinical electroencephalography.* New York: Raven Press, 1973. Pp. 243–268.

Lassen, N. A., Munch, O., & Totty, E. R. Mental function and cerebral oxygen consumption in organic dementa. *Archives of Neurology and Psychiatry,* 1957, *77,* 126–133.

Lavy, S., & Bental, E. The prognostic value of the electroencephalogram in hemiplegia following cerebrovascular accidents. *Confinia Neurologica* (Basel), 1964, *24,* 182–188.

Lavy, S., Carmon, A., & Schwartz, A. Depression of electrical cortical activity in acute cerebrovascular accidents. *Confinia Neurologica* (Basel), 1964, *24,* 349–358.

Leiman, A. L., & Christian, C. N. Electrophysiological analyses of learning and memory. In J. A. Deutsch (Ed.), *The physiological basis of memory.* New York: Academic Press, 1973. Pp. 125–173.

Lelord, G., Laffont, F., Jusseaume, P., & Stephant, J. L. Comparative study of conditioning of average evoked responses by coupling sound and light in normal and autistic children. *Psychophysiology,* 1973, *10,* 415–425.

Lenard, H. G., von Bernuth, H., & Hutt, S. J. Acoustic evoked responses in newborn infants. The influence of pitch and complexity of the stimulus. *Electroencephalography & Clinical Neurophysiology,* 1969, *27,* 121–127.

Lifshitz, K. An examination of evoked potentials as indicators of information processing in normal and schizophrenic subjects. In E. Donchin & D. B. Lindsley (Eds.), *Average evoked potentials, methods, results, evaluations.* NASA SP-191. Washington, D.C.: Government Publications Office, 1969.

Livanov, M. N. In I. N. Knipst (Ed.), *Contemporary problems of electrophysiology of the central nervous system.* Moscow: Academy of Science, 1967.

Lodge, A., Armington, J. C., Barnet, A., Shanks, B. L., & Newcomb, C. H. Newborn infants' electroretinograms and evoked electroencephalographic responses to orange and white light. *Child Development,* 1969, *40,* 267–293.

Lombroso, C. T. The CNV during tasks requiring choice. In C. R. Evans & T. B. Mulholland (Eds.), *Attention in neurophysiology.* London: Butterworths, 1969.

Low, M. D., Borda, R. P., Frost, J. D., Kellaway, P. Surface-negative, slow-potential shift associated with conditioning in man. *Neurology,* 1966, *16,* 771–782.

Luce, R. A., & Rothchild, D. The correlation of EEG and clinical observations in psychiatric patients over 65. *Journal of Gerontology,* 1953, *8,* 167–172.

Lüders, H. Effects of aging on the waveform of the somatosensory cortical evoked potentials. *Electroencephalography & Clinical Neurophysiology,* 1970, *19,* 450–460.

McAdams, W., & Robinson, R. A. Senile intellectual deterioration and the EEG. *Journal of Mental Science,* 1956, *102,* 819–825.

McCallum, C. The contingent negative variation as a cortical sign of attention in man. In C. R. Evans & T. B. Mulholland (Eds.), *Attention in neurophysiology.* London: Butterworths, 1969.

McCandless, G. A. Clinical application of evoked response audiometry. *Journal of Speech Research,* 1967, *10,* 468–478.

McDowell, F., Wells, C. E., & Ehlers, C. The electroencephalogram in internal carotid occlusion. *Neurology* (Minneapolis), 1959, *9,* 678–687.

Mackay, D. M., & Jeffreys, D. A. Visual evoked potentials and visual perception in man. In H. Autrum (Ed.), *Central processing of visual information, Handbook of sensory physiology.* Berlin and New York: Springer-Verlag, 1971.

Maggs, R., & Turton, E. G. EEG findings in old age. *Journal of Mental Science,* 1956, *102,* 812–818.

Magnus, O., Van Leeuwen, W., & Cobb, W. A. *Electroencephalography and cerebral tumours* (Supplement 19, EEG). Amsterdam: Elsevier, 1961.

Marguardsen, J., & Harvald, B. The electroencephalogram in acute cerebrovascular lesions. A report of 50 cases verified at autopsy. *Neurology* (Minneapolis), 1964, *14,* 275–282.

Marsh, J. T., & Worden, F. G. Auditory potentials during acoustic habituation: cochlear nucleus, cerebellum and auditory cortex. *Electroencephalography & Clinical Neurophysiology,* 1964, *17,* 685–692.

Martin, P. Trois aspects EEG des hématomes intracérébraux. *Electroencephalography & Clinical Neurophysiology,* 1953, *5,* 133–136.

Matoušek, M., & Petersén, I. Automatic evaluation of EEG background activity by means of age-dependent EEG quotients. *Electroencephalography & Clinical Neurophysiology,* 1973, *35,* 603–612. (a)

Matoušek, M., & Petersén, I. Frequency analysis of the EEG in normal children and adolescents. In P. Kellaway & I. Petersén (Eds.), *Automation of clinical electroencephalography.* New York: Raven Press, 1973. Pp. 75–102. (b)

Maulsby, R. L., Saltzberg, B., & Lustick, L. S. Toward an EEG screening test: a simple system for analysis and display of clinical EEG data. In P. Kellaway & I. Petersén (Eds.), *Automation of clinical electroencephalography.* New York: Raven Press, 1973. Pp. 45–53.

Meisel, W. S. *Computer oriented approaches to pattern recognition.* New York: Academic Press, 1973.

Mengoli, G. L'EEG nei vecchi. *Rivista Neurologica,* 1952, *22,* 166–193.

Menkes, M., Rowe, J. S., & Menkes, J. H. A twenty-five year follow-up study on the hyperkinetic child with minimal brain dysfunction. *Pediatrics,* 1967, *39,* 393–399.

Merritt, H. H. *A textbook of neurology* (5th ed.). Philadelphia: Lea and Febiger, 1973.

Minskoff, J. G. Differential approaches to prevalence estimates of learning disabilities. In F. F. de la Cruz, B. H. Fox, & R. H. Roberts (Eds.), *Minimal brain dysfunction* (Vol. 205). New York: New York Academy Sciences, 1973. Pp. 139–145.

Miranda, S., & Fantz, M. Recognition memory in Down's syndrome and normal infants. *Child Development*, 1974, *45*, 651–660.

Monod, N., & Dreyfus-Brisac, C. Prognostic value of the neonatal EEG in full-term newborns. In A. Rémond (Ed.), *Handbook of electroencephalography and clinical neurophysiology* (Vol. 15B). Amsterdam: Elsevier 1972. Section X, pp. 89–100.

Monod, N., & Garma, L. Auditory responsivity of the human premature. *Biological Neonatology* (Basel), 1971, *17*, 292–316.

Monod, N., Pajot, N., & Guidasci, S. The neonatal EEG: Statistical studies and prognostic value in full-term and pre-term babies. *Electroencephalography & Clinical Neurophysiology*, 1972, *32*, 529–544.

Morrell, F. Electrophysiological contributions to the neural basis of learning. *Physiological Review*, 1961, *41*, 443. (a)

Morrell, F. Effect of anodal polarization on the firing pattern of single cortical cells. *Annals of the New York Academy of Sciences*, 1961, *92* (3), 860–876. (b)

Morrell, F., & Morrell, L. Computer aided analyses of brain electrical activity. In L. D. Proctor & W. R. Adey (Eds.), *The analysis of central nervous system and cardiovascular data using computer methods.* Washington: National Aeronautics and Space Agency (NASA), 1965, pp. 441–478.

Muehl, S., Knott, J., & Benton, A. EEG abnormality and psychological test performance in reading disability. *Cortex*, 1965, *1*, 434–440.

Muller, H. F., & Grad, B. EEG: Biochemical and behavioral characteristics of elderly psychiatric patients. *Electroencephalography and Clinical Neurophysiology*, 1970, *29*, 410–411.

Muller, H. F., Grad, B., & Kral, V. A. Functional patterns in geriatrics. *Laval Medicine*, 1971, *42*, 180–184.

Mundy-Castle, A. C. Theta and delta rhythm in EEG of normal adults. *Electroencephalography & Clinical Neurophysiology*, 1951, *3*, 477–486.

Mundy-Castle, A. C., Hurst, L. A., & Beerstrecher, D. M. The EEG in senile psychoses. *Electroencephalography & Clinical Neurophysiology*, 1954, *6*, 245–252.

Myklebust, H. R., & Boshes, B. *Minimal brain damage in children.* Final Report to United States Public Health Service, Contract #108-65-142. Washington, D.C.: S.S. Department of Health, Education and Welfare, 1969.

Näätänen, R. Selective attention and evoked potentials. Doctoral dissertation, University of Helsinki, Finland, 1967.

Näätänen, R. Evoked potential, EEG, and slow potential correlates of selective attention. *Acta Psychologica*, 1970, *33*, 178–192.

Naitoh, P., Johnson, L. C., Lubin, A., & Wyborney, G. Brain wave "generating" processes during waking and sleeping. *Electroencephalography & Clinical Neurophysiology*, 1971, *31*, 294.

National Institute of Neurological Diseases and Stroke (NINDS). *Neurological and sensory disabilities* (DHEW Publication No. 73-152). Information Office, NINDS, 1973.

Niedermeyer, E. Migraine and allied disorders. In O. Magnus (Ed.), *Handbook of electroencephalography and clinical neurophysiology* (Part A, Vol. 14). Amsterdam: Elsevier, 1972.

Obrist, W. D. The electroencephalogram of normal male subjects over age 75. *Journal of Gerontology*, Supplement to No. 3, 2nd International Gerontological Congress, 1951, 130.

Obrist, W. D. The electroencephalogram of normal aged adults. *Electroencephalography & Clinical Neurophysiology*, 1954, *6*, 235–349.

Obrist, W. D., & Bissel, L. F. The EEG of aged patients with cardiac and cerebrovascular disease. *Journal of Gerontology*, 1955, *10*, 315–330.

Obrist, W. D., & Busse, E. W. Temporal lobe abnormalities in normal senescence. *Electroencephalography & Clinical Neurophysiology,* 1960, *12,* 249.

Obrist, W. D., Busse, E. W., & Henry, C. E. Relation of EEG to blood pressure in elderly persons. *Neurology,* 1961, *11,* 151–158.

O'Gorman, G. Conference on the psychiatric care of deaf children. *Lancet,* 1962, *11,* 985.

Omae, T., Katsuki, S., Nishimaru, K., Yamaguchi, T., Takeya, Y., Fujushima, M., & Katao, M. Clinical features of cerebral infarction in the Japanese. *Journal of Chronic Disease,* 1969, *21,* 585–606.

Oosterhuis, H. J. G. H., Ponsen, L., Jonkman, E. J., & Magnus, O. The average visual response in patients with cerebrovascular disease. *Electroencephalography and Clinical Neurophysiology,* 1969, *27,* 23–34.

Otero, G. Unpublished doctoral thesis, Centro Nacional de Investigaciones Cientificas, Havana, 1973.

Otero, G., Harmony, T., & Ricardo, J. Polarity coincidence correlation coefficient and signal energy ratio of ongoing EEG activity. II. Brain tumors. *Activitas Nervosa Superior,* 1975, *17,* 120–126. (a)

Otero, G., Harmony, T., & Ricardo, J. Polarity coincidence correlation coefficient and signal energy ratio of ongoing EEG activity. III. Cerebral vascular lesions. *Activitas Nervosa Superior,* 1975, *17,* 127–130. (b)

Otero, G., Harmony, T., and Ricardo, J. La simetria del EEG como metodo diagnostico en las lesiones cerebrales. Paper presented at the symposium on Applications of Computation in the Study of the Nervous System. Havana, Cuba, October, 1975. (c)

Pavy, R., & Metcalfe, J. The abnormal EEG in childhood communication and behavior abnormalities. *Electroencephalography & Clinical Neurophysiology,* 1965, *19,* 414.

Perry, N. W., Jr., & Childers, D. G. *The human visual evoked response: method and theory.* Springfield, Ill.: Charles C Thomas, 1969.

Petersén, I. In P. Kellaway & I. Petersén (Eds.), *Automation of clinical electroencephalography.* New York: Raven Press, 1973. P. 27.

Posner, M. I. Psychobiology of attention. In C. S. Blakemore & M. S. Gazzaniga (Eds.), *The handbook of psychobiology.* New York: Academic Press, 1974.

Posner, M. I., Klein, R., Summers, J., & Buggie, S. On the selection of signals. *Memory and Cognition,* 1973, *1,* 2–12.

Preston, M. S., Guthrie, J. T., & Childs, B. Visual evoked responses (VERs) in normal and disabled readers. *Psychophysiology,* 1974, *11,* 452–457.

Pribram, K. H., Spinelli, D. N., & Kamback, M. C. Electrocortical correlates of stimulus response and reinforcement. *Science,* 1967, *157,* 94–95.

Price, R. L., & Smith, D. B. D. The $P_{3(00)}$ wave of the averaged evoked potential: A bibliography. *Physiological Psychology,* 1974, *2,* 387–391.

Prichep, L. S., Sutton, S., & Hakerem, G. Evoked potentials in hyperkinetic and normal children under certainty and uncertainty: A placebo and methylphenidate study. *Psychophysiology,* 1976, *13,* 419–428.

Ramos, A., Schwartz, E., & John, E. R. Unit activity and evoked potentials during readout from memory. *XXVI International Congress of Physiological Science,* New Delhi, 1974.

Rao, D. B., & Norris, J. R. A double-blind investigation of hydergine in the treatment of cerebrovascular insufficiency in the elderly. *Johns Hopkins Medical Journal,* 1972, *130,* 317–324.

Rapin, I., & Graziani, L. J. Auditory evoked responses in normal, brain damaged and deaf infants. *Neurology* (Minneapolis), 1967, *17,* 881–894.

Rapin, I., Ruben, R. J., & Lyttle, M. Diagnosis of hearing loss in infants using auditory evoked responses. Paper presented at a meeting of the Eastern American Laryngology Rhinology and Otology Society, 1970, Boston.

Redick, R. W., Kramer, M., & Taube, C. A. Epidemiology of mental illness and utilization of psychiatric facilities among older persons. In E. W. Busse & E. Pfeifer (Eds.), *Mental illness in later life.* Washington, D.C.: American Psychiatric Assoc., 1973.

Regan, D. *Evoked potentials in psychology, sensory physiology and clinical medicine.* New York: Wiley-Interscience, 1972.

Rémond, A. Appraisal and perspective of the functional exploration of the nervous system. In A. Rémond (Ed.), *Handbook of electrophysiology and clinical neurophysiology* (Vol 1). Amsterdam: Elsevier, 1971.

Rémond, A. (Ed.), *Handbook of electroencephalography and clinical neurophysiology.* Vol. 15B: *Hereditary, congenital and perinatal diseases*. Amsterdam: Elsevier, 1972.

Renshaw, B., Forbes, A., & Morison, B. R. Activity of isocortex and hippocampus: Electrical studies with micro-electrodes. *Journal of Neurophysiology,* 1940, *3,* 74–105.

Rhodes, L. E., Dustman, R. E., & Beck, E. C. Visually evoked potentials of bright and dull children. *Electroencephalography & Clinical Neurophysiology,* 1969, *26,* 237.

Richlin, M., Weisinger, M., Weinstein, S., Giannini, M., & Morgenstern, M. Interhemispheric asymmetries of evoked cortical responses in retarded and normal children. *Cortex,* 1971, *7,* 98–105.

Rietveld, W. J., Tordoir, W. E. M., Hagenouw, J. R. B., Lubbers, J. A., & Spoor, T. A. C. Visual evoked responses to blank and to checkerboard patterned flashes. *Acta Physiologica & Pharmacologia Neerlandia,* 1967, *14,* 259–285.

Rietveld, W. J., Tordoir, W. E. M., Hagenouw, J. R. B., & van Dongen, K. J. Contribution of fovea-parafoveal quadrants to the visual evoked response. *Acta Physiologica Pharmacologia Neerlandia,* 1965, *13,* 340–347.

Riggs, L. A., & Sternheim, C. E. Human retinal and occipital potentials evoked by changes of the wavelength of the stimulating light. *Journal of the Optometry Society of America,* 1969, *59,* 635–640.

Ritter, W., & Vaughan, H. G., Jr. Averaged evoked potentials in vigilance and discrimination: a reassessment. *Science,* 1969, *164,* 326–328.

Ritter, W., Vaughan, H. G., Jr., & Costa, L. D. Orienting and habituation to auditory stimuli: a study of short-term changes in average evoked potentials. *Electroencephalography & Clinical Neurophysiology*, 1968, *25*, 555–556.

Ritvo, E. R., Ornitz, E. M., Walter, R. D., & Haley, J. Correlation of psychiatric diagnoses and EEG findings: a double-blind study of 184 hospitalized children. *American Journal of Psychiatry,* 1970, *126,* 988–996.

Rohmer, F., Gastaut, Y., & Dell, M. B. L'EEG dans la pathologie vasculaire du cerveau. *Review Neurologie,* 1952, *87,* 93–144.

Rohrbaugh, J. W., Donchin, E., & Eriksen, C. W. Decision making and the P300 component of the cortical evoked response. *Perception & Psychophysics,* 1974, *15,* 368–374.

Rose, G. H., & Ellingson, R. J. Ontogenesis of evoked potentials. In W. A. Himwich (Ed.), *Developmental neurobiology.* Springfield: Charles C Thomas, 1970. Pp. 393–440.

Roseman, E., Schmidt, N. P., & Foltz, E. L. Serial electroencephalography in vascular lesions of the brain. *Neurology* (Minneapolis), 1952, *2,* 311–331.

Roth, W. T. Auditory evoked responses to unpredictable stimuli. *Psychophysiology,* 1973, *10,* 125–138.

Roth, W. T., & Kopell, B. S. P(300) – An orienting reaction in the human auditory evoked response. *Perceptual and Motor Skills*, 1973, *36*, 219–225.

Roth, W. T., Kopell, B. S., Tinklenberg, Jr., J. R., Darley, C. F., Sikora, R., & Vesecky, T. B. The contingent negative variation during a memory retrieval task. *Electroencephalography & Clinical Neurophysiology*, 1975, *38*, 171–174.

Ruchkin, D. S. An analysis of average response computations based upon aperiodic stimuli. *Institute of Electrical and Electronics Engineers Transactions on Bio-Medical Engineering,* 1965, *BME-12,* 87–94.

Ruchkin, D. S. Sorting of nonhomogeneous sets of evoked potentials. *Communications in Behavioral Biology,* 1971, *5,* 383–396.

Ruchkin, D. S., & John, E. R. Evoked potential correlates of generalization. *Science,* 1966, *153,* 209–211.

Ruspini, E. H. A new approach to clustering. *Information Control,* 1969, *15,* 22–32.

Ruspini, E. H. Numerical methods for fuzzy clustering. *Information Sciences,* 1970, *2,* 319–350.

Ryan, H. F. Information content measure as a performance criterion for feature selection. *Institute of Electrical and Electronics Engineers Proceedings on the 7th Symposium on Adaptive Processes,* Los Angeles, 1968.

Ryers, F. W. The incidence of EEG abnormalities in a dyslexic and a control group. *Journal of Clinical Psychology,* 1967, *23,* 334–336.

Saletu, B., Saletu, M., Itil, T. M., & Hsu, W. Changes in somatosensory evoked potentials during fluphenazine treatment. *Pharmakopsychiatry & Neuropsychopharmakology,* 1971, *4,* 158–158. (a)

Saletu, B., Saletu, M., Itil, T. M., & Marasa, J. Somatosensory-evoked potential changes during haloperidol treatment of chronic schizophrenics. *Biological Psychiatry,* 1971, *3,* 299–307. (b)

Saletu, B., Saletu, M., Simeon, J., Viamontes, G., & Itil, T. M. Comparative symptomatological and evoked potential studies with *d*-amphetamine, thioridazine and placebo in hyperkinetic children. *Biological Psychiatry,* 1975, *10,* 253–275.

Salzman, C., Kochansky, G. E., & Shader, R. I. Rating scales for geriatric psychopharmacology–a review. *Psychopharmacology Bulletin,* 1972, *8,* 3–50.

Sammon, J. W., Jr. A nonlinear mapping for data structure analyses. *Institute of Electrical and Electronics Engineers Transactions on Computers,* 1969, *C-18,* 401–409.

Sathananthan, G. L., & Gershon, S. Cerebral vasodilators: A review. In S. Gershon & A. Raskin (Eds.), *Genesis and treatment of psychiatric disorders in the elderly.* New York: Raven Press, 1976. Pp. 155–168.

Satterfield, J. H. Auditory evoked cortical response studies in depressed patients and normal control subjects. In T. A. Williams, M. M. Katz, & J. A. Shield, Jr. (Eds.), *Recent advances in the psychobiology of the depressive illnesses.* U.S. Government Printing Office, DHEW Publication #(HSM) 70-9053, 1972. Pp. 87–98.

Satterfield, J. H. EEG issues in children with minimal brain dysfunction. In S. Walzer & P. H. Wolff (Eds.), *Minimal cerebral dysfunction in children.* New York: Grune & Stratton, 1973. Pp. 35–46.

Satterfield, J. H., Cantwell, D. P., Lesser, L. I., & Podosin, R. L. Physiological studies of the hyperkinetic child. I. *American Journal of Psychiatry,* 1972, *128,* 1418–1419.

Schenkenberg, T., & Dustman, R. E. Visual, auditory and somatosensory evoked response changes related to age, hemisphere and sex. *Proceedings of 78th Annual Convention American Psychological Association,* 1970, 183–184.

Schwartz, E., Ramos, A., & John, E. R. Cluster analysis of evoked potentials in behaving cats. *Behavioral Biology,* 1976, *17,* 109–117.

Shagass, C. Evoked response studies of central excitability in psychiatric disorders. *Proceedings of Colloque Inserm on Average Evoked Responses and Their Conditioning in Normal Subjects and Psychiatric Patients.* Tours, France, September 1972.

Shagass, C., & Schwartz, M. Reactivity cycle of somatosensory cortex in humans with and without psychiatric disorder. *Science,* 1961, *134,* 1757–1759.

Shagass, C., Schwartz, M., & Amadeo, M. Some drug effects on evoked cerebral potentials in man. *Journal of Neuropsychiatry,* 1962, *3,* S49–S58.

Sharpless, S., & Jasper, H. Habituation of the arousal reaction. *Brain,* 1956, *79,* 655–680.

Shelburne, S. A., Jr. Visual evoked responses to word and nonsense syllable stimuli. *Electroencephalography & Clinical Neurophysiology,* 1972, *32,* 17–25.

Shelburne, S. A., Jr. Visual evoked responses to language stimuli in normal children. *Electroencephalography & Clinical Neurophysiology*, 1973, *34*, 135–143.

Shields, D. T. Brain responses to stimuli in disorders of information processing. *Journal of Learning Disabilities*, 1973, *6*, 501–505.

Sheridan, F. P., Yeager, C. L., Oliver, W. A., & Simon, A. EEG as a diagnostic and prognostic aid in studying the senescent individual. *Preliminary Report Journal of Gerontology*, 1955, *10*, 53–59.

Shipley, T., Jones, R. W., & Fry, A. Evoked visual potentials and human color vision. *Science*, 1965, *150*, 1162–1164.

Shipley, T., Jones, R. W., & Fry, A. Intensity and the evoked occipitogram in man. *Vision Research*, 1966, *6*, 657–667.

Sigman, M., & Parmelee, A. H. Visual preferences of four-month-old premature and full-term infants. *Child Development*, 1974, *45*, 959–965.

Silverman, A. J., Busse, E. W., & Barnes, R. H. Studies in the process of aging: EEG findings in 400 elderly subjects. *Electroencephalography & Clinical Neurophysiology*, 1955, *7*, 67–79.

Silverman, L. J., & Metz, A. S. Numbers of pupils with specific learning disabilities in local public schools in the United States: Spring 1970. In F. F. de la Cruz, B. H. Fox, & R. H. Roberts (Eds.), *Minimal brain dysfunction* (Vol. 205). New York: New York Academy of Sciences, 1973. Pp. 146–157.

Slater, E., & Roth, M. Aging and the mental diseases of the aged. In E. Slater & M. Roth (Eds.), *Clinical psychiatry*. London: Gailliere, Tindall & Cassel, 1969.

Smith, D. E., King, M. B., & Hoebel, G. B. Lateral hypothalamic control of killing: Evidence for a cholinoceptive mechanism. *Science*, 1970, *167*, 900–901.

Sneath, P. H. & Sokal, R. R. *Numerical taxonomy: The principles and practice of numerical classification*. San Francisco: Freeman, 1973.

Sokal, R. R. & Sneath, P. H. *Principles of numerical taxonomy*. San Francisco: Freeman, 1963.

Sokoloff, L. Cerebral circulation and metabolism in the aged. In S. Gershon & A. Raskin (Eds.), *Genesis and treatment of psychiatric disorders in the elderly*. New York: Raven Press, 1976. Pp. 45–54.

Sorel, L., Lebrun, P., Leotard, G., Delaite, F., Rucquoy-Ponsar, M., Basecqz, G., & de Biolley, D. Records of a group of 200 normal subjects aged 14–20 years. *Electroencephalography & Clinical Neurophysiology*, 1965, *18*, 633.

Speck, L. B., Dim, B., & Mercer, M. Visual evoked responses of psychiatric patients. *Archives of General Psychiatry*, 1966, *15*, 59–63.

Spehlmann, R. The averaged electrical response to diffuse and to patterned light in the human. *Electroencephalography & Clinical Neurophysiology*, 1965, *19*, 560–569.

Squires, K. C., Hillyard, S. A., & Lindsay, P. H. Cortical potential evoked by confirming and disconfirming feedback following an auditory discrimination. *Perception and Psychophysics*, 1973, *13*, 25–31.

Squires, N. K., Jones, K. C., & Hillyard, S. A. Two varieties of long-latency positive waves evoked by unpredictable auditory stimuli in man. *Electroencephalography & Clinical Neurophysiology*, 1975, *38*, 387–401.

Stein, S. N., Goodwin, C. W., & Garvin, A. A brain wave correlator and preliminary studies. *Transactions of the American Neurological Association*, 1949, *74*, 196–199.

Stevens, J. R., Sachdev, K., & Milstein, V. Behavioral disorders of childhood and the electroencephalogram. *Archives of Neurology* (Chicago), 1968, *18*, 160.

Stotsky, B. Psychotropic drug use in the elderly. In S. Gershon & A. Raskin (Eds.), *Genesis and treatment of psychiatric disorders in the elderly*. New York: Raven Press, 1976. Pp. 229–258.

Straumanis, J. J., Shagass, C., & Schwartz, M. Visually evoked cerebral response changes associated with chronic brain syndromes and aging. *Journal of Gerontology,* 1965, *20,* 498–506.

Strauss, H., & Greenstein, L. The electroencephalogram in cerebrovascular disease. *Archives Neurology & Psychiatry* (Chicago), 1948, *59,* 395–403.

Stuhl, M. L., Cloche, M., & Kartun, M. P. Interêt de l'EEG dans l'étude des insufficiencies cardiaques avec cyanose. *Archives Malaise de Coeur,* 1952, *45,* 921–926.

Surwillo, W. W. Frequency of alpha rhythm, reaction time and age. *Nature,* 1961, *191,* 823–824.

Surwillo, W. W. The relation of simple response time to brain-wave frequency and the effects of age. *Electroencephalography & Clinical Neurophysiology,* 1963, *15,* 105–114.

Surwillo, W. W. The relation of decision time to brain-wave frequency and to age. *Electroencephalography & Clinical Neurophysiology,* 1964, *16,* 514.

Suter, C. *Computer analysis of evoked potential correlates of the critical band.* Technical report. Computer Science Center, University of Maryland, 1968, 68–98.

Suter, C. Principal component analysis of average evoked potentials. *Experimental Neurology,* 1970, *29,* 317–327.

Sutton, S., Braren, M., Zubin, J., & John, E. R. Evoked potential correlates of stimulus uncertainty. *Science,* 1965, *150,* 1187–1188.

Sutton, S., Tueting, P., Zubin, J., & John, E. R. Information delivery and the sensory evoked potential. *Science,* 1967, *155,* 1436–1439.

Suzuki, T., & Taguchi, K. Cerebral evoked responses to auditory stimuli in waking man. *Annals of Otolaryngology* (St. Louis), 1965, *74,* 128–139.

Tamura, K., Lüders, H., & Kuroiwa, Y. Further observations of the effects of aging on the waveform of the somato-sensory C.E.P. *Electroencephalography & Clinical Neurophysiology,* 1972, *33,* 325–327.

Tecce, J. J. Contingent negative variation and individual differences: A new approach in brain research. *Archives of General Psychiatry,* 1971, *24,* 1–16.

Tecce, J. J., & Scheff, N. M. Attention reduction and suppressed direct-current potentials in the human brain. *Science,* 1969, *164,* 331–333.

Terry, R. D., & Wisniewski, H. M. Structural and chemical changes of the aged human brain. In S. Gershon & A. Raskin (Eds.) , *Genesis and treatment of psychiatric disorders in the elderly.* New York: Raven Press, 1976, pp. 127–141.

Thatcher, R. W. A quantitative electrophysiological analysis of human memory. Paper presented at a symposium on: Brain mechanisms and memory processes: An overview. Eastern Psychological Association, Philadelphia, 1974. (a)

Thatcher, R. W. Evoked potential correlates of human short-term memory. *Fourth Annual Neuroscience Convention,* 1974, p. 450. (Abstract) (b)

Thatcher, R. W. Evoked potential correlates of hemispheric lateralization during semantic information processing. In S. Harnard (Ed.), *Lateralization in the nervous system.* New York: Academic Press, 1977.

Thatcher, R. W., and April, R. S. Evoked potential correlates of semantic information processing in normals and aphasics. In R. Riebert (Ed.), *The neuropsychology of language: Essays in honor of Eric Lenneberg.* Plenum Press, New York, 1976.

Thatcher, R. W., & John, E. R. Information and mathematical quantification of brain state. In N. R. Burch & H. L. Altshuler (Eds.), *Behavior and Brain Electrical Activity.* New York: Plenum Press, 1975. Pp. 303–324.

Thompson, L. W. Effects of hyperbaric oxygen on behavioral functioning in elderly persons with intellectual impairment. In S. Gershon & A. Raskin (Eds.), *Genesis and treatment of psychologic disorders in the elderly.* New York: Raven Press, 1975.

Torres, F., & Ayers, F. W. Evaluation of EEG of dyslexic children. *Electroencephalography & Clinical Neurophysiology*, 1968, *24*, 287.

Tueting, P., & Sutton, S. The relationship between prestimulus negative shifts and post-stimulus components of the averaged evoked potential. In F. Kornblum (Ed.), *Attention and performance* (Vol. IV). New York: Academic Press, 1973.

Umezaki, H., & Morrell, F. Developmental study of photic evoked responses in premature infants. *Electroencephalography & Clinical Neurophysiology*, 1970, *28*, 55–63.

U.S. Bureau of the Census, *Statistical abstract of the U.S. 1971 census* (92nd ed.). Washington, D.C., 1971.

Valdés, P., Ricardo, J., & Harmony, T. Utilizacion de la clasificacion automatica en el estudio de los potenciales evocados visuales. *Proceedings of the Symposium on Applications of Computation to the Study of the Nervous System* (Aplicacion de la computacion al estudio del sistema nervioso), National Scientific Research Center of Cuba, Havana, Cuba, October 20–24, 1975.

Van Buskirk, C., & Zarling, V. R. EEG prognosis in vascular hemiplegia rehabilitation. *Archives of Neurology & Psychiatry* (Chicago), 1951, *65*, 732–739.

Van Der Drift, J. H. Cardiac and vascular diseases. In A. Remond (Ed.), *Handbook of electroencephalography and clinical neurophysiology* (Part A, Vol. 14A). Amsterdam: Elsevier, 1972.

Van der Tweel, L. H., Regan, D., & Sprekreijse, H. Some aspects of potentials evoked by changes in spatial brightness contrast. Paper presented at the ISCERG Symposium, 1970, Istanbul.

Vanzulli, A., Wilson, E., & Garcia-Austt, E. Visual evoked responses in the newborn infant. *Acta Neurologia Latin-American*, 1964, *10*, 129–136.

Vaughan, H. G., Jr., & Katzman, R. Evoked responses in visual disorders. *Annals of the New York Academy of Science*, 1964, *112*, 305–319.

Vaughan, H. G., Jr., Katzman, R., & Taylor, J. Alterations of visual evoked responses in the presence of homonymous field defects. *Electroencephalography & Clinical Neurophysiology*, 1963, *15*, 737–746.

Vella, E. J., Butler, S. R., & Glass, A. Electrical correlate of right hemisphere function. *Nature* (London), 1972, *236*, 125–126.

Viglione, S. S., & Martin, W. B. Automatic analysis of the EEG for sleep staging. In P. Kellaway & I. Petersén (Eds.), *Automation of clinical electroencephalography*. New York: Raven Press, 1973. Pp. 269–285.

Walter, D. O., & Brazier, M. A. B. (Eds.), Advances in EEG analysis. *Electroencephalography & Clinical Neurophysiology*, 1969, Suppl. 27.

Walter, W. G. An automatic low frequency analyzer. *Electronic Engineering*, 1943, *16*, 9–13.

Walter, W. G. Spontaneous oscillatory systems and alterations in stability. In R. G. Grenell (Ed.), *Neural physiopathology*. New York: Hoeber, 1962. Pp. 222–257.

Walter, W. G. Specific and non-specific responses and autonomic mechanisms in human subjects during conditioning. *Progress in Brain Research*, 1963, *1*, 395–401.

Walter, W. G., Cooper, R., Aldridge, V. J., McCallum, W. C., & Winter, A. L. Contingent negative variation: an electric sign of sensori-motor association and expectancy in the human brain. *Nature*, 1964, *203*, 380–384.

Wechsler, D. A standardized memory scale for clinical use. *Journal of Psychology*, 1945, *19*, 87–95.

Weinberg, H., & Cooper, R. The recognition index: a pattern recognition technique for noisy signals. *Electroencephalography & Clinical Neurophysiology*, 1972, *33*, 608–613.

Weinberg, H., & Papakostopoulos, D. The frontal CNV: its dissimilarity to CNVs recorded from other sites. *Electroencephalography & Clinical Neurophysiology*, 1975, *39*, 21–28.

Weinberg, H., Walter, W. G., & Crow, H. H. Intracerebral events in human related to real and imaginary stimuli. *Electroencephalography & Clinical Neurophysiology*, 1970, *29*, 1–9.

Weiss, G., Minde, K., Douglas, V., Werry, J., & Sykes, D. Comparison of the effects of chlorpromazine, dextroamphetamine and methylphenidate on the behavior and intellectual functioning of hyperactive children. *Canadian Medical Association Journal*, 1971, *104*, 20.

Wender, P. H. *Minimal brain dysfunction in children*. New York: Wiley-Interscience, 1971.

Weitzman, E. D., Fishbein, W., & Graziani, L. Auditory evoked response obtained from the scalp electroencephalogram of the full-term human neonate during sleep. *Pediatrics*, 1965, *35*, 458–462.

Weitzman, E. D., & Graziani, L. J. Maturation and topography of the auditory-evoked response of the prematurely born infant. *Developmental Psychobiology*, 1968, *1*, 79–89.

Weitzman, E. D., Rémond, A., & Lesèvre, N. Étude chronotopographique des potentiels évoqués auditifs recueillis sur le scalp chez l'homme normal au cours du sommeil et de l'état de veille. *Revue de Neurologie*, 1965, *113*, 253–261.

White, C. T., & Eason, R. G. Evoked cortical potentials in relation to certain aspects of visual perception. *Psychological Monographs*, 1966, *80*,

Wikler, A., Dixon, J. F., & Parker, J. B., Jr. Brain function in problem children and controls: psychometric, neurological, and electroencephalographic comparisons. *American Journal of Psychiatry*, 1970, *127*, 634–645.

Wilkinson, R. T., & Lee, M. V. Auditory evoked potentials and selective attention. *Electroencephalography & Clinical Neurophysiology*, 1972, *33*, 411–418.

Wilkinson, T. R., & Spence, M. T. Determinants of the post-stimulus resolution of the contingent negative variation (CNV). *Electroencephalography & Clinical Neurophysiology*, 1973, *35*, 503–509.

Wishart, D. An algorithm for hierarchical classifications. *Biometrics*, 1969, *22*, 165–170.

Woody, C. D. Characterization of an adaptive filter for the analysis of variable latency neuroelectric signals. *Medical and Biological Engineering*, 1967, *5*, 539–553.

Wu, Y. H., Rayburn, J. W., Allen, L. E., Ferguson, H. C., & Kissel, J. W. Psychosedative agents. 2. 8-(4-Substituted-1-piperazinylalkyl)-8-azaspiro[4.5]decane-7,9-diones. *Journal of Medicinal Chemistry*, 1972, *15*, 477–479.

Zetterberg, L. H. Spike detection by computer and by analog equipment. In P. Kellaway & I. Petersén (Eds.), *Automation of clinical electroencephalography*. New York: Raven Press, 1973. Pp. 227–234.

Zfass, I. S., & Hoefer, P. F. A. Electroencephalographic findings in cases of cerebral vascular lesions. *Electroencephalography & Clinical Neurophysiology*, 1950, *2*, 361–362.

Author Index

Numbers in *italics* refer to the pages on which the complete references are listed.

Y

Z

Subject Index

F

www.ingramcontent.com/pod-product-compliance
Lightning Source LLC
Chambersburg PA
CBHW070601270326
41926CB00013B/2385
* 9 7 8 0 3 6 7 7 5 4 0 2 0 *